Environment and
Services

MITCHELL'S BUILDING SERIES

Environment and Services

8th edition

PETER BURBERRY
MSc, DipArch, RIBA, FCIOB, FRSA

LONGMAN

Addison Wesley Longman Limited
Edinburgh Gate, Harlow
Essex CM20 2JE, England
and Associated Companies throughout the world

First published by B.T. Batsford 1970
Second edition 1975
Third edition 1977
Fourth edition 1979
Fifth edition 1983
Sixth edition 1988
Seventh edition published by Longman Scientific and Technical 1992
Eighth edition published by Addison Wesley Longman 1997
Second impression 1997

British Library Cataloguing in Publication Data
A catalogue entry for this title is available from the British Library

ISBN 0-582-24521-4

Set by 4 in Times 9½/11
Produced by Longman Singapore Publishers (Pte) Ltd
Printed in Singapore

Contents

Preface

The demand for increasing standards of environmental control and provision of services continues to increase as does the sophistication of the techniques and equipment available. Traditionally the design decisions were influenced solely by balancing desirable performance against acceptable cost. The global need for conservation of resources and control of pollution is now well understood and is a new determinant in the environmental and services design.

The construction process, the use of the building and the disposal of waste, both during occupancy and upon demolition, all have major environmental consequences. It is impossible for individuals or building designers to determine the degree to which individual buildings should be constrained. Global, national and local factors have to be considered: The British Government has established a national target for reduction of CO_2 emmissions; new buildings are subject to increasingly stringent standards of insulation and plant performance to reduce energy consumption. However, only a very small proportion of the building stock is renewed each year. If effective environmental improvements are to be achieved involvement of the existing building stock is essential. At present there is legislation to restrict the maximum temperature during the heating season to 19 °C but this is totally unenforced.

Designers, owners and users of buildings have a considerable individual respnsibility not only to design well but also to adopt responsible standards. Some factors are very clear: the energy consumption and use of natural resources reflect the size of the building; new buildings should be as small as possible to achieve their function. Making provisions based on assumptions about the future is a high risk strategy best avoided except for arranging structural members to allow flexibility of future planning. Existing buildings, which form by far the greatest part of the building stock, have been, in the main, neglected by both government and designers. To play their proper part in sustaining the environment they should be as fully utilized as possible. Some existing buildings may be so profligate in their energy use or so inconvenient in their planning that they should be replaced or refurbished. However, the demolition and construction processes impose a major load on natural resources and the environment. Generally, the most environmentally friendly solution will be to continue to use existing buildings for as long as possible. At present public and professional recognition goes almost exclusively to new buildings, almost all of which claim to be energy conserving. There is an urgent need for methods of assessing the relative resource loads which different design strategies place upon the environment.

Whether buildings are new or existing proper environmental control and provision of utility services are as important as ever. The interdependence of the form, fabric and fenestration of the building and the choice and management of the environmental installations is increasingly recognized. With utility services, the economical layout and careful analysis of loads to be met are critical to the efficiency and economy and installation. In all types of environmental and utility installations, ready access for adjustment and maintenance and, as is likely in many buildings, replacement during the life of the building are important design considerations. In the environmental area the importance, for user satisfaction, in many building types of control by occupants rather than automatic systems is being appreciated.

Technical developments have sometimes given rise to unexpected problems. Ventilation limitation to save energy, air conditioning and mechanical ventilation, and the chemical nature of new materials, finishes and processes have caused difficulties which include high concentrations of radon gas in susceptible geographical areas, and a variety of problems often described as the 'sick building syndrome'. Furnishings, ventilation, temperature levels and maintenance routines can lead to concentrations of house mites which can give rise to discomfort and health problems. Air conditioning and water supply systems in centrally heated buildings can allow the rapid growth of the bacteria which give rise to legionnaire's disease.

A recent and welcome design innovation uses skilled understanding of air flow and building form, rather than the use of mechanical plant, to allow a natural ventilation of office buildings which would, until very recently, have been air conditioned.

This new edition of *Environment and Services* continues to give an introduction to the basic principles governing human requirements for health, comfort and activity, and the nature of the physical phenomena involved. It identifies the main utility service installations and the ways in which they operate. In each case the general principles of design are described, together with the standards involved and the associated materials and equipment. It will provide a basic understanding of objectives, methods and standards for environmental and services design. For readers embarking on a professional career in the architectural or building fields it also provides, in many cases, in addition to the general principles, data and methodology for design.

Many detailed items have been brought up-to-date for this edition. In particular there are major revisions to thermal regulations, global environment, natural ventilation in dwellings and large offices, electric lighting, pipes and firefighting. However, since the need for conservation of resources means that existing buildings must be used and maintained effectively, information on equipment likely to be still in use has been retained.

Peter Burberry

Acknowledgements

The author and publishers are grateful to the following individuals and organizations for their assistance during the preparation of this and earlier editions of *Environment and Services*, and for permission to reproduce material from certain books and articles, and to use drawings as a basis for illustrations in this volume:

Peter Stocks, of Cundall, Johnston and Partners, for advice and assistance with natural ventilation for large buildings; Ian Maclean, of Thorn Lighting, for information on lamps and lighting; John Slater, of Fullflow Systems Ltd, for information on syphonic roof drainage.

The *Architects' Journal* for our Table 5.1 (from *Designing for Thermal Comfort* by A. Harvey; *AJ* 20/11/58), and our Fig. 7.47 (adapted from *AJ* 5/3/86 and 27/1/88).

Building Research Establishment for our Tables 3.1 (the Beaufort scale), 5.3 (from BRE CP 14/71), 5.4 (from BRE CP 61/74), 5.7 (from BRE Digest 145), 5.26 (from BRE CP 47/68) and 12.4 & 12.5 (from BRE Digest 34), and our Fig. 5.1 (adapted from *Theoretical and Practical Aspects of Thermal Comfort*). (© Crown copyright reserved.)

HMSO for our Tables 4.1 (from PWBS 30 *The Lighting of Office Buildings*), and our Figs 4.3 & 4.4 (from *Architectural Physics: Lighting*) and 4.14 (adapted from *Sunlight and Daylight*). (© Crown copyright reserved.)

Institution of Heating and ventilating Engineers for our Tables 5.27, 5.28 & 5.29 (from *IHVE Guide 1970*), and our Fig. 5.21 (adapted from *IHVE Guide 1970*).

Extracts from British Standards Institution publications (our Tables 3.3, 4.3, 11.1, 11.2, 11.4 & 11.5, and our Fig. 2.6) are reproduced with the permission of BSI. Complete copies can be obtained by post from BSI Customer Services, 389 Chiswick High Road, London W4 4AL; Telephone: 0181 996 7000.

ENVIRONMENT

_____

1 Comfort and global environment

1.1 Natural control

One of the main objects of building design is to ensure the provision of continuous comfort for occupants in spite of adverse and variable external conditions. In terms of daylighting, air quality, air movement and noise, the natural external conditions are surprisingly similar in all habitable climates and, for much of the time, provide acceptable conditions for human comfort and activity. If these were the only variables, traditional building forms might well have been very similar throughout the world. Rainfall, humidity and thermal conditions vary greatly with location on the earth's surface and it is particularly thermal conditions which have been the prime determinant of the traditional forms of buildings. The need to maintain adequate daylighting and natural ventilation has provided constraints upon design, governing depth and height of rooms and the provision of openings.

Traditionally, the best combinations of materials and forms for local conditions were established slowly by trial and error. Figure 1.1(a)–(d) shows three examples which demonstrate the range of conditions which could be addressed successfully. Figure 1.1(a) and (b) shows the type of courtyard house developed in North Africa in a hot, dry climate with clear skies. At night the building radiates to the clear night sky and cool air accumulates in the courtyard and the ground floor. The fabric is cooled by both convection and radiation. During the day the sun shines on the building, where the white surfaces reflect some of the radiation. The courtyard and the ground floor remain shaded from the sun. Since the climate is dry, a fountain can be used to give evaporative cooling.

Figure 1.1(c) shows an example from a hot climate where heating is never required and where

opportunities to improve comfort are very limited due to high air temperature and relative humidity. The only available option is to encourage air movement as much as possible to help the evaporative cooling of perspiration. This is achieved by using open verandas as living spaces and providing privacy with louvred walls which restrict air flow as little as possible.

Figure 1.1(d) shows a typical English cottage of the Middle Ages, when fuel was difficult to obtain and the variable climate often cold and damp. The thick walls and the roof of thatch provide massive insulation. The windows are small. The interior therefore fluctuates in temperature relatively slowly and does not reach the extremes of the external conditions. A slow-burning fire of wood or peat burns continuously in winter, which is much more economical of energy than rapid variations of heat output. The relatively high thermal capacity of the chimney and walls, together with the substantial insulation, enable comfort conditions to be maintained in both winter and summer. Figure 1.1(e) gives a section through a classroom wing of a school at Wallasey. The structure is massive and provided with substantial external insulation. The large south-facing windows have an unobstructed prospect. They have two layers of glazing, widely spaced with windows that can be opened in various configurations to control ventilation. The electric lighting is by tungsten lamps. Although it has now been equipped with central heating, the classroom block performed to the reasonable satisfaction of the occupants for several years without additional heat sources.

Effects of environmental plant

The development of mechanical and electrical plant for heating, cooling, ventilation and lighting has made it

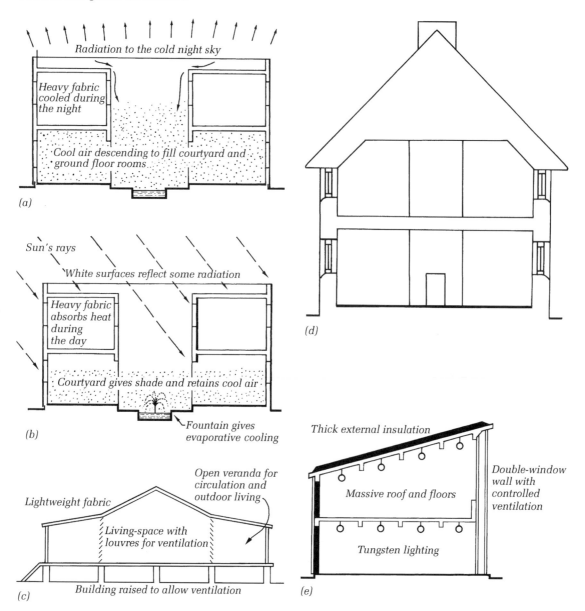

Figure 1.1   Control of environment through construction: North African courtyard house – (a) night-time and (b) during the day; (c) a dwelling for a hot and humid climate; (d) an English country cottage; (e) St George's School, Wallasey.

possible to provide closer control of internal temperature and comfort than can be achieved by natural means. In technologically advanced countries almost all buildings employ plant of this type. Designers are no longer constrained by the traditional forms. Comfort can be maintained in buildings with shapes and constructions which would have been uninhabitable in the days before mechanical plant. Geographical variations in building form have diminished.

Parallel technological development has also resulted in the use, in many buildings, of mechanical and particularly electrical equipment which may require a more stringent control of environmental conditions than is required for human occupancy alone.

Pollution of external environment and exhaustion of fuels

Until very recently environmental design in most buildings was directed simply to the provision of comfort for occupants or preservation of contents. Little, if any, thought was given to the depletion of resources or the pollution of the environment. With advancing standards for environmental control, the use of fuel in and for buildings has increased very rapidly, creating problems of supply and external environmental quality. Smoke from the partial combustion of coal in open fires caused increasing air pollution problems in cities and, ultimately, after statistics showed that some four thousand people had died as the result of a London fog, the Clean Air Act was introduced to prohibit this type of heating in urban areas. Although there was a temporary shortage of fuel at the end of the war in the mid-1940s, it was not until political problems with the continuity of oil supplies in the 1970s that concern began to arise about the possible exhaustion of fossil fuels. This concern has been overtaken by the increasing certainty that carbon dioxide emissions are giving rise to a greenhouse effect, apparently causing global warming and acid rain. Other activities contribute to carbon dioxide emissions but buildings are the main contributor. Thus, for current and future building environmental design, the impact of the building on the external environment has to be considered as well as the traditional inward-looking approach.

Other effects of building operation on the external environment

Although fuel-related problems enjoy most current interest, there are other ways in which building design decisions affect the external environment. Several problems can occur in the immediate vicinity of the building. Ensuring that daylight and sunlight for surrounding buildings are not prejudiced by new construction has been part of town planning control for many years. More recently the effects that large buildings can have on wind speeds at ground level have been appreciated and standards established. Health hazards can result from the use of air-conditioning with wet cooling towers. If these towers are not rigorously maintained, passers-by can be infected with legionnaires' disease. In the global context flue gases – the same gases that produce global warming – can also give rise to acid rain with consequent damage to buildings and vegetation. Felling of timber is reducing the areas of forest; this is affecting the local climate and may have serious consequences for the whole planet. Many modern insulating materials use chlorofluorocarbons (CFCs) in their manufacture. The CFC gases, and other halogens, rise in the atmosphere and destroy the ozone layer which protects the earth against ultraviolet rays. The process of construction generates substantial waste. Rehabilitation and particularly demolition generate very large volumes of waste. It is becoming more and more difficult to dispose of this waste without damage to the environment.

Design standards for global factors

Designers of buildings cannot be expected to determine the contribution which individual buildings should make to global or local environmental conservation. Targets have to be established by governments taking into account the international situation.

In the United Kingdom the means of government action range from legal requirements through recommendations supported by financial grants if the work is carried out, to hopeful exhortations. The UK government has legislated to stop the manufacture and use of CFCs. It does not directly control the use of energy. In this context action takes the form of improving the thermal performance of existing and particularly of new buildings in the anticipation that appropriate energy savings will result. Although the situation is changing rapidly, the following notes give an idea of the approaches currently adopted for energy conservation and carbon dioxide emissions in the building field:

- *Existing buildings*
  - There is a legal requirement which limits temperatures in a wide range of building types to 19 °C; it is not enforced and has had no observable effect.
  - Financial assistance is offered to householders who install insulation in roof spaces.
- *New buildings*
  - Comprehensive Building Regulations specify standards of insulation, ventilation and some aspects of plant efficiency.
  - An advisory service offers designers a free consultation and more detailed advice at a subsidized rate.

To offer general assistance and encouragement for energy conservation, the government has established the Energy Efficiency Office, which supports a wide range of investigation and information provision. In particular the publication of a very useful range of booklets under the heading of the Best Practice Programme. The series provides data on comparative performance for many building types, descriptions of successful techniques, general information on energy use and management, and descriptions of new approaches to design and management.

For a more comprehensive assessment of environmental issues, including global, local and internal aspects, the government, through the agency of the Building Research Establishment, has established the BREEAM system (Building Research Establishment environmental assessment method). This method has versions for dwellings and several types of commercial buildings. In each case a series of relevant factors are identified and assessed at the design stage. A certificate is issued indicating, for each factor, whether satisfactory performance has been achieved. There is no overall assessment. The scheme is voluntary and the developer pays the cost of the assessment. From the developer's point of view, a satisfactory certificate should enhance the value of the project.

The factors taken into account vary to suit the particular building type. Among other items they can include

- carbon dioxide emission
- ozone depletion
- acid rain
- timber from sustainable sources
- use of recycled materials
- efficient waste collection
- water economy
- loads on transport systems

In addition to current actions, the government is attempting to gather information and develop an informed strategy for sustainable development. The general aims were published in 1994. In 1996 a very comprehensive document was issued, *Indicators of Sustainable Development for the United Kingdom*, identifying and quantifying the trends over the last 25 years for a wide range of environmental issues. According to its predictions, if the current pattern is sustained, the mean temperature in the United Kingdom will increase by 5 °C over the next 100 years, known reserves of oil will be exhausted in 12 years and reserves of gas in 25 years. In spite of its direct importance for energy consumption, use of resources and pollution, the building field is not specifically identified as a coherent element. It indicates trends in transportation, energy use, land use, water resources, climate change, ozone layer depletion, acid deposition and air quality. It is very much to be hoped that a practical policy can result.

In the field of building, the most effective government action has been directed towards new buildings. The building stock is renewed very slowly, and if sustainability is to be achieved, it must involve existing buildings. Remedial measures in existing buildings are often difficult and involve investment of energy and resources. An area not yet explored is to increase the density of occupancy of buildings; although administrative problems would have to be solved, greater density does offer a very effective energy-saving measure.

1.2 Health

Developing medical expertise, especially during the nineteenth century, led to an appreciation that conditions in buildings, particularly housing could give rise to disease. At first the conditions identified were gross examples of lack of hygiene: polluted water, exposed excrement, damp, infestation, fumes and cross-infection due to inadequate ventilation. New standards along with new building and engineering solutions represented a major triumph of nineteenth-century endeavour. Improved working and housing environments achieved a dramatic improvement in public health.

It is seldom appreciated that the virtual disappearance of several major killing diseases owes more to building standards than to drugs or medical treatment. Even in the twentieth century the dramatic reduction in the incidence of tuberculosis appears to have resulted from improved housing standards. The current increase in its incidence is attributed to inadequate accommodation.

The main health factors associated with buildings are infestation, temperature, ventilation, lighting, dampness, sanitation, food storage and the removal of waste. Most of these were regarded as adequately covered by current standards and the universal use of refrigerators has rendered the special ventilated cupboard for food storage redundant. However, amounts of waste have increased, robust dustbins have been replaced by plastic sacks and the rat population is increasing. There is cause for concern in this area. Concern to minimize energy consumption has led to a considerable reduction in ventilation rates and, in dwellings, to condensation problems. Toxic vapours can come from paints and adhesives. Dust from building materials, furniture and finishes, and germs from badly maintained services may all present comfort or health hazards, hazards made more critical by reduced ventilation. Electromagnetic fields are relatively new and increasing in their extent. Any effect on health is vigorously denied. But regardless of their effect on people, they should still be considered by designers because some types of electronic equipment need to be protected from electromagnetic fields.

Some time ago, in commercial buildings, particularly those equipped with air-conditioning or mechanical ventilation, complaints of discomfort and even illness became frequent. So frequent, in fact, that the problem became described as 'sick building syndrome'. A great deal of research was carried out to identify the causes. Many buildings were found to have conditions that were disadvantageous to comfort and health, but no specific cause could be identified. However, it did become apparent that fewer complaints occurred where the environmental plant, particularly the control systems, had been properly related to the requirements of the building as well as the needs and comfort of the occupants; once installed, they were also rigorously maintained. A very high standard of cleaning and freedom from dust also helped to lower the number of complaints.

The origins of building legislation lie in the fields of health and safety, and the results have been highly successful. But up to now little attempt appears to have been made to evaluate the possible consequences to health of innovations in design and construction. In a situation where any unrecognized problem will be incorporated into a very large number of buildings, which inevitably have a long life, it is very important that the effects of innovations should be assessed before health problems become apparent.

1.3 Design

Until the nineteenth century there were few, if any, mechanical installations in buildings. Environmental design was based on experience. During the 1800s constructional innovations enabled greater spans and heights, and mechanical installations for providing heat, light, ventilation and communication were progressively employed. The present pattern of building professions was established in the mid nineteenth century. At first the new mechanical installations were deployed in designs conceived within a traditional context. The architectural design was, in effect, completed before the mechanical installations were designed to fit. However, the development of mechanical installations progressed to a point where satisfactory environmental conditions and circulation could be provided in shapes and forms of buildings which could not otherwise have been made practicable. In this situation the overall design has to be conceived as one whole solution rather than as a succession of separate solutions. There are notable examples of buildings where integrated solutions were successfully realized, but these successes are far from universal. The tradition of the architectural profession is to generate an overall concept for the building. For this to be complete in contemporary terms, it must include the environmental and mechanical aspects. The building professions still have the opportunity to improve their capability for integrated design.

2 Moisture

Water in its free liquid form must be totally excluded from the interiors of buildings. And its presence as vapour in the air, or as moisture content of building materials, must be controlled to within acceptable limits. Moisture in buildings comes from three main sources: precipitation as rain or snow; damp rising from the ground through building materials by capillary action or as vapour; and from occupants and processes (such as cooking and washing). In addition, condensed droplets in the air (fog) can be carried by wind to wet the surfaces of the building and to penetrate joints and materials. External relative humidity is likely to affect internal relative humidity levels. This can influence both thermal comfort and the risk of condensation, a major problem in many buildings.

2.1 Precipitation

Precipitation falling on to roofs, walls and paved areas must be prevented from damaging the fabric of the building and penetrating to the interior. This involves the selection of external finishes that are impervious or, if they can absorb water, are not subject to deterioration and are effectively isolated from contacts which could allow moisture migration to the interior. Other Mitchell's Building Series (MBS) volumes deal with properties of materials and detailing to prevent moisture ingress or degradation of materials. This book deals with the drainage of water from the surfaces of the building.

Rain

It is conventional to think of rainfall in terms of the volume of water falling during the course of the year. This relates well to perception of areas of the country regarded as relatively wet or dry, and to agricultural needs. For the drainage of surface water, however, it is the maximum intensity of rainfall which is the critical factor for the design of buildings.

Short periods of heavy rainfall have long been of interest to meteorologists. However, the original methods for recording them were far from accurate. The observer had first to notice that a spell of unusually heavy rain had begun, before recording the time and the level in the rain gauge. When the heavy rain eased, the time and the new level were then recorded. The results were of limited use. Simple recording gauges began to come into use and, in 1935, Bilham of the Meteorological Office published a formula for predicting frequency, duration and intensity of heavy rainfall from data based on 12 stations with recording gauges. This represented a major step in the understanding of heavy rainfall and was used as a basis for design for many years. In the late 1950s more accurate recording gauges came into general use, capable of identifying rainfall amounts minute by minute. During the same period Gumbell developed and validated a frequency distribution which was able to display data on the frequency and intensity of a variety of natural phenomena, and also to predict the probability of more extreme values outside the range of measured records. On the basis of these developments the Meteorological Office has been able to calculate the probable frequency and duration of heavy rainfall over the whole of the British Isles. The results demonstrate that, contrary to what might have been expected, the south and east are more subject to intense rainfall than the north and west, where annual rainfall is higher.

The Building Regulations for England and Wales prescribe a simple standard for design of a rainfall rate of 75 mm/h. This corresponds to a rainfall of 2 min duration once a year in the south-east and lower intensities in the north-west. BS 6367 gives advice for determining design rates for buildings where more rigorous safety standards are required. For most buildings, however, the standards in the Building Regulations will be appropriate. In any design for surface water it is essential to remember that the design rainfall might be exceeded. Although, statistically, this may be only once in a hundred years, there is no guarantee it will not take place next week. Gutters and outlets to downpipes are inevitably somewhat obstructed by leaves and solid matter, thereby substantially reducing their discharge capacities. It is not uncommon for plastic bags to be blown on to roofs and into gutters. Perhaps the most critical time for overflow from gutters is when the building is new and debris from construction has not been cleared. So, although gutters and downpipes should be designed to give the discharge capacities required by the regulations, the critical factor in design must be to ensure that any overflow will not cause damage to the building or its contents.

Rainfall rarely falls vertically. Allowance must be made for the effects of wind. It is not possible to do this on a precise analytical basis. Airflow patterns around buildings are complex. Indeed, in large and tall buildings the wind flow on the upper part of a façade facing the wind can be upwards and flashing details will require careful thought.

Nominally flat roofs will be provided with falls to outlets; this avoids ponding, reduces the consequences of leakage and ensures the appropriate pattern of drainage. However, strong winds may reverse the nominal flows, so gutter and downpipe capacities should take this into account. The amount falling on pitched roofs and vertical walls will be influenced by wind velocity and direction. For practical design, BS 6367 recommends that a value of 26° to the vertical and 90° to the face of the wall or roof should be adopted in order to calculate the run-off to gutters. Vertical rainfall must be assumed for flat roofs. Figure 2.1 shows how the design rate for gutters and downpipes can be determined.

Snow

Compared with heavy rainfall, the melting of snow does not normally place such an acute load on the gutters and drains of individual buildings, so it is not a critical problem from this point of view. The MBS title *Finishes* deals with the precautions desirable to prevent the penetration of wind-driven snow into buildings and damp penetration due to melting. Care

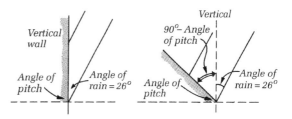

Figure 2.1   Calculation of design rainfall for surfaces at an angle:
Design rainfall rate = normal rate × sin (90° − angle of rain)
= 75 mm/h × sin (90° − 26°)

must be taken in gutter layout to avoid damage to the building from overflows during snow thaw, and snow boards may be needed to prevent snow falls on to glazed areas or entrances.

Design of buildings and surrounding features in relation to snow drifting is not normal in this country.

2.2 Moisture migration through the building fabric

Movement of moisture through the materials of which the building is constructed can occur through capillary action or by diffusion of vapour through the pores of the material. The sources of moisture which must be considered in this context include not only precipitation on external surfaces but also damp from the ground on which the building stands. Roofs are traditionally made of impervious materials but walls and ground floors are not often made in this way. Since, in the case of walls, the exposure to moisture is intermittent it is possible to use thick solid walls of porous materials which absorb moisture into their external faces then allow it to evaporate without giving rise to serious internal penetration. This type of wall generally has to be very much thicker than is required for structural needs and in modern buildings, economy dictates that means, other than thickness, have to be used to limit moisture penetration.

Two methods of control are normally employed. Either a barrier of impervious material, or an air gap, is placed to intercept the movement. Flashings (capillary movement only), damp-proof courses, damp-proof membranes (capillary and vapour diffusion) and vapour barriers (vapour only) are examples of impervious materials being used as barriers. Cavities, provided they are not bridged by moisture-transmitting features, are a very effective means of arresting capillary movement. They are not necessarily effective against vapour diffusion since, unless well ventilated, it will be possible for the vapour to pass across. Traditional suspended ground floors are a good

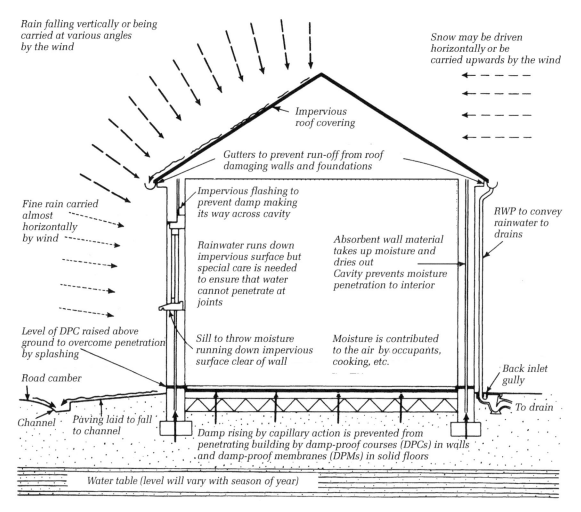

Rain falling vertically or being
carried at various angles
by the wind

Snow may be driven
horizontally or be
carried upwards by the wind

Impervious
roof covering

Gutters to prevent run-off from roof
damaging walls and foundations

Fine rain carried
almost
horizontally
by wind

Impervious flashing to
prevent damp making
its way across cavity

RWP to convey
rainwater to
drains

Rainwater runs down
impervious surface but
special care is needed
to ensure that water
cannot penetrate at
joints

Absorbent wall material
takes up moisture and
dries out
Cavity prevents moisture
penetration to interior

Level of DPC raised above
ground to overcome penetration
by splashing

Sill to throw moisture
running down impervious
surface clear of wall

Moisture is contributed
to the air by occupants,
cooking, etc.

Road camber

Back inlet
gully

To drain

Channel

Paving laid to fall
to channel

Damp rising by capillary action is prevented from
penetrating building by damp-proof courses (DPCs) in walls
and damp-proof membranes (DPMs) in solid floors

Water table (level will vary with season of year)

Figure 2.2 How to control moisture penetration in buildings.

example of a cavity formed to prevent capillary and vapour movement. To be effective in preventing vapour reaching the other side, the cavity has to be adequately ventilated. For a period, solid ground floors were more widely used than suspended floors. Recent increases in insulation requirements, coupled with the need to deal effectively with associated leakage and damp-proofing problems, have led to a return to the use of suspended floors.

2.3 Condensation: background

Until the 1960s condensation was an unusual occurrence in buildings in England and Wales and comparatively rare in Scotland. The main problems arose in sheeted roofs, particularly in industrial buildings. Since that time, however, the incidence of condensation, particularly in housing, has increased spectacularly, and large numbers of dwellings and other buildings have been affected. The severity of condensation varies from the comparatively minor inconvenience of condensation upon windows, through black mould forming on internal surfaces to decay of finishing and structural materials, and ultimately to structural failure. Many of the reasons for the situation can be relatively simply identified. They include changes in building design and construction and also in patterns of occupancy. Although the changes in occupancy arose because of influences not related to building, most of the changes in building design and construction were consciously made for reasons which, taken in isolation, seemed well merited but which, in the overall context of design and use, had highly undesirable consequences. The main changes are summarized below. They are not the only reasons for condensation and it is not necessary for all these factors to be present for condensation to result. All the developments described contribute either to an increase in the level of internal humidity or to a reduction of the structural temperature.

1. Ventilation of habitable rooms

In 1965 the new Building Regulations relaxed the requirement that every habitable room should be provided with either a flue or a special ventilator. It had been considered that, since many occupants obstructed the flow of air through the ventilators, there was little point in providing them. Subsequent history has indicated clearly that many ventilators did serve a useful purpose in providing ventilation and dissipating internal moisture. The 1990 edition of Building Regulations Approved Document F1 reintroduced a requirement for background ventilation, which was developed in 1995 (see Chapter 3).

2. Changes in window design

These fall into two categories. Window manufacturers regarded it as important that windows should not allow large quantities of air to leak into and out of the rooms they served, and made considerable efforts to improve the airtightness of windows in the closed position. Standards were established and testing procedures were developed to improve designs and to test finished products. During winter conditions the result was inevitably a further diminution of natural infiltration of air into rooms.

During the same period architects, in particular, showed a very marked preference for windows without night ventilators. The night ventilator, a small top-hung opening light, could be left open in all weather conditions, and presented less security risk than a large window. With large opening lights, even when fitted with special stops to give limited opening, it is very difficult to provide acceptable control of ventilation during cold weather. The night ventilator, although not ideal from this point of view, was very much smaller and situated at a high level so that draughts resulting from limited opening were not so noticeable. The practical effect of this change in window selection is also to reduce rates of winter ventilation.

3. New building materials

The application of new building materials has continued since the major effort at introducing

prefabricated housing after the war. New materials range from lightweight cladding and insulation to massive concrete construction. Such materials often have very different thermal and vapour properties from the traditional ones. Frequently the significance of these changes was not appreciated and new materials were used without any analysis of the consequences for condensation risk. Lightweight construction of sheeted materials is more liable to suffer from condensation due to moist internal air making its way through joints and gaps. Heavyweight construction, such as dense concrete panels, may have high moisture resistance thereby reducing the rate of dissipation of internal moisture. The high thermal capacity of this construction may, in the case of intermittent heating, produce slow warming and lower the surface temperature at critical periods.

4. Changes in heating installations

The traditional form of domestic heating, the open fire, or its more efficient development, the openable stove, operated in a way which automatically minimized condensation risk. Substantial quantities of air were drawn through the room and passed up the flue, thereby ensuring significant levels of internal ventilation. A large proportion of the heat output was in the form of radiation, which warmed internal surfaces. The fire or stove could not be switched on and off easily. Heating was inevitably provided over a period of several hours and the substantial brickwork which contained the fireplaces stored heat which limited overnight temperature drops in the interior.

Modern domestic central heating systems offer the possibility of much more rapid control, and in the case of warm air heating the response of the system is almost instantaneous. Economy of running encourages intermittent operation, which can result in the fabric of the building remaining relatively cold, thereby substantially increasing the condensation risk.

Lack of proper heating installations or inability to run a central installation because of its cost frequently results in the use of free-standing, flueless oil- or gas-fired appliances. Since water is one of the major products of combustion, the use of this type of apparatus can be disastrous from the condensation viewpoint.

5. Patterns of occupancy and operation

Many, probably even a majority, of dwellings are now unoccupied during most of the day. This encourages the adoption of intermittent patterns of operation of the heating system and also, for security reasons, leads to windows being kept tightly shut. The low level of ventilation and the lack of warming of the fabric of the building contribute significantly to condensation risk.

The use of washing-machines and particularly clothes-drying machines, which are rarely arranged to extract the moist air to the outside, contribute to the problem.

6. Future problems

It is important to bear in mind that major changes in the thermal behaviour of buildings are still taking place. Very much higher standards of insulation are now being called for in new buildings. This has important consequences for condensation. With highly insulated dwellings, the heat loss through the fabric is dramatically reduced and ventilation becomes the dominant factor in energy conservation. With increasing fuel prices there can be no doubt that ventilation rates will be reduced as much as possible. Excessive reduction will, however, inevitably lead to condensation problems. It seems possible that, in the future, minimum rates of ventilation in dwellings will be governed by condensation considerations rather than the previous standards.

2.4 Condensation: general principles

It is not possible to postulate a set of simple design rules which, if observed, will avoid condensation problems. Calculation procedures form an essential check in the development of design solutions but cannot generate them. Since the avoidance of condensation involves many different aspects of planning, construction and the environment, it is essential that designers should be familiar with the physical mechanisms which give rise to condensation and should bear these principles in mind from the outset of design so that critical problems are avoided.

It is well known that the capacity of the air to contain water vapour varies with temperature and that, if air containing a given quantity of moisture is cooled to a temperature below that which enables the air to retain its original quantity of vapour, then water will be precipitated. This phenomenon can most usually be observed where air comes into contact with cold surfaces and moisture is deposited as dew or condensation. In nature this circumstance arises because, in clear weather, surfaces lose their heat by radiation to the cold sky and become much colder than the air surrounding them. It is rare to observe this phenomenon on the outside of buildings. Walls are usually protected from exposure to the cold sky and the

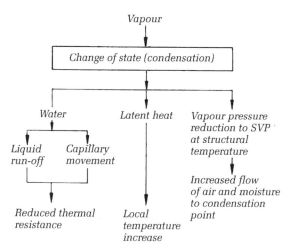

Figure 2.3   Condensation phenomena.

substantial thermal capacity of most building materials will limit the temperature drop. An exception would be a thin single-skin roof where the material would be rapidly cooled and condensation could take place both outside and inside. To explain condensation in buildings, one must look for other mechanisms. The primary factor in building condensation is that, except in specially conditioned buildings, moisture levels are invariably raised above the levels of those prevailing outside. Significant quantities of moisture are given off by occupants by sweating and transpiration from the lungs. In buildings such as offices and schools there may be little other source of moisture. In dwellings, however, cooking, washing, clothes drying, and the use of unflued heating appliances combine to give substantial quantities of vapour input. There are many industrial processes which either generate vapour directly, or give rise to exposed areas of liquid or wetted surfaces from which evaporation can take place. The final level of moisture within the building will be a balance between the rate of input and the rate at which the moisture is carried away by ventilation. It is clear, therefore, that every effort should be made to reduce the amount of vapour which is input into the building. Unflued gas and oil appliances should be avoided wherever possible and apparatus having major vapour output should have its own extraction arrangements. To minimize condensation, ventilation rates should be as high as possible; but to conserve energy they should be as low as possible. Determining standards for ventilation and providing methods of control are urgent problems as yet unsolved.

Whatever steps are taken to reduce internal humidity levels, they will inevitably be higher than those outside and if walls had no insulating value, there would inevitably be condensation inside them. Single glazing forms a good example of this circumstance where, particularly at night, when ventilation rates inside are relatively low and external temperatures also low, condensation is the rule rather than the exception. The original Building Regulations requirements for some measure of insulation in walls arose from a need to control condensation. Notice that, when condensation takes place on walls, it usually does so at the top of the wall in the corners at window openings and at low level on the ground floor. These are all positions where heat can escape more readily, surface temperatures are lower and careful detailing of insulation is required if condensation is to be avoided. The problem is further complicated by the capability of modern heating installations for intermittent operation and the desire of occupants to achieve economy by using this facility to the full. If heating is very intermittent the inner surfaces of walls may remain at comparatively low temperatures and be subject to condensation; whereas if continuously heated, they might escape the problem. In the same way that standards for ventilation will undoubtedly have to be revised to take account of modern problems, the design and control of heating installations and the amount of thermal capacity provided in buildings will need to be considered in relation to control of condensation in intermittently heated conditions.

Surface condensation, and the black mould which inevitably accompanies it, is quickly and easily visible. It damages decorations but does not usually result in major structural deterioration. Condensation which takes place, unobserved, within the thickness of the elements of construction frequently gives rise to structural failure; indeed, sagging of structural members may be one of the first indications that condensation is taking place. The problem arises either from diffusion of moisture through materials until it meets a cold zone or from currents of moist air entering the cavities and voids. Single homogeneous elements do not suffer accumulation of water. Problems arise because of multiple layers and the differential performance of building materials in relation to thermal and vapour resistance. It is very common for thermal insulating materials to have a very low vapour resistance. Thus it is possible to apply an internal lining of insulation to a wall which previously was not troubled by condensation and discover that, because the wall surface is now at a much lower temperature but vapour can still penetrate readily to it, condensation now takes place at the interface of the materials. Clearly it is necessary in

designing any element of an external envelope to ensure this circumstance will not occur. Ideally, from the condensation viewpoint, the concentration of vapour resistance should be towards the inner face of the wall and the concentration of thermal resistance outside this.

Although the diffusion of moisture through materials can give rise to serious condensation problems, a more common mechanism for condensation is the movement of moist air from within the building through cracks in the construction into cavities and particularly into ceiling voids. Large quantities of moisture can be transported in this way and rapid deterioration of organic or ferrous materials can result. Figure 2.3 shows the phenomena involved in the condensation process. Figure 2.4 shows the patterns of heat, vapour and airflow through building elements and their relation to condensation risk.

Until recently there was little useful guidance on condensation from official or semi-official sources. The revision in 1989 of BS 5250 : 1989 *Control of condensation in buildings* has changed this situation. The standard is a very valuable reference and contains useful information on principles, design and calculation for walls, floors and roofs. A wide range of typical constructions are shown and desirable design features and condensation risk discussed.

2.5 Condensation: calculations

General

Since condensation as a major building problem is relatively recent, the extent and sophistication of prediction techniques and of data to use in them is limited. There is, however, a range of basic design procedures available to designers and these should certainly be employed in appropriate situations where there may be a risk of condensation.

The techniques are mainly based on steady-state conditions but they permit peak and mean rates of condensation to be predicted. For many building applications such calculations are entirely appropriate. Many roofs are of low thermal capacity and may be entirely effectively analysed on the steady-state basis. In many other cases where high thermal capacity materials are used it is the accumulation of moisture over a period of time which is critical rather than the peak rate of deposition and this problem cannot be adequately tackled by steady-state methods.

Dynamic condensation predictions – taking into account the varying internal and external temperatures and vapour pressures and thermal and vapour

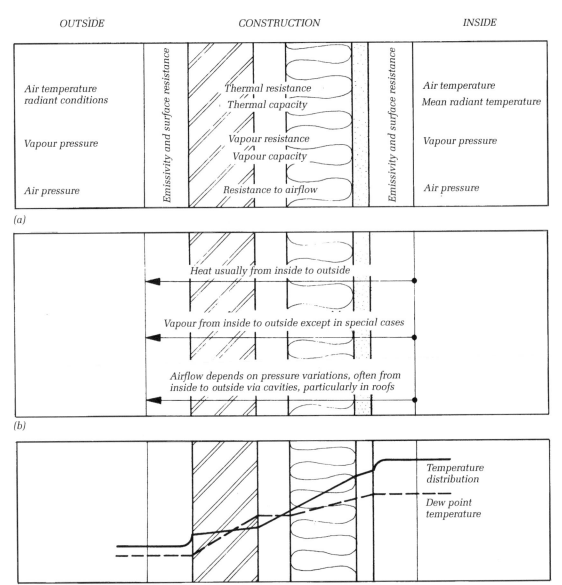

Figure 2.4   Patterns of heat, vapour and airflow in elements of construction and their relation to condensation risk: (a) basic conditions; (b) resulting flows; (c) resulting temperature and vapour pressure distribution (VP expressed as dew point temperature). Note that rain can affect the thermal properties of a construction.

resistances and capacities of materials – are difficult to carry out, even by large computers, and will only be employed in very critical circumstances. Figure 2.5 shows graphical output from a hybrid computer which takes into account all the variables described and predicts the rate and duration of condensation. Steady-state conditions are assumed for design calculations.

Condensation calculations involve:

- knowledge of relevant physical phenomena, terminology and units
- information on external conditions
- information on internal conditions
- data on properties of building materials
- knowledge of patterns of moisture movement
- standards to be achieved
- methods of analysis

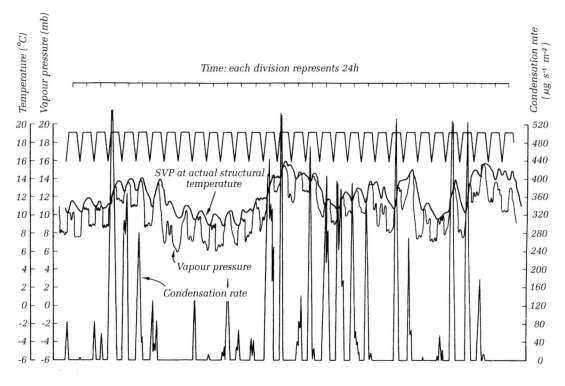

Figure 2.5 Graphical output from a computer analysis.

Physical phenomena, terminology and units

Moisture properties of air

Air can contain water vapour. The amount which can be contained varies with temperature. At 0 °C 1 kg of dry air can contain 3.7 g of moisture. At 20 °C/ this capacity increases to 14.4 g.

There are several ways in which the moisture content of air may be defined.

Mixing ratio (or moisture content or absolute humidity) The weight of water contained in a kilogram of dry air. (Units kg/kg or g/kg.) It enables amounts of water to be summed to give consequent overall mixing ratios.

Vapour pressure The pressure exerted by the molecules of vapour contained in air. (Units kPa.) Enables the rate of diffusion of vapour through materials of construction to be estimated. Millibars were originally used, 1 mb = 0.1 kPa.

Dew point temperature Temperature at which a sample of air with given moisture content becomes saturated. (Units °C.) Used in calculations.

Saturation vapour pressure Vapour pressure which would be given by saturated air at a specific temperature. (Units kPa.) Used in calculations.

Relative humidity Quantity of water contained in air expressed as a percentage of the maximum which could be contained in air at that temperature. (Units per cent RH at temperature in °C.) Vapour pressure is the basis of relative humidity values:

$$RH (\%) = \frac{\text{actual vapour pressure}}{\text{saturation vapour pressure at given air temperature}} \times 100$$

Relative humidity varies both with moisture content and with temperature, which means it is not a convenient quantity for use in calculations. It is, however, important in moisture studies since physiological reactions of people and the moisture content of building materials are governed by relative humidity, not by any of the other factors related to absolute humidity. It is also the most easily measured of the moisture phenomena described above and much of the meteorological and other data used for condensation estimation uses RH as the measure of moisture quantity.

Percentage saturation   Similar, and for most practical purposes, identical with relative humidity. (Units percentage saturation at temperature in °C.) Based on mass of moisture in atmosphere rather than vapour pressure.

$$\text{Saturation (\%)} = \frac{\text{mass of moisture in air}}{\substack{\text{mass of moisture required to} \\ \text{saturate air at given} \\ \text{temperature}}} \times 100$$

Used particularly for air-conditioning calculations where quantities of moisture are to be input or removed from air.

The psychrometric chart (Fig. 2.6) relates the various moisture phenomena described and enables dew points and saturation vapour pressures to be determined. The values are affected by changes in barometric pressure but condensation predictions are not sufficiently precise to require these variations to be taken into account and one psychrometric chart meets all requirements.

Use of psychrometric chart   The chart (Fig. 2.6) expresses the relationships between the quantity of water in the atmosphere (expressed as a mixing ratio), the vapour pressure, the relative humidity (and dew point) and the wet and dry bulb temperature.

With normal meteorological or environmental data on dry bulb temperature and relative humidity, the appropriate vapour pressure can be determined for condensation calculations (e.g. 20 °C 60% RH gives 1.4 kPa vapour pressure).

The variation of relative humidity with temperature for air with a given quantity of water can be established. (Consider in the case of the above a reduction in air temperature to 15 °C, while the vapour pressure and mixing ratio remain constant. As a result of the temperature change the relative humidity changes to 83%.)

Dew point temperature may be predicted, e.g. taking the same case, 100% RH (dew point) will occur at 12.5 °C dry bulb.

The effect of adding moisture to the air may also be established by adding the vapour introduced to the mixing ratio. Consider external air at 0 °C 90% RH warmed on entering a dwelling at 20 °C and with 3.4 g of moisture per kilogram of dry air added. The mixing ratio of 0 °C 90% RH is 3.4 g/kg. Adding 3.4 g/kg produces a total mixing ratio of 6.8 g/kg. This corresponds to an RH of 46% at 20 °C and a dew point of 8.5 °C.

The actual moisture content of air in buildings depends on the original content externally and the additional moisture introduced by occupants and processes. Meteorological data can be used to establish a design value for external conditions (in England very moist conditions often exist). This information is usually in terms of wet and dry bulb temperatures, or dry bulb temperature and relative humidity. In either case a point representing the conditions can be found on Fig. 2.6 and the quantity of moisture in the atmosphere per kilogram of dry air read off from the right-hand scale.

Latent heat   If air is cooled below the temperature at which it is saturated with moisture, the excess moisture will be precipitated. In building applications the air is normally cooled by contact with cold building materials and the excess moisture is then deposited as water on the surface or in the interstices of the material.

In changing from water to vapour a considerable quantity of heat is absorbed and the same quantity of heat is given up when vapour changes to water. To vaporize 1 g of water, at whatever temperature, requires 2.3 kJ of heat. This is several times as much heat as would be required to raise the water from freezing point to 100 °C. (The specific heat capacity of water is 4.2 J/g K.) The heat required to vaporize liquid is normally supplied from sources not forming part of the building. Cooking, washing and sweating are examples. The heat given up in condensing will be contributed to building materials. It will raise the temperature of the materials.

In a material of thermal capacity equivalent to wood wool, the condensation of 1 g of moisture over an area of 1 m² could increase the temperature of a layer some 4 mm thick by 1 °C. This assumes, however, that there is no heat loss. Over a period of 1 h some 43 J pass through 1 m² of walling with a normal temperature difference of 20 °C and a *U* value of 0.6. Thus abnormally high rates of condensation of over 10 g/m² h would be required for the latent heat of condensation to become a significant factor in this example.

Moisture properties of building materials

Building materials, in addition to their thermal properties, have analogous moisture properties. Vapour will pass through materials at a rate depending on the resistance of the material and the vapour pressure difference between the two sides of the material. As with heat flow and temperature difference, the rate of vapour flow through a given sample of material is directly proportional to the

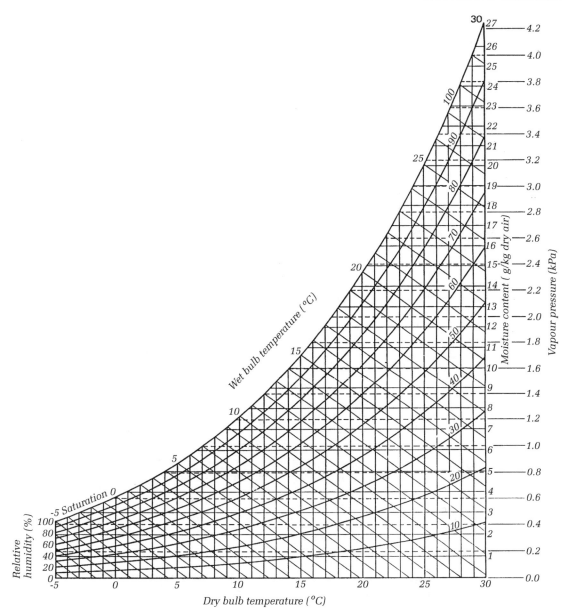

Figure 2.6 A psychrometric chart plots temperature, relative humidity and dew point. (Reproduced courtesy of the British Standards Institution).

vapour pressure difference. Thermal effects can influence vapour flow but these are ignored in condensation calculations, as are the effects of surface resistance to vapour flow. The significant vapour properties are as follows.

Vapour diffusivity Rate of vapour transfer through a material resulting from a vapour pressure difference.

(Units g m/MN s.) Note that this represents the rate of transfer in grams per second through 1 m^2 of material 1 m thick under a vapour pressure difference of 1 meganewton.

Vapour resistivity The reciprocal of diffusivity and, in effect, the resistance of a 1 m^2 sample 1 m thick. (Units MN s/g m.)

Vapour resistance   The resistance of a given thickness of material to vapour flow. (Units MN s/g.)

Resistivity × thickness (m) = resistance

The resistance of composite roofs, etc., is expressed in these terms.

Moisture content   A building material can contain moisture absorbed into its volume and adsorbed into thin films on internal surfaces of pores. Apart from contributions by condensation, the quantity contained depends more upon the surrounding relative humidity than upon vapour pressure. The main effect of moisture content is to influence the thermal properties. (Units per cent by weight.) Section A3 of the *CIBSE Guide* deals with the effects of moisture on thermal properties.

Vapour capacity   Only used in very sophisticated calculations. Complex phenomena are involved which are difficult to quantify. Usually taken as the vapour capacity of the air spaces in the material. (Units g/kg or g/m^3.)

Properties of materials

Appendix 1 gives the vapour resistivities of typical building materials. Two factors should be borne in mind when using the tables.

- The information about vapour resistance is limited and less precise than that for thermal properties. Condensation is a problem of relatively recent significance and prediction techniques which could use accurate resistance data are very recent indeed. In addition the measurement of vapour resistance is difficult and laborious. The traditional method is shown in Fig. 2.7. It involved weighing the moisture passing through a sample under the influence of high and low moisture levels on opposite sides. A single determination could take several weeks. Figure 2.8 shows a more elaborate testing apparatus which gives more rapid results.
- When considering condensation it is important to bear in mind that, unlike thermal performance where an average result is acceptable, the lowest values of thermal and vapour resistance should be used as the design standard. If this is not done, some parts of the construction may give rise to condensation, even though the average level is satisfactory. Cold areas will be unacceptable in terms of condensation, even if the average performance is adequate.

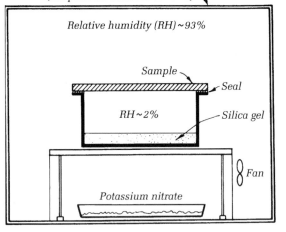

Figure 2.7   Weighing method of measuring vapour resistance.

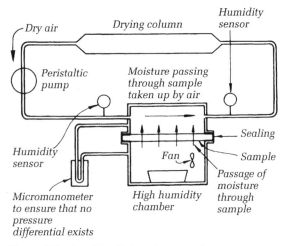

Figure 2.8   A rapid technique for measuring vapour resistance.

External design conditions

Standard

External conditions must be known in order to determine internal conditions and to estimate condensation risk. In the United Kingdom, it has become conventional to use mean conditions over a 60 day period. For such calculations BS 5250 : 1989 recommends an external air temperature of 5 °C and an external vapour pressure of 0.83 kPa (95% RH). Table 2.1 enables comparison with means at Heathrow over various periods.

Table 2.1 Mean temperature at Heathrow

Period	Air temperature (°C)	Vapour pressure (kPa)
30 day mean (Jan)	3.4	0.65
60 day mean (Jan, Feb)	4.2	0.67
210 day mean (winter)	8.0	0.82

These, or similar local data might be used for analysis of mean conditions in circumstances where some transitory excess is acceptable. No allowance is made for radiation to the night sky since this is more than compensated by daytime solar gain.

Short cold periods

It is not unusual for mean temperatures well below 5 °C to exist for short periods and it is very desirable to check whether this could give rise to problems, particularly if condensation takes place in parts of the construction where it will be retained in the form of ice until a general thaw takes place. A range of temperatures and durations should be considered, to explore the sensitivity of the construction, and weighed against the consequences of condensation taking place.

Lightweight construction

In lightweight construction, and particularly where sheeted roofs are involved, significant deposits of moisture may occur within very short periods, even overnight. In such cases it is important to take into account not only the air temperature but also the effect of heat loss by radiation. In cold weather when the sky is clear, radiation from the earth's surfaces to the sky will reduce external surface temperatures. The condensation consequences can readily be observed by the deposition of moisture from the air either in the form of dew or hoarfrost. In the case of thin sheets, the internal surface will be effectively at the same temperature as the outer surface. To take account of this phenomenon it is usual to assume that the surfaces of roofs can be cooled to 5 °C below air temperature. For walls the temperature depression is taken at 3 °C.

Ventilated cavities

Where sheeted roofs have ventilation from the external air it will be appreciated that radiant heat losses described above can reduce the surface temperatures of both the external and the internal surfaces of the external cladding to below the dew point temperature of the surrounding external air. Condensation can occur on external surfaces in a similar way to dew or hoarfrost on grass, bushes or trees. On the external surface this is not likely to be of any significance. However, if there is a void under the sheeting which is ventilated to the external air then condensation may form, and even overnight, may give rise to sufficient deposition to cause drips which may stain ceilings. In sustained very cold weather ice may build up on the underside and release substantial quantities of water when the thaw takes place. In situations where this may happen, precautions should be taken either to avoid the deposition of moisture or to ensure that drips from the cladding cannot cause damage.

Internal conditions

The internal conditions directly affecting condensation are internal temperature, usually taken as the air temperature, and the vapour pressure. The vapour pressure will be determined by the balance between vapour inputs and ventilation rate.

The Building Research Establishment has identified three general types of building use and recommended appropriate levels of increased internal moisture levels. They are shown in Table 2.2.

Table 2.2 Moisture levels recommended for design by BRE

Building types	Moisture level increase		Design internal moisture levels based on 0 °C and 90% RH external conditions
	Mixing ratio (g/kg)	Vapour pressure (kPa)	
Shops, offices, club rooms, public buildings, dry industry process	1.7	0.25	7.8
Dwellings[a]	3.4	0.55	10.8
Catering establishments or wet industrial processes	6.8	1.07	16.0

[a] This recommendation appears to have been adopted by BS 5250.

Table 2.3 Belgian recommended internal vapour pressures

Building types	Mean annual vapour pressure (kPa)
Garages, warehouses	1.1–1.2
Offices, shops, schools, public buildings	1.2–1.4
Dwellings, hospitals	1.4–1.5
Humid or badly ventilated buildings, e.g. lavatories, swimming-pools, kitchens	>1.5

It is interesting to compare these categories and standards with those in use in Belgium, shown in Table 2.3.

The Belgian figures are significantly higher than those recommended by BRE. It seems that the explanation of the difference is more likely to be in different ventilation rates than any other factor.

Figure 2.9 shows the variation of internal and external moisture levels for a dwelling in the United Kingdom. Note how they confirm the BRE figures.

In spite of legislation prescribing a maximum of 19 °C, it is usual to take 20 °C as the internal temperature design level for peak conditions, except where it is known that higher or lower temperatures will be the case. For calculations over a period of time, lower temperatures will be used to take account of night setbacks in temperatures and intermittency of use.

For dwellings, BRE have established that mean internal temperatures may vary from 13.5 °C for a poorly insulated dwelling with no central heating to 19.5 °C for a fully centrally heated, well-insulated dwelling. Commercial, industrial and educational buildings tend to have lower mean temperatures than dwellings because of the relatively short daily heating period and lack of weekend heating. Estimates will have to be made for each particular case.

In some cases the general design values described above may be considered inappropriate and it will be desirable to take into account the special circumstances of a particular case either because of different vapour inputs or ventilation rates. Tables 2.4 and 2.5 give typical moisture inputs for domestic activities. BS 5250 gives a combined total of 14.4 kg as the daily amount of moisture input into a typical five-person household.

Where moisture input can be viewed as a steady rate of input, the resulting internal mixing ratio can be simply determined from the following formula (ac = air changes):

$$\text{Internal mixing ratio (g/kg)} = \frac{\text{moisture input rate (g/h)}}{\text{ventilation rate (ac/h)} \times \text{volume (m}^3)} + \text{external mixing ratio (g/kg)}$$

Figure 2.10 is based on this formula; it shows how the increase (over outside content) in the moisture content of air in a room varies with varying input rates and rates of ventilation.

Another useful formula gives the ventilation rate required to achieve a desired internal mixing ratio.

$$\text{Ventilation rate (m}^3\text{/h)} = \frac{\text{moisture input (g/h)}}{\text{desired internal mixing ratio (g/kg)}} - \text{external mixing ratio (g/kg)}$$

Wetted surfaces or water areas, particularly in industrial buildings and swimming-pools, make a major contribution to moisture in the air.

All the considerations of internal moisture levels so far described relate to steady-state conditions with regular moisture input and ventilation rates. There will undoubtedly be circumstances when sudden moisture inputs or changing ventilation rates will need to be considered either from the viewpoint of condensation risk or to limit relative humidity to the 70% limit for comfort.

Varying moisture inputs

In some cases, such as lecture rooms or theatres, where the moisture load varies with the pattern of occupancy, it may be desirable to make an estimate of the peak moisture levels. Apart from condensation risks, moisture levels in rooms of this type may well rise above the 70%, regarded as the maximum for comfort. Information about moisture levels is therefore needed

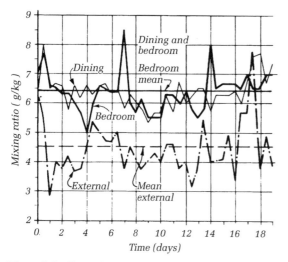

Figure 2.9   External and internal mixing ratios for a dwelling.

Table 2.4   Approximate moisture input from a variety of sources

Source	Approximate quantity
Wetted surfaces[a]	Evaporation rate (g/h) = 100 × area of wetted surface (m²) × (SVP at water temperature − vapour pressure of surrounding air)
Occupants	
At rest	40 g/h
Moderate activity	50 g/h
Very active	200 g/h
Typical domestic daily activities[b]	
Cooking Electricity	2 kg/day
Gas	3 kg/day
Dishwashing	0.4 kg/day
Bathing and showers	0.3 kg/day per person
Clothes washing	0.5 kg
Clothes drying, internal or unvented drying	1.5 kg/day per person
Flueless heaters	
Oil	100 g/KWh
Natural gas	150 g/KWh
Propane and butane	125 g/KWh

[a] Suggested by Neil Millbank of the BRE.
[b] An approximation for a complete average household may be made on the basis of 6 kg plus 1 kg per occupant for daily input. Table 3.3 gives typical ventilation rates.

Table 2.5   Clothes drying: moisture contributed to the atmosphere

Item	Dry weight (g)	Wet (but after spin drying) (g)	Moisture given out in drying (g)
Medium towel	240	450	210
Bath towel	550	1080	530
Sheet 1: synthetic, double size	910	1260	350
Sheet 2: single, cotton	1010	1950	940
Shirt, cotton	280	450	370

to aid decisions about ventilation rates as well as condensation precautions.

A simple graphical method may be used to give an answer sufficiently accurate for design purposes. Figure 2.11 shows this method. The period being investigated is divided into convenient time units, in this case half-hours. In each half-hour period the total weight of water present and input into the room is found and divided by the total weight of air present and input. The result is the mean mixing ratio, from which

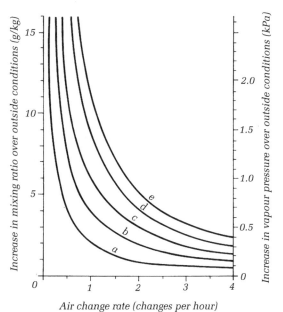

Figure 2.10   Moisture equilibrium for room of volume 20 m³ with different rates of moisture input and ventilation: (a) moisture input from 1 occupant = 50 g/h; (b) from 2 occupants = 100 g/h; (c) from 3 occupants = 150 g/h; (d) from 4 occupants = 200 g/h; (e) from 5 occupants = 250 g/h.

the vapour pressure, or the relative humidity at any given temperature can be established.

The example given is a classroom 50 m² in floor area, 3.5 m high. The internal temperature is to be 20 °C. For daytime occupancy in term time an external temperature of 8 °C with 100% relative humidity presents a possible selection for worst conditions. For an occupancy of 40 pupils, a room of this volume should have a ventilation rate of 1120 m³/h or $6\frac{1}{2}$ changes per hour (see Chapter 3). But in a building of this type, with window-controlled natural ventilation, it is unlikely this rate would be achieved during the winter, particularly if blackout were used. Three air changes might be a reasonable assumption. Occupancy is scheduled to be between 0930 and 1030, 1100 and 1230, 1400 and 1630. Data required for the calculation is as follows:

- Volume of classroom 175 m³, weight of air 222 kg.
- Volume of three air changes 525 m³, weight of three air changes 666 kg.
- Moisture input from occupants (4) at 50 g/h = 2 kg/h.
- Mixing ratio for 100% RH at 8 °C = 6.5 g/kg, weight of water in 175 m³ of air = 1.44 kg.

The calculation is made in the rows *a–e* in the top part of Fig. 2.11 and plotted graphically.

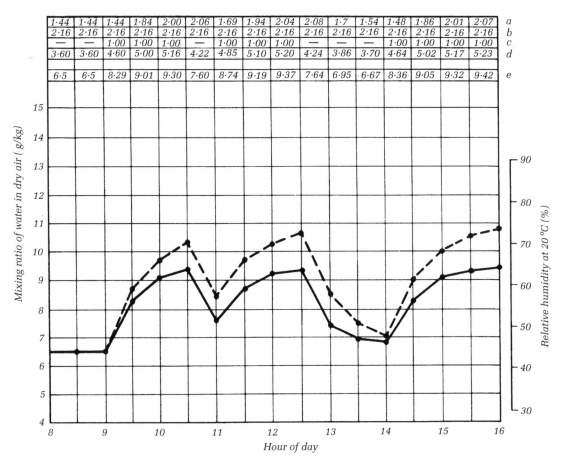

1·44	1·44	1·44	1·84	2·00	2·06	1·69	1·94	2·04	2·08	1·7	1·54	1·48	1·86	2·01	2·07	a
2·16	2·16	2·16	2·16	2·16	2·16	2·16	2·16	2·16	2·16	2·16	2·16	2·16	2·16	2·16	2·16	b
—	—	1·00	1·00	1·00	—	1·00	1·00	1·00	—	—	—	1·00	1·00	1·00	1·00	c
3·60	3·60	4·60	5·00	5·16	4·22	4·85	5·10	5·20	4·24	3·86	3·70	4·64	5·02	5·17	5·23	d
6·5	6·5	8·29	9·01	9·30	7·60	8·74	9·19	9·37	7·64	6·95	6·67	8·36	9·05	9·32	9·42	e

Figure 2.11 Graphical prediction for variation of moisture level with occupation.

1 **Row a**: Enter weight of water in the room air (mixing ratio × weight of air in room). (Initially this will be based on the same mixing ratio as the external air. For subsequent calculations it will be the mixing ratio resulting from the last half-hour multiplied by the weight of air in room.)

2 **Row b**: Enter weight of water introduced by ventilation (external mixing ratio × weight of air introduced).

3 **Row c**: Enter weight of water introduced by occupants.

4 **Row d**: Sum total weight of water present and input during half-hour period.

5 **Row e**: Enter resulting mean mixing ratio calculated by dividing weight of water from row d by the weight of air present and input during half-hour period.

6 **Row a**: Using mixing ratio from step 5 multiplied by weight of air in room, enter weight of water in the room air.

7 etc.   Repeat for each time period.

The effects of two and three air changes are shown in the diagram.

As in most other aspects of environmental design, the wishes and behaviour of occupants vary considerably. The internal standards described above envisage normal use with some margin of safety. In the non-domestic field extreme occupancy loads arise from particular uses such as wet industrial processes, hydrotherapy pools and cold stones. In these cases the conditions to be overcome will be known and can be taken into account in design. In dwellings, however, the occupants will not normally be known and standard conditions must be used. Condensation is likely if particular occupants reduce ventilation rates to low levels or indulge in activities which input large amounts of moisture to the interior. It is impracticable to design every dwelling to cater for rare extreme cases. It is desirable, however, to make clear what consideration standards have been used as part of the normal presentation of the design.

Patterns of moisture movement

Airborne movement within rooms

Moisture diffuses relatively slowly through still air. The main mechanisms of movement in the air in rooms will be buoyancy and turbulence.

Moist air is less dense than dry air and consequently has a tendency to rise. Upper corners of rooms, which are often colder than lower areas in dwellings, are therefore exposed particularly to rising currents of moist air from below. The moisture-induced buoyancy will be reinforced by thermal buoyancy effects.

The air in most rooms is in general turbulent motion as a result of air infiltration and air movements due to ventilation. Whether this effect will be more marked than buoyancy and ensure a more uniform vapour distribution depends on the particular case. The degree of movement resulting from buoyancy or turbulence will govern the amount of moisture which can be deposited on exposed inner surfaces.

Airborne movement between rooms

Stack effects are important elements in air movements in buildings and during winter substantial volumes of air will rise even in two-storey dwellings. This encourages concentration of moisture and is particularly critical in dwellings where most of the moisture input is on the ground floor while the first floor may be less well heated. The problem is not confined to habitable rooms. Air can pass into roof spaces through cracks, openings for services and hatches and ducts. It is typical for nearly half the air lost from dwellings as ventilation to make its way into the roof space. Condensation can result, particularly where the ceiling is insulated and the temperature in the roof space particularly low.

Diffusion through materials

Where a vapour pressure difference exists across a layer of building material molecules of moisture will make their way through the interstices and there will be a transfer of moisture from the higher to the lower vapour pressure zone.

The rate of flow of vapour can be determined from the following equation:

$$\text{Rate of vapour flow (g/m}^2\,\text{s)} = \frac{\text{difference in vapour pressure (kPa)}}{\text{vapour resistance of material (MN/s g)}} \times 10,000$$

Diffusion and air movement

Many practical constructions of external walls and roofs involve several layers of different materials and have cavities between some of the layers. In these cases moisture may make its way to the cavity, not only by diffusion through the materials themselves, but also as a result of air pressure differences which produce currents of air and its contained moisture. In roof spaces, particularly, the amounts of moisture making their way into cavities by this means are likely to be much greater than the amounts that penetrate by diffusion. Figure 2.12 shows graphically the greater moisture levels which may occur in cavities as a result of air currents conveying vapour from interior to cavity.

The transport of moisture from the interior into cavities in construction is often the critical factor in condensation, and ventilation of cavities from the exterior, particularly roof spaces, is an important design feature in preventing condensation. It is very desirable therefore to be able to estimate the effects of cavity ventilation on condensation risk. To be able to do this it is necessary to estimate the temperature and vapour pressures which will prevail in the cavity. Few designs attempt this calculation, although it is a critical factor and must be taken into account when designing to avoid condensation.

Standards

It would be unnecessary, and very expensive to attempt to design buildings generally to avoid condensation in all circumstances. It is perfectly acceptable for some materials to have an increase in moisture content for a limited period, in other cases it may be undesirable to have any deposition of moisture at all.

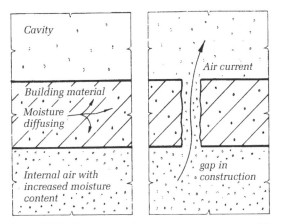

Figure 2.12 Comparison of moisture movements caused by diffusion through materials and air movement through gaps.

BS 5250:1989 identifies several different cases:

- Nuisance condensation: surface condensation that is not harmful.
- Interstitial condensation that is not harmful.
- Harmful condensation: interstitial or surface condensation that could cause mould or degradation of materials, impair thermal performance or be detrimental to health.

The following sections identify quantitative standards and estimation techniques for the control of harmful condensation.

Internal surfaces

Glazing Where single glazing is used, it is very difficult and would impose unacceptable limits on environmental conditions if it were attempted to avoid condensation at all times. Provided adequate arrangements are made to drain water from sills without involving deterioration of materials, condensation on the glass may be accepted at peak times. Condensation on window frames themselves is difficult to contain and should be avoided by selection of materials and designs that ensure the inner surfaces do not fall below dew point temperature.

Solid walls The critical element for solid walls is not visible quantities of water deposited on or running down the surface. Except in the case of glazed or impervious surfaces, which do not encourage such growth and are easily cleaned, the essential condensation problem for walls is the growth of black mould. This growth can take place as a result of imperceptible quantities of condensation or even in conditions of high humidity. A standard which has been used for some time is to ensure that room air will have a monthly mean relative humidity not greater than 70%. On the other hand, normal variations of external and internal conditions are very likely to ensure the transitory deposition of water in cases where the relative humidity is in the vicinity of 70% and this standard provides some element of safety factor. It is important to remember that surface temperatures near corners and at lintels and other cold bridges may be several degrees Celsius below the normally calculated surface temperature, which assumes a slab extending indefinitely in all directions. Figure 2.13 shows how internal surface temperatures can vary in an actual building.

Surface condensation

If the internal surface temperatures and the internal vapour pressures are known, the psychrometric chart (Fig. 2.6) can be used to determine whether

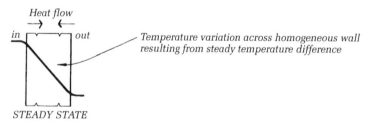

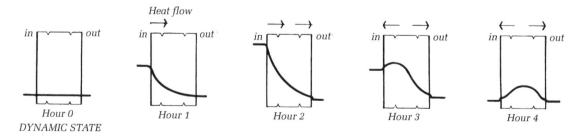

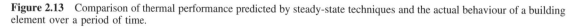

Figure 2.13 Comparison of thermal performance predicted by steady-state techniques and the actual behaviour of a building element over a period of time.

condensation is taking place. The risk of condensation increases at corners or cold bridges. Diagrams show typical temperature distributions. It is possible to make an estimate of particular corner conditions by taking the internal surface resistance at three times its normal value and the external surface resistance at zero. Some approximate methods for estimating the effects of cold bridges are given in Chapter 5. Recent studies have demonstrated that mould growth can occur without actual condensation if very high relative humidities are sustained. The current UK standard is to keep the internal relative humidity below 70% in well-insulated buildings. This assumes that air in contact with the internal surfaces will be cooler than the main body of air in the room and that the relative humidity at the surface will be higher than 70%. Some European authorities identify a monthly mean RH of 80% at the surface as the design standard.

BS 5250 : 1989 suggests a method for assessing surface condensation risk either in individual rooms or in whole buildings. The method involves establishing the average internal temperature and the average internal vapour pressure over the period being considered. The average internal relative humidity can then be established from the psychrometric chart (Fig. 2.6). In many cases, both of proposed and existing buildings, the temperatures will be known. If this is not the case, and information on energy consumption is available, the formula for average internal temperature can be used (see below). Average internal vapour pressures will rarely be known. They can be estimated as follows:

Average internal temperature (°C) =
    average external temperature (°C) +

$$\frac{\text{rate of heat input (W)}}{[\text{sum of areas} \times U\text{s}] + [0.33 \times \text{ac/h} \times \text{vol (m}^3)]}$$

Average internal VP = average external VP +

$$\frac{\text{moisture input (kg/day)}}{0.191 \times \text{ ac/h} \times \text{vol (m}^3)}$$

Typical values appropriate for use in the above formulae are:

Average external air temperature   5 °C
Average external vapour pressure   0.83 kPa
U values   Calculate or take from Appendix 2
Number of air changes   Fig. 5.22
Moisture input   Table 2.6
Average rate of heat input   As described above

Table 2.6 Moisture generation rates for dwellings

Dwelling type	Moisture generation
Normal well-ventilated occupancy	3.5 kg/day for first occupant 0.5 kg/day for each additional occupant
Limited ventilation and high moisture input	2 × above values
Inadequate ventilation and very high moisture input	2.7 × above values

Within building elements

On impervious surfaces This problem is typified by sheeted roofs with profiled metal sheet covering. Condensation forming on the inner surface of the metal sheeting will not be absorbed. If the sheeting is of aluminium or steel adequately protected against deterioration, the transitory deposition of moisture may not be regarded as serious. However, the impervious surface will only be able to hold a limited amount of water; drops will form when this threshold is exceeded, and depending on the slope and configuration of the sheeting, they will either fall or run down the slope. Impervious, smooth, flat surfaces may well contain over 10 g/m^2 of moisture before any drips occur. Even if the sheeting slopes at quite shallow angles, drops may form and run down until they reach a supporting member. At this point concentrated dripping can occur. If the slope is a long one, the quantities dripping can be considerable, even if only small amounts are deposited per square metre. It is very difficult to assess a specific amount of deposition at which trouble can be anticipated since surfaces vary greatly in the amount of moisture they can contain before runs or dripping occur. It is important to consider this aspect of performance in all cases where condensation can be deposited on the impervious sheets.

It is interesting to note that asbestos cement sheets, once very widely used for sheeted roofs, were able to contain and absorb condensation deposited overnight or at periods of high internal moisture concentration and allow it to evaporate later in the diurnal cycle whereas the same conditions with metal sheeting could present condensation problems because of reduced absorption and greater liability for drops to run and coagulate.

Analysis of condensation risk should be based on peak values, and accumulation of moisture can be allowed up to the point where dripping would be anticipated.

Organic materials, particularly timber Organic materials are particularly affected by moisture.

Excessive moisture contents can reduce strength, induce decay and encourage rot. The conditions inside elements of construction are suitable for dry rot if high moisture levels are sustained. Impregnated timber or boards manufactured from timber or other organic materials may in some cases have increased resistance to moisture-induced decay but their properties should be checked before any relaxation of condensation standards is allowed. Sustained relative humidities of over 85% can produce moisture levels in timber over the 20% threshold for development of rot. Moisture does not have to be uniformly distributed in timber to cause decay. Water trapped between a timber face and impervious sheet materials can be the cause of trouble.

In the case of organic materials it is desirable to design to avoid any condensation risk. Such a standard is also likely to prevent sustained relative humidities in cavities or interstices of over 85%.

Analysis of condensation risk should be based on peak values. The desirable standard is to avoid condensation risk.

Porous and fibrous, non-organic materials Deposition of moisture in materials of this sort not subject to moisture-induced deterioration is acceptable up to limits which will ensure that adjacent materials (e.g. plaster) are not affected and prevent significant increases in thermal conductivity which could give progressively worse condensation conditions. Although there are no recognized design standards, it is sensible to ensure that the overall thermal resistance of the building element containing the material subject to condensation is not reduced by more than 5%. Figure 2.14 enables the effect of moisture content on the conductivity of masonry materials to be established.

Condensation risk analysis

Two separate problems are involved. One is the estimation of surface condensation risk and the second is the estimation of interstitial condensation.

Surface condensation

Internal and external conditions must be established and the temperature of the inner surface of the wall calculated. If the steps involved in columns *a–f* of Table 2.7 are completed, the surface temperature between the innermost layer and the internal surface resistance will be established. The psychrometric chart (Fig. 2.6) can be used to establish whether condensation will be present. A point representing the internal air temperature and relative humidity is

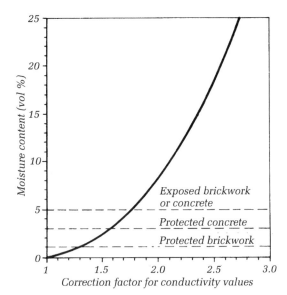

Figure 2.14 Effect of moisture content upon thermal conductance of masonry materials.

selected on the chart then projected horizontally to the left to cut the vertical line representing the internal surface temperature. If this point falls outside the diagram, condensation will take place.

The risk of condensation increases at corners or cold bridges. Figure 5.11 shows typical temperature distributions. It is possible to estimate corner conditions in the way described above by the internal surface resistance at three times its normal value and the external surface resistance at zero. Section 5.7 provides details of the prediction method for cold bridges.

Interstitial condensation

In the past a very simple method was used for the prediction of interstitial condensation. It consisted of comparing the structural and dew point temperatures at each interface in a wall or roof. A risk of condensation was indicated where the structural temperature fell below the dew point temperature. The method was very simple to carry out manually and did give some indication of where condensation might take place. However, it did not give any indication of the amount of condensation, and the method used for estimating the dew point temperatures was inaccurate.

In many constructions it is not possible to eliminate condensation totally and it is desirable to be able to estimate and control the amount of moisture which can accumulate over critical periods. BS 6229 : 1982 suggested maximum limits for accumulation and BS 2560 : 1989, now the definitive standard, gives

Table 2.7 Condensation risk: example calculation

a	b	c	d	e	f	g	h	i	j	k	l	m	n	o	p	q
Component	\Thermal\						\Moisture\									
	Thickness (m)	Resistivity ($m\,K\,W^{-1}$)	Resistance ($m^2\,K\,W^{-1}$)	Temperature drop (°C)[a]	Interface temperature (°C)[a]	Saturated vapour pressure (kPa)	Resistivity (MN s/g m)	Resistance (MN s/g)	Nominal vapour pressure drop (kPa)	Interface nominal vapour pressure (kPa)[a]	Fixed vapour pressures (kPa)[a]	Vapour resistance between FVPs (MN s/g)	Vapour pressure drops (kPa)	Actual vapour pressure (kPa)[a]	Vapour flow (g/h m²)	Moisture deposition (g/m² per 60 days)[a,b]
Exterior																
					5.0	0.87				0.78	0.78			0.78		
Ext. surface	–	–	0.06	0.3			–	0	–			–	–		–	
					5.3	0.88				–	–			0.78		
Brickwork	0.114	1.1	0.13	0.5			40	4.6	0.16						0.11	
					5.8	0.92				0.94	0.92			0.92		55
Cavity	0.05	–	0.18	0.7			5	0.3	0.01			0.3	0.01		0.15	
					6.5	0.97				0.95	–			0.93		0
Insulation	0.05	30.0	1.50	6.4			8	0.4	0.01			0.4	0.02		0.15	
					12.9	1.48				0.96	–			0.95		0
Concrete block	0.1	3.5	0.35	1.5			30	3.0	0.11			3.0	0.12		0.15	
					14.4	1.64				1.07	–			1.07		0
Plaster	0.013	2.0	0.03	0.1			60	0.8	0.03			0.8	0.04		0.15	
					14.5	1.65				1.1	1.1			1.1		–
Int. surface	–	–	0.12	0.5			–	0	–			–	–		–	
					15.0	1.70				1.1	1.1			1.1		
Interior																
Total	0.327		2.36	10.0				9.1	0.32							55

[a] Values at interfaces.
[b] Note that 55 g of moisture deposition over a 60 day period on the outer brick leaf of the cavity wall is not a significant condensation risk.

1 Establish the inside and outside temperatures and vapour pressures and calculate the overall differences

	Temperature (°C)	RH	Vapour pressure (kPa)
External	5	95	0.83
Internal	15	65	1.1
Differences	10		0.27

2 Enter the names and thicknesses of all layers.   *a,b*
3 Enter the thermal and vapour resistivities for any solid materials.   *c,h*
4 Enter the thermal and vapour resistances for any cavities, vapour barriers and surfaces.   *d,i*
5 Calculate the thermal resistances for any solid layers: Thickness × resistivity = resistances [$b \times c = d$]   *d*
6 Total the thermal resistances.   *d*
7 Calculate the temperature drop across each layer:

$$\text{Difference} = \frac{\text{resist of layer}}{\text{total resistance}} \times \text{overall temperature difference} \left[\frac{d \times \text{temp. diff.}}{\text{total } d}\right]$$ *e*

8 Sum the temperature drops and check that the total is equal to the temperature difference between inside and outside.   *e*
9 Establish the temperatures at any interfaces by cumulative totals of the temperature differences.   *f*
10 Using the psychrometric chart (Fig. 2.6) calculate the interface saturated vapour pressures corresponding to the interface temperatures.   *g*
11 Enter the vapour resistivities for each layer.   *h*
12 Calculate the vapour resistances for solid layers: Thickness × resistivity = resistance [$b \times h = i$]   *i*

13 Total the vapour resistances.   *i*
14 Calculate the nominal vapour pressure difference for each layer.   *j*

$$\text{Difference} = \frac{\text{resist of layer}}{\text{total resistance}} \times \text{overall VP difference} \left[\frac{i \times \text{VP diff.}}{\text{total } i}\right]$$

15 Establish the nominal vapour pressures at any interfaces by cumulative totals of the vapour pressure differences.   *k*
16 Calculate the fixed vapour pressures, as described above and shown in Figs 2.15 and 2.16.   *l*
17 Enter layer vapour resistances and calculate total vapour resistance between fixed vapour pressure points.   *m*
18 Calculate actual vapour pressure drops between fixed vapour pressures for each separate layer: *n*

$$\frac{\text{Layer vapour resistance} \times \text{VP diff. between fixed points}}{\text{Total vapour resistance between fixed points}} \left[\frac{i \times l}{m}\right]$$

19 Establish the actual vapour pressures at any interfaces by cumulative totals of the actual vapour pressure drops.   *o*
20 Calculate the vapour flows for each layer: *p*

$$\frac{\text{Actual vapour pressure differences across layer}}{\text{Vapour resistance of layer}} \times 0.36 \left[\frac{n \times 0.36}{i}\right]$$

Where the vapour resistance of a layer is zero, the flow will be the same as for the adjacent inner layer.
21 Calculate the 60 day vapour deposition at each interface:

(Flow through layer − flow through next outer layer) × 1440   *q*

more detailed standards for accumulation and a more accurate prediction method. Some experts already use the principles adopted by BS 5250, but almost all current manual methods and the great majority of computer programs will have to be replaced.

The critical element in the calculation method is shown in Figs 2.15 and 2.16. Figure 2.15 shows the curves representing nominal and saturated vapour pressures which have been used conventionally for calculation. In reality, the actual vapour pressure cannot rise above the saturated vapour pressure. The value of these curves is therefore to identify fixed actual vapour pressure locations which can then be used to establish the actual vapour pressure distributions that would exist in reality.

The fixed points are established in two regions:

- *At external or internal surfaces*
  Fixed vapour pressure (FVP) is the lower of either saturated vapour pressure (SVP) or the actual internal or external vapour pressure.
- *At interfaces between layers*
  If the saturated vapour pressure (SVP) is lower than the nominal vapour pressure (NVP) then the SVP is the fixed value.

The vapour pressures at points between the fixed values must be recalculated using the fixed points as a basis.

Figure 2.16 shows a curve representing the actual vapour pressure distribution calculated as described. Table 2.7 gives a worksheet to assist in carrying out this calculation and shows a typical calculation for a wall. The steps required in the calculation are also given in the table.

This technique automatically provides an analysis of moisture accumulation. In the example given, the accumulation is for a period of 60 days. It is possible to determine the accumulation for other periods by taking appropriate mean values for external and internal conditions.

In the context of design, where many possibilities must be explored, this type of analysis is best undertaken by computer. This makes it possible to include facilities such as the effect of ventilated cavities, which would not be practicable by manual methods.

Figures 2.17 to 2.22 use tables and charts to show how the condensation risk in a roof can be influenced by vapour barriers, external insulation and cavity ventilation. In steady-state analysis, condensation will invariably take place at interfaces rather than over a zone and determination of the actual vapour pressure is

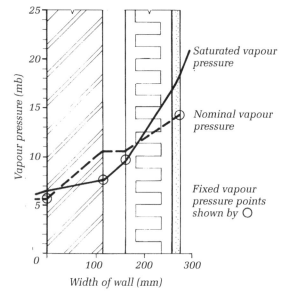

Figure 2.15 Nominal and saturated vapour pressures.

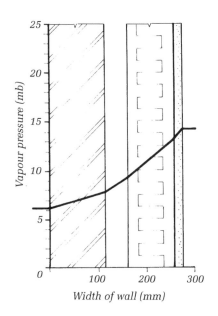

Figure 2.16 Actual vapour pressure distribution.

essential to the estimation of amounts of moisture deposited. Except for organic materials, which must be kept dry, it is the quantity of moisture which accumulates that is the critical element rather than the existence of a condensation risk. The quantities used in the diagrams are as follows:

THERMAL PROPERTIES AND *U* VALUE

Ref	1 Element	2 Thick (mm)	3 Resistivity (mK/W)	4 Resistance (m²K/W)
0	Ext. SR			0.04
1	Felt	6	2.00	0.01
2	Woodwool	50	10.00	0.50
3	Cavity	150	0.00	0.18
4	Insulation	100	30.00	3.00
5	Plasterboard	13	7.00	0.09
6	Int. SR			0.10
	Totals	319		3.92
	U value of construction		0.25	W m⁻² K

Ref: Roof

VAPOUR PROPERTIES

Ref	1 Element	2 Thick (mm)	3 Resistivity (MNs/gm)	4 Resistance (MNs/g)
1	Felt	6	10 000	60.0
2	Woodwool	50	15	0.8
3	Cavity	150	5	0.8
4	Insulation	100	7	0.7
5	Plasterboard	13	15	0.2
	Totals	319		62.4

Ref: Roof

BOUNDARY CONDITIONS

Internal		
Internal Air Temperature	15.0	C
Internal Vapour Pressure	1.0	kPa
Int. Dew Point Temp.	7.3	C
Int. Relative Humidity	60	%
External		
Ext. Air Temperature	5.0	C
Ext. Rad. Allowance	0.0	C
Ext. Vapour Pressure	0.8	kPa
Ext. Dew Point Temp.	4.1	C
Ext. Relative Humidity	94	%
Moisture Deposition		
Accumulation during	60	Days

Ref: Roof

Figure 2.17 Basic construction of the roof, its thermal and vapour properties, and the boundary conditions.

TEMPERATURE AND MOISTURE DISTRIBUTION

Face	Elements	Temp (ºC)	DPT (ºC)	RH (%)	V.P. (kPa)	Flow (g/hr)	Depos (g/60d)
	Ext. S.R.	5.0	4.1	94	0.8		
0/1	Ext. S./Felt	5.1	4.1	93	0.8	0.00	
1/2	Felt/Woodwo	5.1	5.1	100	0.9	0.12	167
2/3	Woodwo/Cavity	6.4	6.4	100	1.0	0.12	0
3/4	Cavity/Insula	6.9	6.9	100	1.0	0.12	0
4/5	Insula/Plaste	14.5	7.2	62	1.0	0.12	0
5/6	Plaste/Int.S.	14.7	7.3	61	1.0	0.12	
	Int. S.R.	15.0	7.3	60	1.0		

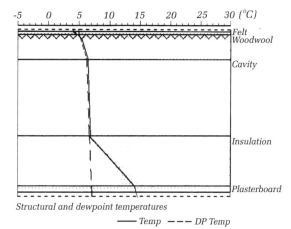

Structural and dewpoint temperatures

——— Temp   – – – DP Temp

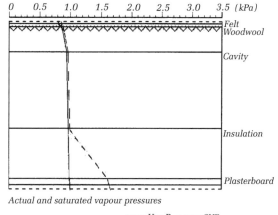

Actual and saturated vapour pressures

——— Vap P   – – – SVP

Figure 2.18 Basic performance of the roof. Moisture is accumulating beneath the roof weathering, but 167 g over a 60 day period in wood wool slabs would not normally be regarded as unacceptable.

TEMPERATURE AND MOISTURE DISTRIBUTION

Face	Elements	Temp (°C)	DPT (°C)	RH (%)	V.P. (kPa)	Flow (g/hr)	Depos (g/60d)
	Ext. S.R.	5.0	4.1	94	0.8		
0/1	Ext. S./Felt	5.1	4.1	93	0.8	0.00	
1/2	Felt/Woodwo	5.1	5.1	100	0.9	0.00	0
2/3	Woodwo/Cavity	6.4	5.1	91	0.9	0.00	0
3/4	Cavity/Insula	6.9	5.1	89	0.9	0.00	0
4/5	Insula/Vapour	14.5	5.1	53	0.9	0.00	0
5/6	Vapour/Plaste	14.5	7.3	62	1.0	0.00	0
6/7	Plaste/Int.S.	14.7	7.3	61	1.0	0.00	
	Int. S.R.	15.0	7.3	60	1.0		

TEMPERATURE AND MOISTURE DISTRIBUTION

Face	Elements	Temp (°C)	DPT (°C)	RH (%)	V.P. (kPa)	Flow (g/hr)	Depos (g/60d)
	Ext. S.R.	5.0	4.1	94	0.8		
0/1	Ext.S./Felt	5.1	4.1	93	0.8	0.00	
1/2	Felt/Woodwo	5.2	5.2	100	0.9	0.45	636
2/3	Woodwo/Cavity	6.6	6.6	100	1.0	0.02	−0
3/4	Cavity/Insula	7.1	7.1	100	1.0	0.02	0
4/5	Insula/Plaste	14.5	7.2	62	1.0	0.02	0
5/6	Plaste/Int.S.	14.8	7.3	61	1.0	0.02	
	Int. S.R.	15.0	7.3	60	1.0		

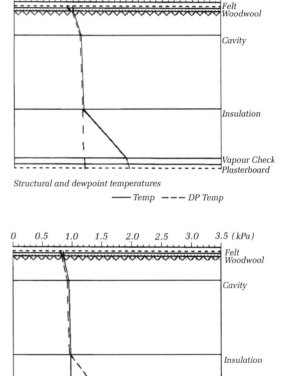

Structural and dewpoint temperatures

——— Temp   ——— DP Temp

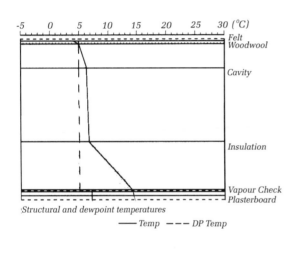

Structural and dewpoint temperatures

——— Temp   ——— DP Temp

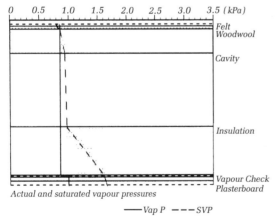

Actual and saturated vapour pressures

———Vap P   ———SVP

Actual and saturated vapour pressures

———Vap P   ———SVP

Figure 2.19 Effect of a vapour control layer above the ceiling. Moisture deposition is effectively eliminated, but it is important to bear in mind the practical problems which make it difficult to achieve a continuous membrane.

Figure 2.20 Effect of one air change per hour in the roof cavity caused by roof air making its way through gaps in the ceiling. Acute condensation may occur if gaps in the ceiling allow internal air to reach the outer layers of the construction. Here the amount of moisture deposition is clearly unacceptable. This mechanism is one of the major causes of condensation.

TEMPERATURE AND MOISTURE DISTRIBUTION

Face	Elements	Temp (°C)	DPT (°C)	RH (%)	V.P. (kPa)	Flow (g/hr)	Depos (g/60d)
	Ext. S.R.	5.0	4.1	94	0.8		
0/1	Ext.S./Felt	5.1	4.1	93	0.8	0.00	
1/2	Felt/Woodwo	5.1	4.1	93	0.8	0.00	0
2/3	Woodwo/Cavity	6.3	4.1	86	0.8	0.00	0
3/4	Cavity/Insula	6.7	4.1	83	0.8	0.00	0
4/5	Insula/Vapour	14.5	4.2	50	0.8	0.00	0
5/6	Vapour/Plaste	14.5	7.3	62	1.0	0.00	0
6/7	Plaste/Int.S.	14.7	7.3	61	1.0	0.00	
	Int.S.R.	15.0	7.3	60	1.0		

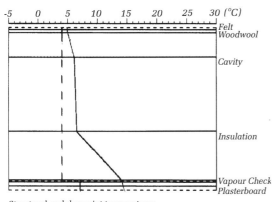

Structural and dewpoint temperatures

—— Temp − − − DP Temp

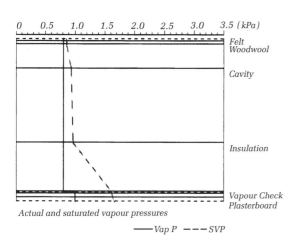

Actual and saturated vapour pressures

——Vap P − − −SVP

Figure 2.21 Effect of three air changes per hour of external air in the cavity. A flow of external air through a cavity can dissipate moisture diffusing through the ceiling. This principle is widely used to control condensation.

Thermal conductivity, λ — W/m K
Thermal resistance, R — m K/W
Vapour resistivity — MN s/g m
Vapour resistance — MN s/g
Vapour pressure — kPa
Vapour flow and vapour deposition — g/m^2 h
Vapour accumulation — g/m^2
— 60 days

The example used is a roof with continuous weathering on screeded wood wool with a cavity, insulation and a suspended ceiling of plasterboard panels. The external and internal conditions used and the duration are those suggested in BS 5250: 1989.

In each case a table sets out the conditions in the roof and the amount of moisture deposition. Charts show the temperature and vapour pressure distributions. Condensation is indicated by a zigzag line.

Special condensation problems

Vapour control layers The penetration of vapour into walls and roofs can be reduced by the provision of a vapour barrier near the internal surface. Vapour barriers are sheets of material of high vapour resistance but negligible thickness and thermal resistance. The minimum vapour resistance which is normally regarded as reaching vapour barrier standards is 15 MN s/g but most materials actually used have resistances of 250 MN s/g or more (Table 2.8). In theory vapour barriers should give an excellent performance in reducing condensation risk, but experience in the United Kingdom is very disappointing. Gaps between sheets, holes, tears and sometimes the omission of whole sections can allow moisture penetration, particularly by air currents between gaps. Ventilation from the external air of any cavity or the cold side of the vapour barrier can aid in the dissipation of any

Table 2.8 Performance of vapour barriers

Barrier	Vapour resistance (MN s/g)[a]
Two coats emulsion paint	45
Two coats flat oil paint	95
Two coats gloss paint	280
Aluminium foil	4000
Impregnated paper	30
Polythene	
0.05 mm	20
0.10 mm	250
0.15 mm	400
Bituminous felt, single	100
Metal-backed felt	500

[a]The figures apply to intact barriers with sealed joints.

BOUNDARY CONDITIONS

Internal		
Internal Air Temperature	15.0	C
Internal Vapour Pressure	1.0	kPa
Int. Dew Point Temp.	7.3	C
Int. Relative Humidity	60	%
External		
Ext. Air Temperature	5.0	C
Ext. Rad. Allowances	0.0	C
Ext. Vapour Pressure	0.8	kPa
Ext. Dew Point Temp.	4.1	C
Ext. Relative Humidity	94	%
Moisture Deposition		
Accumulation during	60	Days
Cavity Ventilation		
Layer Number	3	
Ventilation Rate	3	ac/hr
Percentage Ext. Air	100	%
Pattern of ventilation	Fully Mixed	
Ref: Roof		

TEMPERATURE AND MOISTURE DISTRIBUTION

Face	Elements	Temp (°C)	DPT (°C)	RH (%)	V.P. (kPa)	Flow (g/hr)	Depos (g/60d)
	Ext. S.R.	−5.0	−0.2	98	0.6		
0/1	Ext.S./Felt	−4.7	−4.7	100	0.4	0.00	***
1/2	Felt/Woodwo	−4.7	−4.7	100	0.4	0.23	3
2/3	Woodwo/Cavity	−1.5	−1.5	100	0.5	0.23	0
3/4	Cavity/Insula	−0.3	−0.3	100	0.6	3.27	37
4/5	Insula/Plaste	18.8	10.1	57	1.2	3.27	0
5/6	Plaste/Int.S.	19.4	12.1	63	1.4	3.27	
	Int.S.R.	20.0	12.0	60	1.4		

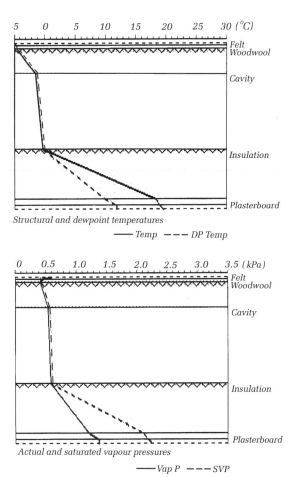

Structural and dewpoint temperatures

——— Temp – – – DP Temp

Actual and saturated vapour pressures

———Vap P – – –SVP

moisture which passes the barrier, but unless very high standards of work can be assured, the performance of vapour barriers must be regarded as suspect.

In the case of flat roofs it is important to remember that the bituminous felt or other finish is normally a very good vapour barrier indeed. It is, however, on the cold side of the construction and moisture from the interior is likely to be able to penetrate to the underside of the felt and condense there. Figure 2.16 shows this problem and how it can be reduced by cavity ventilation and by well-installed vapour barriers.

Another approach to the problem is the so-called upside-down roof, where slabs of weatherproof and water-resistant insulation are laid on top of the weatherproof covering, then held in position by paving slabs or gravel. A properly designed roof of this type will raise the temperature of the underside of the felt to a level which will overcome condensation.

Cold bridges The calculation of temperature distribution across elements of construction is normally carried out for the central areas of the element remote from corners, lintels or other configurations of construction which may cause higher rates of heat loss and consequently lower temperatures.

If cold bridges remain in an otherwise well-insulated construction, much of the potential energy saving will be lost. If minimum internal temperatures and ventilation rates appropriate to the general standard of the construction are employed in the interests of energy conservation, condensation is likely to form on the surfaces of the cold bridges. Thus internal temperatures or ventilation rates will have to

Figure 2.22 BS 5250 : 1989 gives realistic design values for a 60 day winter period and moisture diffusion is a relatively slow process, but more extreme conditions can prevail for short periods, so they should be considered in condensation design. In this case the temperature is 20 °C, the internal relative humidity 60% and the external temperature is 0 °C, but an allowance of −5 °C is made to take account of radiation to the night sky.

Table 2.9 Condensation risk: checklist of design factors

Factor	Means of reducing condensation risk
Moisture input	Moisture input from occupants cannot be controlled. For ablutions, washing, drawing and cooking, and for flueless heaters, change the process to generate less moisture (e.g. electric cooker instead of gas); reduce the wetted area surrounding wet activities; select apparatus with a built-in exhaust; extract moisture-laden air near the point of moisture input.
Internal air and environmental temperature	Although RH varies with air temperature it is absolute humidity which governs condensation and this is not affected by temperature changes. Internal temperature, however, governs the structural temperature in walls, floor and roofs. Increases in air temperature are not normally possible since temperatures are governed by comfort considerations. If heating is intermittent, however, more continuous heating will raise structural temperatures and thereby reduce condensation risk.
Ventilation of rooms or cavities in walls and roofs	Increased ventilation reduces levels of moisture in the air, a very effective way of reducing condensation risk. Unventilated cavities present a barrier to heat flow but not to vapour; they therefore increase condensation risk in the outer skin. Small amounts of ventilation can be effective in reducing moisture concentration without excessive heat loss.
Structural temperatures of walls, floors and roofs	Increases in structural temperature reduce condensation risk. This can be achieved by increasing the internal air temperature, making heating more continuous, or by providing insulation towards the outer face of the wall. Insulation on the inner face, if not associated with a vapour barrier or inherently vapour-resisting itself, increases condensation risk. Wall surface temperatures in external corners can be as much as 3 °C less than those in the centre of the wall and cold bridges in construction may also give areas where structural temperatures are very much lower than on the wall generally. Cupboards and fitted furniture can also reduce wall temperatures without affecting vapour pressure.
Vapour pressure in walls, floors and roofs	Reducing vapour pressure will reduce condensation. Increased ventilation achieves this. Vapour barriers towards the inside surface can limit the diffusion of vapour and thereby reduce vapour pressure within the thickness of elements of construction. Note that problems arising from inadequate jointing of sheet-type vapour barriers are not infrequent. (Flat roofs, which have an effective vapour barrier on the cold side, present a critical condensation problem since the internal vapour pressure will be maintained all the way through the construction and condensation on the cold underside of the finish is very common. The most effective remedies are either to provide insulation outside the vapour barrier or to provide ventilation in the roof cavity. The movement of relatively moist internal air into roof and wall cavities through holes for services, unsealed joints in sheeted materials, and cracks or defects in construction is a major cause of condensation. Where such holes cannot be avoided the vapour pressure in cavities must be assumed to be the same as that prevailing internally.
Pipe and trunking temperatures in cold rooms, etc.	In winter mains water will often enter buildings at temperatures of about 5 °C. Condensation on exposed pipes is likely in many places, particularly kitchens and bathrooms. Cold pipes on trunkings conveying chilled water or cold air may be at temperatures below the dew point for considerable periods and will inevitably be subject to condensation. Vapour barriers on the outside of the insulation can take place but the effect cannot be avoided.

be maintained at a level higher than is generally needed, simply to avoid problems arising at the cold bridge.

Condensation in pitched roofs Condensation in flat roofs with impervious coverings has long been recognized as a major problem. In England and Wales, however, there has been very little trouble with condensation in pitched roofs. In Scotland, where weather conditions are more adverse and roofs are usually lined with boarding, some problems have arisen. Recent developments have changed this situation and, if care is not exercised in both new buildings and schemes of rehabilitation, acute condensation problems can arise in pitched roofs.

Much of the air lost by ventilation in dwellings enters the roof space through cracks and pipe openings. This conveys moisture from the interior into the roof space. In traditional construction large quantities of fresh air entered the roof space at the eaves and the moisture was dissipated. If substantial thicknesses of insulation are installed in roofs, they usually have little effect on the penetration of air from the building below; but by blocking the eaves, they can reduce the inflow of fresh air to the roof and allow moisture levels to rise unacceptably. And the insulation prevents heat from reaching the roof space, so in a well-insulated building the roof space is now very much colder than in the past. The increased risk of condensation is apparent.

The problem is most readily overcome by ensuring that air can still enter the roof space from the eaves. Battens may be used to hold the insulation away from the underside of the slates or tiles. Short sections of plastic pipes may be used or special plastic eaves-ventilating strips employed.

The Building Regulations now prescribe minimum areas for vents in both flat and pitched roofs. Approved Document F2:1992 specifies the requirements.

Any roof void above an insulated ceiling shall be ventilated:

(a) Pitched roofs (over 15 degrees) [flat ceiling]
    Through ventilation by vents on each side equivalent to a 10 mm wide continuous opening.
(b) Pitched roofs where ceilings follow the pitch
    A continuous cavity with a minimum width of 50 mm and vents at eaves equivalent to at least a 25 mm wide opening plus ridge vents equivalent to a 5 mm wide continuous opening.
(c) Roofs with pitch of less than 15 degrees
    A continuous cavity with a minimum width of 50 mm and vents equivalent to at least a 25 mm wide continuous opening on two opposite sides. Where the span exceeds 10 m or the roof is not rectangular, vents totalling 0.6% of the roof area.

The requirements of the regulations will also be met by constructors in accordance with CSB 250:1989.

Notice that the vents required are at eaves level. Ventilators in the ridge are sometimes used but the high wind velocities and pressure differences prevailing in this location can give rise to high rates of ventilation, so the eaves position is much to be preferred.

Design to minimize condensation
Although essential as a check on design and to demonstrate compliance with proper standards, calculation techniques do not provide a suitable approach to design. It is not satisfactory to design without regard to condensation, then to apply remedial measures as a result of unsatisfactory numerical checks. Table 2.9 summarizes factors which should be borne in mind during design.

3 Ventilation and air quality

3.1 Wind

The movement of air, which takes place smoothly well away from the earth's surface, is affected by the topography of the surface. Velocity is reduced towards the ground; direction of movement is affected by the varied relief of mountains hills and vegetation. A turbulent airflow results, which is not realistically represented by steady velocity values. Admiral Beaufort devised a scale in 1806 which is used to classify winds at sea; this has proved so useful that it has been extended to cover land conditions and, recently at the Building Research Establishment, conditions for human comfort. Table 3.1 shows the Beaufort scale and the associated conditions for comfort. Ground areas where wind speeds exceed 5 m/s for significant periods are likely to be thought uncomfortable for human occupation; speeds greater than 10 m/s are positively unpleasant; and speeds over 20 m/s can be dangerous because of the risk of being blown over and the risk of injury from broken branches and trees, even falling pieces of building.

The traditional town gave substantial protection from high winds, and discomfort was only rarely experienced. In recent years the construction of high buildings has caused a dramatic change. These buildings present a large area to the wind, producing considerable air pressure differentials and consequent increases in air velocity, particularly at the corners of buildings. Figure 3.1 shows a typical pattern of pressure distribution. The pressures themselves are not directly noticeable to people walking outside, but Fig. 3.2 shows the resulting velocity patterns, which do affect comfort. Note that particularly high velocities occur at the leading edges of the building, both top and sides, where high and low pressure zones approach closely. Buildings often have open areas at ground level, and Fig. 3.2 shows that high velocities of air movement are inevitable since these openings allow airflow between the high pressure area facing the wind and the low pressure areas on the other side. This feature of design usually presents air movement problems for people walking there.

The juxtaposition of high buildings with lower ones gives rise to special problems. Figure 3.3 shows how a standing vortex may form between the high and low building, giving very high velocities at ground level. Downwind of the high building downcurrents may develop; they are unlikely to be noticeable directly but they can carry smoke downwards from chimneys. Courtyards in buildings with blocks of different heights can suffer from the same problem.

These effects can multiply the original wind velocity. This contrasts with the traditional town pattern, where velocities experienced at ground level were generally reduced. Pedestrians in new developments can therefore discover wind velocities four or five times as great as those to which they are accustomed. There have been examples where new layouts were quite unacceptable for pedestrians and ground areas have had to be roofed over to enable them to be used. Town planning authorities are conscious of the problem and many now require promoters of major developments to demonstrate that unacceptable velocities will not occur in pedestrian areas. Although general principles of air movement, some of which have been described, are well understood, and can be borne in mind during design, there are no simple drawing-board means of predicting velocities. Schemes to be studied will have to be tested in a

Table 3.1   Beaufort scale

Beaufort number	Windspeed (m/s)	Description	Sea condition	Land condition	Comfort
0	0–0.5	Calm	Mirror-like surface	Smoke rises vertically	No noticeable wind
1	0.5–1.5	Light air	Ripples	Smoke drifts	
2	1.6–3.3	Light breeze	Small wavelets	Leaves rustle	Wind felt on face
3	3.4–5.4	Gentle breeze	Large wavelets scattered white horses	Wind extends flags	Hair disturbed, clothing flaps
4	5.5–7.9	Moderate breeze	Small waves, frequent white horses	Small branches in motion, raises dust and loose paper	Hair disarranged
5	8.0–10.7	Fresh breeze	Moderate waves, many white horses, chance of spray	Small trees in leaf begin to sway	Force of wind felt on body
6	10.8–13.8	Strong breeze	Large waves and white foam crests	Whistling in telegraph wires, large branches in motion	Umbrellas used with difficulty Difficult to walk steadily Noise in ears
7	13.9–17.1	Near gale	Sea heaps up and foam begins to be blown in streaks	Whole trees in motion	Inconvenience in walking
8	17.2–20.7	Gale	High waves spindrift, and foam	Twigs broken from trees	Progress impeded Balance difficult in gusts
9	20.8–24.4	Strong gale	High waves with toppling crests Spray affects visibility	Slight structural damage (chimney pots and slates)	People blown over in gusts
10	24.4–28.5	Storm	Very high waves with long overhanging crests Surface white with foam visibility affected	Seldom experienced inland; trees uprooted Considerable structural damage	

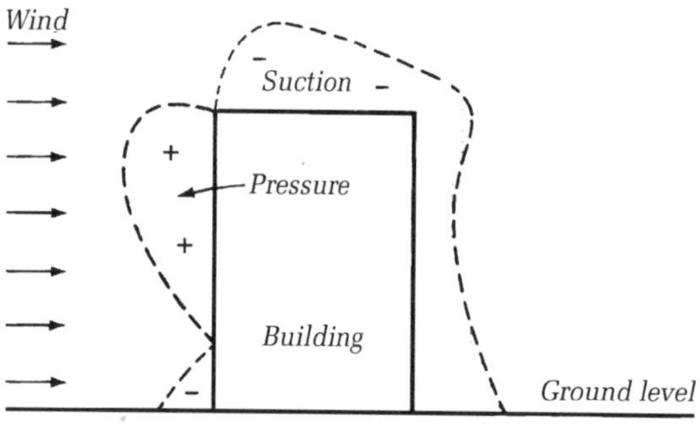

Figure 3.1   A typical pattern of pressure distribution.

wind tunnel. Suitable tunnels exist at the National Physical Laboratory (NPL) and at some universities. Normal architectural models can often be used. It is important, however, to ensure that the tunnel used can model not only the velocity gradient of the wind but also the pattern of turbulence.

In addition to influencing form and siting, wind must be taken into account in some aspects of the internal planning of buildings. It is clearly desirable to avoid locating doors near corners where velocities will be high. (Dustbins, similarly, should not be located at high velocity positions.) Sometimes entrance areas link doors on different façades, as shown in Fig. 3.2, where pressures are very different. Unless revolving doors are used there may be substantial inconvenience. Exhausts and intakes for ventilation systems must also be considered in relation to pressures round the building. In recent years the velocities of airflow and the total resistance of ventilation systems have increased, thereby reducing the effect of wind, but it is still important to ensure the exhaust and inlet are not subject to pressure differentials that might produce excessive or reduced flow. Positions on opposite faces of the building should be avoided for the reasons shown in Fig. 3.4. One solution is to place both exhaust and intake on the same façade. If this is done it is desirable, particularly in high buildings, to have the intake below the exhaust, otherwise the natural buoyancy of the usually warm exhaust coupled with the updraught on the face of the building may convey exhaust air to the intake. A new solution is to have the

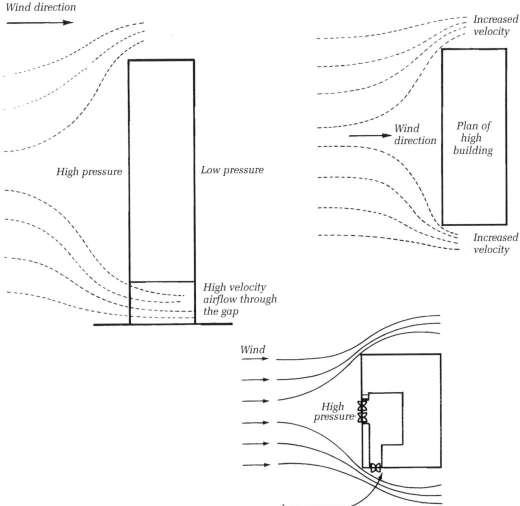

Figure 3.2 Airflow patterns round large buildings showing high velocity at parapets and through gaps in buildings.

exhaust at roof level, since this is normally a low pressure zone. The intake may then be on any façade. Figure 3.5 shows these arrangements.

Chimneys carry effluent gases up to a level at which they will be adequately diluted before they reach ground level or other buildings. The chimney itself should therefore project beyond the zone immediately affected by the building. If it is high enough the velocity at which the flue gases leave the chimney may significantly aid proper dispersion. In the case of small buildings, past performance of chimneys in similar circumstances is the best possible guide, although differences in topography and surrounding buildings must be taken into account. Wind tunnel tests should be carried out when siting chimneys in circumstances

without precise precedent, and in all cases where large chimneys are involved.

The structural loading implications of wind are outside the scope of this book. Vibrations resulting from wind also require specialist consideration. It is, however, worth mentioning briefly that the variations in velocity during turbulent flow produce pressure variations. If very rapid measurements are taken, it is possible to show that the resultant power varies with frequency. There is a risk of damage if part of the cladding or structure of a building has a natural frequency that corresponds with a high power from the wind. Vibration can also be imparted to buildings by vortex shedding at the sides of the airstream. Vortices are rotating masses of air shed alternately from each

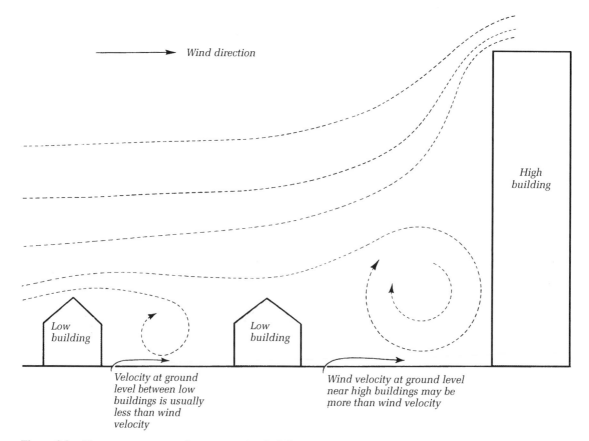

Figure 3.3   Air movement patterns between two low buildings and between a low building and a high building.

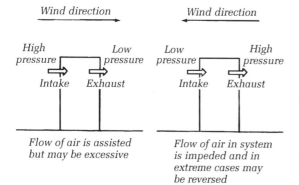

Figure 3.4   Bad position for ventilation intake and exhaust.

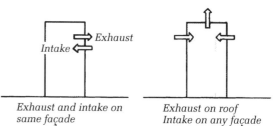

Figure 3.5   Good position for ventilation intake and exhaust.

side; the result is vibration of chimneys and towers, even parts of buildings themselves. Metal chimneys are clearly very much at risk from this phenomenon. The spiral strakes now seen round the top part of chimneys of this type represent a very successful solution developed by the National Physical Laboratory (Fig. 3.6).

3.2 Ventilation

Air change and movement within buildings, or ventilation, is a vital aspect of design which has

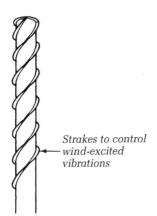

Figure 3.6 Strakes on a metal chimney.

Strakes to control wind-excited vibrations

governed form throughout history. Acceptable electric light sources and electric fans have now given a situation where natural ventilation is not essential in every case, but the majority of buildings are still naturally ventilated.

Two factors control natural ventilation. One is the pressure variations due to wind. The other is the 'stack' effect which results from warm air in the building rising and being displaced by colder external air (Fig. 3.7). In low buildings with small rooms it is easy to achieve acceptable levels of ventilation, and to control both wind and stack effects using windows with controllable opening lights. In countries with more severe climates than England, such as Scandinavia, special controlled vents are often incorporated into window frames because operation of the normal opening light would give only crude control. Ventilation of this sort is designed on the basis of experience supplemented by regulations which call for opening lights equivalent to a percentage (in England and Wales 10% was traditional) of the floor area and for adequate space round buildings to allow air movement.

In high buildings, above about 10 floors, wind and stack effects become very marked. In winter the temperature difference between the cold external air and the air inside the building will be markedly different, often by as much as 20 °C. Under these conditions the warm air inside the building tends to rise through stairwells, lift shafts, ducts and any other openings, making the top of the building excessively hot and the bottom unacceptably cold. Flats do not suffer from this problem because of the relatively effective subdivision into individual dwellings, but high buildings of other types have to be mechanically ventilated to overcome the problem.

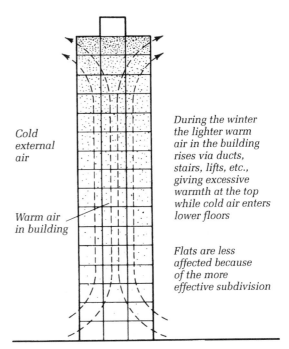

Cold external air

Warm air in building

During the winter the lighter warm air in the building rises via ducts, stairs, lifts, etc., giving excessive warmth at the top while cold air enters lower floors

Flats are less affected because of the more effective subdivision

Figure 3.7 The stack effect in a tall building.

Ventilation criteria

The basic human requirement for fresh air in order to gain an adequate supply of oxygen is well known. In buildings, however, this never forms a standard for ventilation since there would be acute discomfort long before any danger to life arose. In fact, ventilation standards are based on keeping various types of contamination of the air or overheating to acceptable levels.

Additional ventilation considerations

In summer the conventional provision of opening windows meets the needs of effective natural ventilation in the majority of buildings except where external noise levels prevent window opening. In winter the problem of providing a limited ventilation rate without discomfort is more difficult.

Traditionally buildings were far from airtight. Flues or ventilators were usually present as well as windows, and always in habitable domestic rooms. Windows did not seal tightly and there were many other passages for air leakage. Natural air change rates when all windows and doors were closed could still be significant and in practice often met the need for a minimum level of air change in winter conditions. In addition to the basic

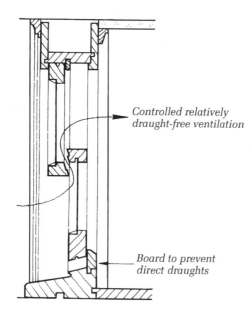

Figure 3.8   Traditional sliding sash window.

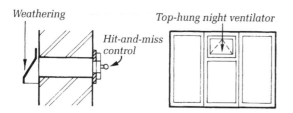

Figure 3.9   Two types of vent.

infiltration, special window arrangements were often made to assist in controlled winter ventilation. Large opening lights, even when opened only a little, gave excessive ventilation rates at levels which caused draughts. Figure 3.8 shows a traditional sliding sash window with a draught board which enables controlled winter ventilation. Figure 3.9 shows a night ventilator. Night ventilators were almost universal in domestic buildings from the 1920s through to the 1940s, when sliding sashes were no longer popular. The small window at high level was relatively controllable and airflows were not close to occupants. The small size enabled night ventilators to be opened without complete loss of security.

Changes in architectural and building fashion have led to the abandonment of special winter ventilation provisions in windows, and Building Regulations progressively relaxed the requirement for flues or ventilators in habitable rooms from 1965 to 1990, a

period during which many millions of dwellings were built. Building product manufacturers try to produce well-sealed components. The need for energy conservation causes owners and occupiers to reduce ventilation rates as much as possible in order to save energy. Some of the factors which have contributed to the reduction of ventilation rates in buildings are as follows:

- Abandonment between 1965 and 1990 of the requirement for flues or vents in habitable rooms.
- More tightly sealed windows.
- Lack of night ventilators in windows.
- Dense airtight construction in some cases.
- Weather stripping.
- Security dictating that windows must be kept closed.
- Flueless heating systems or balanced flues not drawing air from the interior.

The reduction in ventilation rates produces increased concentrations of air pollutants; new materials, products and equipment contribute new types of pollution to the internal atmosphere (Table 3.2). This makes it a difficult problem to establish standards for ventilation. For energy conservation, ventilation rates should be low, whereas for health and comfort they should be high.

In the past, except for special circumstances, standards for ventilation have been based on the control of odour. This is still an important factor for the determination of ventilation rates, but the nature of air pollutants has changed. More frequent bathing reduces the concentration of body odour but many manufactured items incorporated both in the building fabric and in the form of furniture may give off polluting substances such as formaldehyde. Although the actual levels depend very much upon how the machines operate, it is likely that dishwashers and clothes dryers, if not ventilated to the exterior, contribute more moisture to the internal air of dwellings than the traditional processes. Many consumer products, particularly aerosols and cleaners, contribute to pollution in the air. In larger buildings air-conditioning systems and even shower sprays can house and distribute biological pollutants. Instances of airborne infection from spray chambers and shower heads have received wide publicity. The higher temperatures prevailing in centrally heated buildings encourage some forms of biological pollution. Figure 3.10 shows typical sources of domestic air pollution and with Table 3.3 give a more comprehensive schedule of pollutants, sources and limits of concentration.

Table 3.2 Indoor air: pollutants and their sources

Contaminant	Typical sources										Typical and allowable concentrations[a]	
	Smoking	Combustion	Metabolic activity	Furnishings	Building materials	Aerosols	Soil	Moulds and static water	Equipment	External environment	Approximate cuttent limits for concentration[b]	Typical concentrations in buildings[b]
Odour			✓									
Particulates	✓	✓			✓		✓				260	30
Moisture		✓	✓						✓		>30% RH >70% RH	
Mites				✓				✓				
Micro-organisms and allergens			✓	✓				✓		✓		
Radon					✓		✓				10 mSv/year[c]	0.4 mSv/year[c]
Asbestos					✓				✓			80 fibres/m^3 [d] >400 fibres/m^3 [e]
Manmade fibres				✓	✓							
Carbon monoxide	✓	✓									10,000–40,000	1000–10,000
Methane							✓					
Hydrogen sulphide							✓					
Oxides of nitrogen	✓	✓									200	
Sulphur dioxide		✓									365	
Carbon dioxide		✓	✓								0.5%	0.1%
Formaldehyde		✓		✓	✓						60–3000	7
Various chemicals including fluoro-carbons and hydrocarbons, ammonia, etc.				✓		✓					160	–
Ozone											235	

[a] Design standards, where they exist, vary greatly from country to country and are subject to change as a result of research. Allowable levels are also greatly affected by the duration of exposure.
[b] mg/m^3 of air (except where indicated).
[c] Millisieverts per annum.
[d] No asbestos panels.
[e] With asbestos panels.

In the case of long-established pollutants where the level of input has not changed significantly, the effect of reduced ventilation is to give higher concentrations in the internal air. This affects tobacco smoke, combustion products from flueless heaters and gas cookers, radon from masonry materials and foundations, moisture and biological pollutants.

In most cases pollutants are introduced directly into the atmosphere. In some cases they arise as a result of conditions which might not themselves be regarded as contributing to contamination. The spores from moulds growing as a result of surface condensation form a good example. Current medical opinion regards these spores as potential allergens capable of affecting a

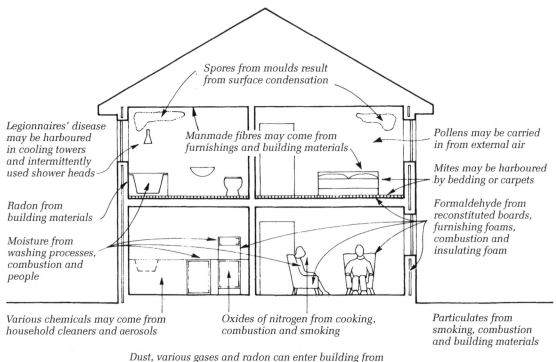

Spores from moulds result from surface condensation

Legionnaires' disease may be harboured in cooling towers and intermittently used shower heads

Manmade fibres may come from furnishings and building materials

Pollens may be carried in from external air

Mites may be harboured by bedding or carpets

Radon from building materials

Formaldehyde from reconstituted boards, furnishing foams, combustion and insulating foam

Moisture from washing processes, combustion and people

Various chemicals may come from household cleaners and aerosols

Oxides of nitrogen from cooking, combustion and smoking

Particulates from smoking, combustion and building materials

Dust, various gases and radon can enter building from the subsoil

Figure 3.10   Sources of internal air pollution.

significant proportion of building occupants. The condensation giving rise to them, however, is a result of a combination of moisture in the air and cold wall or ceiling surfaces. It can occur when the moisture levels would be regarded as normal and with wall temperatures which would not usually be regarded as relevant to air pollution (page 41). Water in air-conditioning systems and even the small quantities retained in shower heads can be the cause of biological pollution, particularly if the water is maintained at a comfortably warm temperature. In both these cases the effect of the arrangements in the buildings is to enable contamination present in the atmosphere in very small quantities to multiply and become concentrated in the buildings. It demonstrates, however, that in designing for high standards of air quality, attention needs to be paid to a wide range of building features.

In principle there are several ways of approaching the problem of indoor air pollution control:

- Remove the source of pollution.
- Provide local air extract at pollution sources.
- Provide sufficient general ventilation to dilute and remove the pollution.
- Purify the internal air.

Masking by acceptable scents has been used in the past but cannot be recommended as a method of solution. And no single measure can solve the whole range of air pollution problems. Removal of the source of pollution can apply to unflued fixed appliances, washing-machines or cookers, but it cannot apply to occupants smoking, furnishing foams, chipboard floors or oil heaters.

The provision of general ventilation is the most widely applicable of the possible control strategies. But high ventilation rates are incompatible with energy conservation and closely controlled ventilation is needed. This is difficult to achieve.

Current ventilation design and standards

There are two basic approaches to the provision of ventilation. With *natural ventilation* air is admitted into each room by windows or large vents leading directly to the external air; the rate of air change is governed by wind and buoyancy effects. Sometimes the rate of air movement is assisted by extractor fans. With *mechanical ventilation* or *air-conditioning* air is drawn in then exhausted by fans; the air is distributed through a network of ducts. In cases of this sort the air will usually be warmed in winter and may be cooled in summer; there may also be humidity control.

Natural ventilation is the norm and its design is the responsibility of the architect. Mechanical or air-conditioning systems are used when conditions preclude natural ventilation. The following situations require mechanical ventilation or air-conditioning:

- Internal rooms.
- Large, closely populated rooms where the distribution of natural ventilation would be inadequate (BS 5925 : 1991).
- Rooms where the volume per occupant is too low for efficient ventilation (BS 5925 : 1991).
- Rooms where close control of the environment is required; typical aspects are temperature and relative humidity.
- Where natural ventilation cannot be provided because of external air pollution or noise.
- In tall buildings where wind and stack effects render natural ventilation impracticable.
- Extract ventilation (or an excess of extract over input) may be required to deal with fumes or smells from cooking or other special processes.

For successful systems of this type the architect must take into account the accommodation required for plant and the needs and characteristics of the system in the overall design of the building. The design of the installation itself will normally be the work of the building services engineer.

In the 1950s, the planning and servicing advantages of internal bathrooms in dwellings were recognized and comparison made with continental examples where natural ducted ventilation for both supply and exhaust to such rooms was accepted. Experiments carried out at the Building Research Station led to a report which supported the use of such systems. In fact, mechanical extract has been adopted as the norm for internal bathrooms because of the simplicity, economy and reliable performance of extractor systems.

The system of natural ventilation

Modern natural ventilation is not simply a matter of having an openable window of given size. It involves the provision of background ventilation for periods of low occupancy, rapid ventilation to deal with temperature and air pollution during occupancy, and the removal of local air pollution at source rather than allowing it to escape to the general atmosphere. These aims have to be achieved while maintaining the security and weather resistance of the building.

Background ventilation

A low and controllable level of ventilation is required at all times and weather conditions, without causing draughts, compromising the security of the building or admitting rain. The traditional air brick, if provided with a damper, met this requirement. At present a 'trickle' ventilator consisting of a narrow vent at the head of the window with an adjustable grille to give control is the most usual method. Vertical sliding sashes can also be used, provided they can give positive and finely adjustable control of the upper sash. Security considerations preclude their use at ground level.

Rapid ventilation

For summer conditions or when the room is overheating or subject to transitory excess levels of pollution or smell, a means of rapid air change is required. This is normally met by openable windows.

Supplementary extract

Where the basic ventilation is by natural means but there are localized sources of pollution, such as cookers or bathrooms, it is sensible to extract air at the source of pollution rather than to allow the pollution to escape and increase the whole rate of ventilation to achieve a less effective result. Extraction can be achieved by fans drawing air from the locality of the source and discharging it externally by a duct. Recently a system of natural extraction, passive stack ventilation (PSV), has been validated and is now included in the Building Regulations. It involves an inlet terminal near the source of pollution serving a duct insulated to prevent condensation. Ideally the duct will be vertical, but it can have a section at 45°. It should not terminate below ridge level. The system is illustrated in Fig. 3.11, which shows a typical solution for the natural ventilation of a dwelling.

Natural ventilation of internal rooms and large buildings

The depth limit of rooms which can be effectively ventilated from one side has normally been placed at about 7 m. Authoritative recommendations vary between 6 m and 10 m, which might be doubled for ventilation on opposite sides. This imposes close limits on the form of buildings and the size of offices. And these limits have become unacceptable to many clients and architects for both practical and visual reasons. Air-conditioning could overcome the problem but would involve considerable initial expense and running costs, and would not be compatible with the

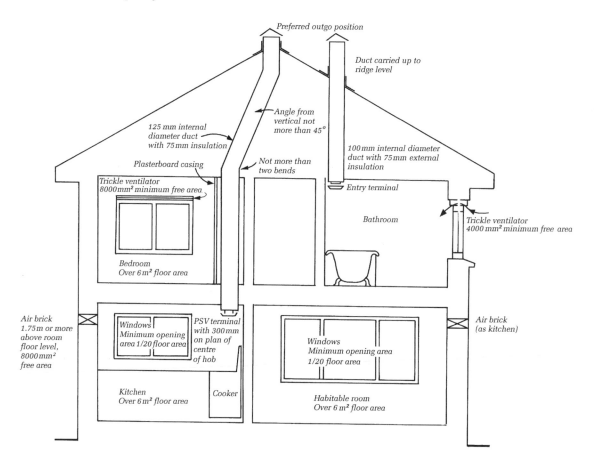

Figure 3.11   Natural ventilation system for a dwelling. Background ventilation: first floor, trickle ventilators; ground floor, air bricks. Rapid ventilation: all floors, windows. Extract ventilation: passive stack ventilators (PSVs); extractor fans could have been used instead.

wish to conserve energy. Several large buildings have been built using internal atria or even ventilation stacks to provide adequate movement of air. They have given satisfactory service. Very careful attention must be given to the detailed design of air inlets, heat emitters, blinds and dampers. Many of the building occupants will have no immediate means of controlling their own environment and the strategy for control and briefing of occupants must be thorough. Figure 3.12 shows some of the features of a building of this type.

Heat recovery, thermal storage, mixed-mode operation and zoning

Heat recovery

Most air-conditioning and mechanical ventilation systems, in their nature, provide the opportunity to

recover some of the heat from air being exhausted. Pages 164–9 show some of the methods. Systems are available to apply the technique to dwellings. In Sweden it is said that the great majority of new dwellings have extractor ventilation systems with heat recovery. And when considering this, it is important to appreciate several differences between the two countries. Sweden's winter temperature lows are more consistent than in the United Kingdom, so opening windows in kitchens and bathrooms is rarely feasible during winter. The building regulations in Sweden therefore require substantial air vents to be carried through the roof, unless mechanical ventilation is provided. Thus the cost of mechanical ventilation is significantly offset by savings in venting; the UK regulations (Table 3.3) do not have the same requirement, so there is no saving to be had. Background infiltration through the fabric of the

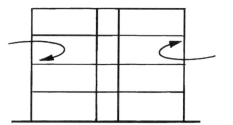

CONVENTIONAL OFFICE BLOCK

*Each office is ventilated
by a window on one side*

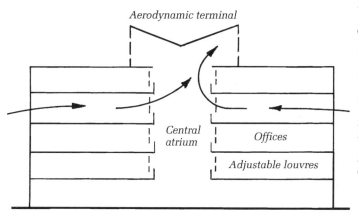

Aerodynamic terminal

DEEP-PLAN OFFICE

Can be closed at night to conserve energy

Provides stack effect at low wind speed

*With higher wind speeds aerodynamic
design or selective closure of vents
maintains stack effect*

*External windows are openable with
minimum trickle control*

Central
atrium   Offices

Adjustable louvres

*Internal louvres control different stack
effects between floors*

Figure 3.12   Natural ventilation strategies.

Table 3.3   Typical ventilation standards from BS 5925 : 1980

Building type	Recommended rate (l/s per person)	Minimum rate (l/s per person)
Commercial	8	5
Private offices Residences Hotel bedrooms	12	8
Restaurants Conference rooms Bars	18	12
Boardrooms	25	18

Spaces where vent rate is not based on occupancy

Corridors	1.3 ac per hour
Domestic kitchens	10 ac per hour
Industrial kitchens	20 ac per hour
Toilets	10 ac per hour

above freezing point and relative humidities very high, it is by no means certain that the ventilation rates used successfully in Sweden would be adequate to avoid condensation. Experience with systems to overcome aircraft noise indicates that complaints about fan noise would be likely.

Thermal storage
Thermal storage in the building structure (sometimes described as free cooling) can reduce cooling loads on air-conditioning plant. As temperatures rise, a massive structure will absorb some of the heat, reducing the peak temperature. The effects can be enhanced by ensuring ventilation at night to cool the structure. Opening windows may not be practicable and some systems draw cool night air through ducts in specially designed concrete floor slabs. Room air can be circulated through the floor slab during the day; or when fresh air is required and the outside temperature is high, outside air can be used instead.

Mixed-mode operation
It is possible to utilize both natural and mechanical ventilation. Mixed-mode ventilation supplements natural ventilation with a mechanical system to give cooling when necessary.

building is carefully controlled in Sweden, often with a complete polythene envelope. The amount of moisture present in the air at very low temperatures is very small, thus reducing the condensation risk. In the UK climate, where typical winter temperatures are usually

Zoning

Thermal installations are normally zoned for control purposes; this applies to mechanical ventilation and air-conditioning systems. But zoned systems allow more than just the type of control to be varied; it is also possible to adjust them according to the local need in the building.

Building Regulations

After a long period when legal requirements and conventional practice progressively reduced standards, in 1995 the government re-established rules for effective ventilation. The current standards are set out in Building Regulations 1991, Approved Document F: 1995. The approved document sets out general aims for control of levels of vapour and pollutants together with a safe and weatherproof supply of fresh air. It describes acceptable solutions rather than defining ventilation standards.

The specific solutions contained in the approved document are confined to the ventilation of rooms and sanitary accommodation located on external walls in dwellings and non-domestic buildings. Storage and circulation space are not included. The approved document refers generally to BS 5925 : 1991, BS 5720 : 1979, BS 5250 : 1989, and the *CIBSE Guide* Parts A and B as suitable guides for circumstances not explicitly covered.

The recommendations for dwellings and non-domestic buildings are very similar. They call for a basic level of background ventilation, facilities for providing rapid ventilation when required and means of extracting air near cooking apparatus and from bathrooms.

Control of air quality and ventilation

At first thought it is surprising there are no general provisions in building regulations giving direct control of ventilation rates and, except for industrial buildings, no control of air quality. It is even more surprising that, since ventilation is one of the major aspects of heat loss from buildings, it has not been controlled from the energy conservation viewpoint. The contrast with thermal insulation of the building's fabric, which has been the subject of frequent and increasing stringent statutory requirements, is very dramatic.

The explanation is a simple one. It is impossible for designers to specify the form, materials and work standards of the building in order to achieve a specified natural ventilation rate, and it is very expensive, complicated and time-consuming to measure ventilation rates to see whether specified standards have been achieved.

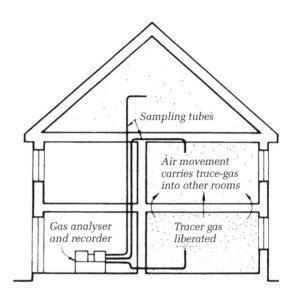

Figure 3.13   Use of tracer gas.

The equipment involved is expensive, the procedures required are elaborate and, since ventilation rates are subject to variation as a result of wind and stack effects, the measured value at a given instant may not be representative of general conditions.

Measurements are based on the rate of decay of the concentration of a small quantity of tracer gas mixed with the air of a room. From this rate of decay the rate at which room air is being replaced can be calculated. Nitrous oxide is one of the most popular tracer gases. It is relatively innocuous and, since it is possible to detect a few parts of nitrous oxide in a million parts of air, it is only necessary to use very small quantities. On being released the gas must be mixed with the room air. Oscillating fans are usually employed. Samples are then drawn from selected points by tubes leading to the gas analysis and recording equipment. It is possible to measure rates of growth of concentration on other rooms due to internal movements as well as the rate of decay in the room where the gas was liberated. One problem with the method is that it is impossible to tell whether the air which has entered the room comes from outside or from other rooms. This may be overcome by the simultaneous use of several different tracer gases released into the different spaces under investigation. Figure 3.13 shows the method. Tracer gas may be liberated continuously at a controlled rate and continuous determinations made. External wind and temperature conditions are always measured at the same time as tracer gas measurements are made.

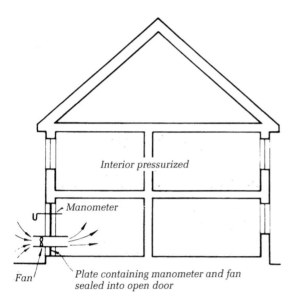

Interior pressurized

Manometer

Fan　*Plate containing manometer and fan sealed into open door*

Figure 3.14　Pressurization testing.

The tracer gas technique is expensive and time-consuming, and the results depend upon wind and temperature conditions. It is not practicable to employ it in general tests of building performance nor to use it as a basis for legislative standards. A very much simpler and quicker method is in the course of development and is widely used in Scandinavia, particularly in dwellings which are to have a simple mechanical extractor system. It is independent of wind and weather, and does not require expert scientists to execute it. The method relies on pressurizing the interior of the building, with all windows and doors shut. A panel containing a fan and measuring equipment is sealed into the open front door and a pressure of 55 Pa is maintained. The volume of air input and the ventilation rate are then determined. Three air changes would represent a very well-sealed building. The procedure is repeated with negative pressure. Figure 3.14 shows the arrangements. This method does not predict rates of natural ventilation, but it does enable the basic airtightness to be checked, so that uncontrolled infiltration will not dominate the proper provisions for ventilation.

In modern buildings there is an increasing number of cases where natural ventilation does not give satisfactory conditions; several situations call for mechanical ventilation.

Air movement

Usually associated with means for ventilation, although it could be separated. Some degree of air movement is essential for feelings of freshness and comfort. Desirable speeds vary with temperature and conditions. In domestic buildings and other similar situations, a velocity of 0.10–0.33 m/s is considered reasonable.

Fumes, smells, products of combustion

Where fumes are likely to be offensive or injurious they should be removed by extraction at source. It is neither satisfactory nor economical to allow the fumes to enter the whole volume of the room. Hoods or extractors should be placed to capture the fumes as soon as they are emitted. High velocities are often required to ensure the fumes are entrained; this may lead to high ventilation rates in associated rooms unless the source of fumes is enclosed and provided with its own air supply as sometimes occurs in fume cupboards.

Bacteria

Special precautions may be required in hospitals or other places where bacterial concentrations could be critical. The precautions may consist of specially high air change rates and minimization or avoidance of recirculation. In high risk areas air may be disinfected by bactericidal sprays or ultraviolet light. In the case of the operating theatres or sterile areas mechanical ventilation may be used to produce special patterns of air movement so that bacteria are swept away from areas where they might do harm.

Excess heat

High rates of ventilation may be used to remove excess heat (1 m^3 can convey 1.3 kJ for each Celsius degree of temperature difference between the heated control and the outside air). The method of calculation is described in Chapter 7. Radiant heat cannot be controlled in this way. Where high levels of electric lighting are provided, it is often necessary to extract air through the light fittings, so the heat from them is removed at source and does not escape into the room.

Relative humidity

It is usually considered that relative humidities of 30–70% are acceptable from the viewpoint of comfort and health, and little effort is normally made to control humidity except in air-conditioned buildings. Levels as high as 70% would, in almost all cases, require substantial vapour inputs within the building and would almost certainly be associated with condensation and mould growth if the vapour inputs could not be controlled. One of the best ways to reduce high levels of relative humidity is to increase ventilation rates. Very low levels can be achieved during the winter when, on cold dry days, low levels of external humidity can produce even lower levels when the air is heated. If air at 0 °C and 70% RH is raised to 20 °C the RH will drop to about 18%. Relative humidities of this order can give rise to complaints of dry throats and will cause woodwork to shrink and crack.

In some industrial buildings high levels of relative humidity are caused by or required for special processes. In the provisions of the Factories Act 1961 the minister may make regulations covering ventilation rates. In the case of factory rooms where humidification takes place, the first schedule of the act appears to indicate that relative humidities of up to 85–87% are admissible. Special precautions will therefore be required against condensation.

3.3 Recently recognized problems

Sick building syndrome

In recent years there have been increasing complaints of acute discomfort in office buildings, particularly those equipped with air-conditioning. Headache, eyeache, sore throat, breathing difficulty, cold-like symptoms and lassitude were usual complaints. Medical investigations have confirmed that these symptoms are real.

Many surveys have been carried out and extensive and detailed measurements made of environmental conditions existing in a wide range of buildings. No clear identification of the cause of the problem emerged. This is not surprising in view of the large number of variables involved in each of the environmental aspects, which include thermal, aural, visual, hygienic and medical considerations.

A fundamental problem was confirmed by the surveys. It is the very large degree of discrepancy between the designed performance of heating and air-conditioning systems and the way in which they actually operated. This is not surprising. To set such complex systems into proper operation a period covering perhaps two annual cycles would be needed. The pattern of building contracts was established in a period when the balancing of a simple gravity hot-water heating system was relatively simple and could be said to have been properly tested by the time of practical completion of the building. This is not the case with modern installations, but the expectation of the great majority of building contracts remains unchanged. A very similar situation exists in terms of maintaining performance. Many modern buildings have very sophisticated environmental installations but

few are considered large enough to justify the employment of the professionally qualified staff who would be needed to maintain operational efficiency.

There is clearly a vital need for standards of installation, commissioning and operation of environmental systems to be established and incorporated into building contracts and legislation. The sophistication of systems installed should not be incommensurate with the quality of operation which can be provided. This would result in a major improvement in comfort and energy conservation, although it could not necessarily eliminate all sick building symptoms.

One factor which has been identified as playing an important part in contributing to sick building symptoms is cleanliness. Sick building complaints have been reduced by a rigorous process of cleaning. This includes interiors of ducts, particularly soft furnishings and carpets. It appears that the constant internal conditions of warmth and humidity, limited fresh air, absence of direct sunlight and universal use of carpets, provides an ideal environment for house mites. Powdered excrement from these mites becomes airborne and gives rise to allergic reactions in some people. It may also produce less well-defined but nevertheless distressing symptoms in many others.

Apart from contaminants in industrial buildings there are no regulations which directly control air quality in buildings. The increasing size, complexity and artificiality of environments in modern buildings call for much closer attention to standards, design and operation.

Radon

Radon is a radioactive gas which is given off by many materials in the earth's crust. Some building materials give off small quantities. The gas decays into particles called the 'daughters of radon'. These particles can be inhaled and become lodged in the lungs, where they discharge radiation. In mines in strata giving out large quantities of radon, very high air concentrations can result, and the resulting incidence of cancer was recognized as a problem before the causes were understood.

Recently, it has been appreciated that undesirable concentrations of radon can occur in buildings. A relatively small proportion comes from building materials themselves. Brick, stone, concrete or plaster can emit radon, particularly if they are exposed and undecorated or wet. However, the most significant source is diffusion of gas through the soil and movement into buildings through cracks and fissures in the soil and in the ground floor. The resulting concentrations give rise to concern in some areas, particularly where the underlying rock is granite. The Department of the Environment gives advice to local authorities where precautions are considered necessary, and the local authorities advise builders and residents where appropriate.

Conserving energy by reducing ventilation rates will increase the concentration of radon.

Reduction in radon levels in buildings can be achieved by increased ventilation or by the use of electrostatic precipitators. High rates of ventilation are required for this method to be effective and the running expense would be high unless mechanical ventilation systems and heat recovery were used. Both this method and electrostatic precipitation, although technically possible, are of a degree of complexity and cost which is likely to preclude them from general use. The most realistic approach, and one which is particularly suitable for new buildings, is to provide a continuous membrane to the ground floor; this will cut off or substantially reduce the penetration of radon gas. Such a membrane (1200 gange polyethylene) should be fully supported, carefully laid with sealed joints and with careful sealing at any service entries, cavities and through walls.

In the case of suspended floors, ventilation should be provided for the void, together with any insulation required for the floor above. Consideration should be given to cold bridges and condensation risk. The ventilation rate may be increased by the provision of a fan. In the case of a dwelling with a solid floor a large diameter (100 mm) subfloor depressurization pipe is currently recommended. This should terminate internally at a central radon sump formed by loose bricks and slabs to ensure the pipe entry is kept clear. A fan to discharge air clear of any openings in the building should be provided. Local authorities will advise on the level of precaution appropriate to their area.

Legionnaires' disease

Bacteria of the family *Legionellaceae* are widespread in soil and surface sources of water. All water systems in buildings must be regarded as potential breeding grounds for the bacteria. The first recognition of disease from this source in buildings was in 1976 at a convention in a US hotel. Since then many outbreaks of the disease have been recognized, several in the United Kingdom. Although there can be little doubt that many unrecognized outbreaks of infection did occur in the past, the complexity of water-containing services and higher temperatures prevailing in modern buildings undoubtedly increase both the risk and severity of any outbreaks.

Spread of infection requires that aerosol-sized droplets of infected water become suspended in the air and are subsequently breathed into the lungs. Sprays of water from showers or cooling towers forming part of air-conditioning plants generate large quantities of aerosols but surface agitation of water in sanitary appliances in use, or when they are filled from taps, can also contribute aerosols to the atmosphere.

The bacteria are so widespread that it is impossible to eliminate them completely. However, the level of occurrence in nature does not give rise to major disease. An effective strategy in buildings is to prevent rapid multiplication of the bacteria. The optimum temperature for rapid multiplication is 37 °C. Below 20 °C multiplication is slow and above 46 °C it ceases. In water supply systems it is important to ensure that cold pipes and storage cisterns are not warmed by their surrounding services to 20 °C or over. It may be appropriate to insulate cold-water pipes so the flow of fresh cold water is not warmed to an unsatisfactory temperature. It is advantageous if the water is not allowed to remain static. Adequate flow reduces the increase of temperature and ensures that chlorination of the mains water is effective in maintaining sterile conditions in pipes. Cisterns should not be larger than 24 h flow. They should be covered and convenient for cleaning. Hot-water systems should be designed and operated to ensure that all sections of pipework receive water regularly at a temperature higher than 46 °C, and calorifiers should be designed and operated to ensure there is not a lukewarm area at low level. Cooling towers clearly represent a major risk. Control of temperature is not practicable. Where they are required they should be sited as far away from the building as possible downstream of the prevailing wind. The design should include drift eliminators to reduce aerosol escape. A rigorous programme of cleaning, disinfection and testing is essential, and the procedures should be monitored and kept up to date with the best current practice. Prosecution by the Health and Safety Executive may result from inadequate maintenance.

4 Daylighting

The sun is important in building design for several reasons. It is the source of natural lighting; the heat from its rays may be utilized to advantage or give rise to severe discomfort through overheating; and its penetration into buildings, at least in northern latitudes, is a strongly felt need for many people. Until recently the health-giving properties of the sun's rays were thought important but medical opinion has changed considerably on this issue and the germicidal effects of sunshine in interiors are not now considered significant in relation to modern cleaning methods. Another aspect of building design which is strongly affected by the sun is the shading of areas round buildings which can affect both the vegetation and the human enjoyment of adjacent open spaces. The critical effect of these factors can be seen in almost all buildings. The depth, shape and spacing of buildings is largely governed by daylighting considerations. In some types of buildings, particularly those with large glazing areas, acute discomfort due to glare or overheating can occur unless careful design consideration is given to the penetration of the sun's rays.

In past ages when the seasons influenced everyone's life to a very great extent, knowledge of the sun's behaviour was almost universal. Nowadays artificial lighting, central heating, food preservation, roads passable in all weathers and occupations largely indoors make us very much less conscious of the basic pattern of the sun's movement. In view of the importance of this factor in modern building design it is worth devoting careful attention to the way in which the sun's rays behave. There are two cases to consider: light from the overcast sky and direct rays of the sun.

4.1 Light from the overcast sky

In some parts of the world where clear skies are the rule it is the direct rays of the sun which must be taken into account for daylighting analysis, but in Britain and many other areas of the world critical daylighting conditions arise when the sky is covered by clouds, and it is this condition which must be taken into account in design. The variation in illumination received from the sky varies very greatly according to the movement of clouds and sun. The assumption made as a basis for daylighting analysis is of a heavily overcast sky giving a total unobstructed illumination (on a horizontal plane) at ground level of 5000 lx. This condition is described as the *standard overcast sky* and measurements show that between 0800 hours and 1700 hours the illumination from the sky will be 5000 lx or more for 85% of the time.

The luminance of the sky is not uniform all over. It is some three times as bright at the zenith as at the horizon. An equation defining a luminance distribution which has been proved to represent conditions in many parts of the world was proposed by Moon and Spencer and adopted in 1955 by the Commission Internationale de l'Eclairage (CIE) as a standard for use in design.* It is sometimes called the Moon and Spencer sky and sometimes the CIE sky. Figure 4.1 shows a section through the imaginary hemisphere of the sky surrounded by a band, the width of which varies according to the luminance distribution of the CIE sky.

* CIE sky: originally in daylighting estimation the sky was assumed to have uniform brightness. A more realistic distribution of brightness which reduces oversizing of rooflights and undersizing of low windows and which has been established as occurring in many parts of the world is defined by the formula:

$$B_\theta = B_z \frac{1 + 2\sin\theta}{3}$$

where θ is altitude
B_θ is the sky brightness at altitude θ
B_z is the sky brightness at the zenith

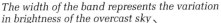

The width of the band represents the variation in brightness of the overcast sky

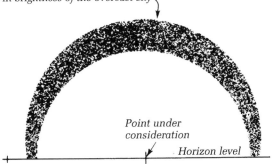

Point under consideration

Horizon level

Figure 4.1 Graphical representation of the brightness variation in the overcast (CIE) sky.

The variation of outdoor illumination makes it unrealistic to use specific values of illumination as standards for the design of interiors and the problem is overcome by the use of a ratio, known as the *daylight factor*, which expresses the illumination received at a point indoors as a percentage of the illumination received simultaneously by an unobstructed point out of doors. The daylight factor is made up of three components. The direct light from the sky falling on the point in question (the sky component), the reflected light from external surfaces (the externally reflected component), and light received by reflection from the internal surfaces of the room (the internally reflected component). It is important to bear in mind that in the early stages of daylighting analysis the importance of the reflected components and the variation of sky luminance was not fully appreciated and that the term *daylight factor* will sometimes be found used as synonymous with sky component.

Table 4.1 shows some recommended minimum standard of daylighting for dwellings and offices.

The daylight factors for dwellings are to be based on the following assumptions of reflection factors:

Walls	40%
Floor	15%
Ceiling	70%

Daylight factors are normally measured or estimated in terms of the light falling on a horizontal plane 0.85 m above ground or at the level of the working plane if this is different from 0.85 m.

The intensity of light falling on a horizontal surface is not necessarily a complete measure of satisfactory lighting. But in rooms enjoying daylighting from windows, much of the light penetrates almost horizontally; so if the working plane is adequately illuminated, the walls will be well lit, giving a pleasant interior. The general direction of the light also gives a pleasant definition to the form of three-dimensional objects. These conditions are not met in top-lit rooms where, if the same standard is applied, the walls will be poorly lit and a gloomy environment achieved. This problem is normally overcome by requiring the daylight factor to be very much higher than would be the case with side-lit rooms. Top-lit factories should have a minimum daylight factor of 5%. The roof glazing required to achieve this may, if not protected from the sun, give rise to acute overheating. The traditional north-light roof is a response to this problem. Calculations of the glare from side-lit windows are possible but they are rarely made in practice.

A wide variety of techniques is available for daylighting analysis in buildings. They range from simple tables giving values of daylighting for specific cases through many types of numerical or graphical analysis up to the measurement and often the visual evaluation of lighting conditions in models made specially for daylighting study.

The most useful techniques for designers are likely to be the average daylight factor, the BRS sky component protractors, the BRE internally reflected component nomograms, the Waldram diagram and model analysis. Each of these methods has its particular application. The following paragraphs describe the methods and provide data for their application.

4.2 Average daylight factor

The average daylight factor procedure can be applied at the early stages of design with useful accuracy. The concept is very recent. It was initiated by Lynes and has been developed at BRE. Previous practical methods can be applied to check and develop window design by trial and error but they do not lend themselves to generation of solutions.

Usually the desired daylight factor will be known. To obtain the necessary window size it is necessary to know the dimensions of the room, the internal reflection factor and the transmission factor for light through glass, allowing for dirt and glazing bars; it is also necessary to assume a location for the centre of the window. The vertical angle which the visible sky subtends from the window centre must be established. The required window area can be determined from the formula:

Table 4.1   Minimum daylighting standards[a]

Room	Daylight factor (%)[b]	Penetration	Daylight area
Kitchen	2	Cooker, sink and preparation table should be placed within the daylighted area	50% of area with minimum 5 m^2
Living-room	1	Half the depth facing the main window	7.5 m^2
Bedroom	0.5	Three-quarters the depth facing the window	6 m^2
Offices	1	4 m	–
Drawing-offices	5	Over whole area of office	–

[a] Recommended minimum standards of daylighting for dwellings from CIBSE. Recommended minimum standards of daylighting for offices from PWBS 30 *The lighting of office buildings*, HMSO, 1952.
[b] Measured on working plane.

$$\text{Window area (m}^2) = \frac{\text{daylight} \times \text{internal reflection factor}}{\text{window light transmission factor}}$$

where

Internal reflection factor (typically 0.4) =
   total internal surface area (m^2)
   $\times$ (1 − area-weighted surface reflectance)^2

Window light transmission factor =
   glass transmission $\times$ dirt and glazing bar factor
   (typically 0.8)   (typically 0.7)

If the window area is known, the formula can be used to obtain the daylight factor.

At BRE the method has been extended to rooflights. BRE Information Paper 12/94 *Average daylight factor* describes the complete method.

4.3 BRS sky component protractors

General principles

The source of light is the overcast sky. Minimum design standards are based on an illumination of 5000 lx from the whole unobstructed sky. The brightness of the sky is assumed to vary from zenith to horizon in accordance with the formula adopted by the CIE.

Light reaching any point within a room will come from three sources:

- Directly from the sky via windows or other opening (sky component); see Fig. 4.2.
- By reflection from landscape and buildings outside (externally reflected component); see Fig. 4.3.
- By reflection from other surfaces in the room (internally reflected component); see Fig. 4.4.

The measure of daylighting is the daylight factor which gives as a percentage the relationship between the light actually falling on a point within a building and the light which would fall on that point from the unobstructed sky. It is expressed as a percentage.

The daylight factor comprises the internally reflected component plus the externally reflected component and the sky component, if any.

The BRE protractors (second series) compute the sky component and enable the externally reflected component to be established.

The sky component protractors perform three functions:

- They establish the proportion of the whole sky which contributes light directly to the point under consideration.
- They make allowance for the effect of glazing.
- They make allowance for the angle of incidence of light with the working plane.

The protractors are designed to be used on drawings.

Plans and sections are required for the rooms concerned, including details of windows and of any external obstructions and their relationships to the window.

Since the behaviour of light is consistent, irrespective of scale, any conveniently sized drawings may be used. Plans and sections need not be to the same scale.

There are five protractors for use (available from government bookshops):

	Protractor no.
Vertical glazing	2
Horizontal glazing	4
30° glazing	6
60° glazing	8
Unglazed apertures	10

Other protractors are available based on a sky of uniform brightness.

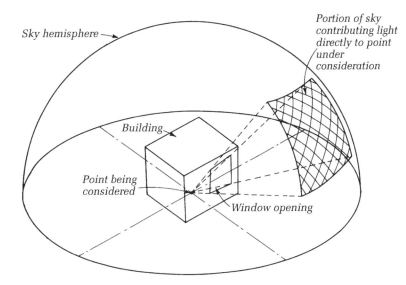

Figure 4.2 Principles of daylight penetration directly from the sky through a window.

Methods of analysis

The information required about daylighting will usually be in one of two forms:

- Values of daylighting at one or more particular spots known to be critical.
- The general distribution of daylighting across the working plane in a room; this will involve the establishment of daylighting values at a grid of points covering the working plane.

The following items should be borne in mind in laying out the grid:

- A rectangular grid is usually convenient. Spacing will depend on the size of the room and closeness of analysis required. Generally 0.75–1.0 m spacing is convenient in rooms up to 200 m².
- Space adjacent to walls may sometimes be known to be occupied by cupboards or circulation and need not be considered. Where this is not the case it will often be convenient to fix grid points near the wall one half grid space away.
- Work may be saved if grid points are arranged symmetrically about the centreline of the window (or the main window if there are more than one) and in lines parallel to the window. Providing there are no symmetrical obstructions, values need only be computed for half the points, and values on section will be identical for each line.
- For easy identification of points, grid lines in one direction should be labelled A, B, C, etc., and those in the other direction 1, 2, 3, etc.

Detailed use of protractors

Using a protractor on a section

Consider vertical glazing. On a section of the room normal to the window, mark the working plane and the point under consideration. Draw lines from the grid point under consideration to the highest and lowest limits at which the sky can be seen from the grid point. Place the base of the sky component (long windows) along the working plane and the centre over the grid point. Read off the value as shown in Fig. 4.5. The lowest limit may be the sill, or the skyline of the landscape or surrounding buildings if this can be seen above the sill from the point being examined.

The resulting value is the sky component which would be received at the point from an infinitely long window having the section indicated. A correction must be made for the limited length of a normal window. And using the normal degree-scale protractor, note the mean altitude of the visible sky.

Using a protractor on a plan

On a plan of the room draw lines from the grid point to the edges of the visible sky. Place the correction factor side of the protractor with its base parallel to the window and its centre over the point under consideration. Using the main altitude found, choose the appropriate concentric scale and read the correction factors for each side of the window. Add them to obtain a final correction factor where readings are on opposite sides of the centre zero. Where both readings are on the same side, subtract the smaller from the larger.

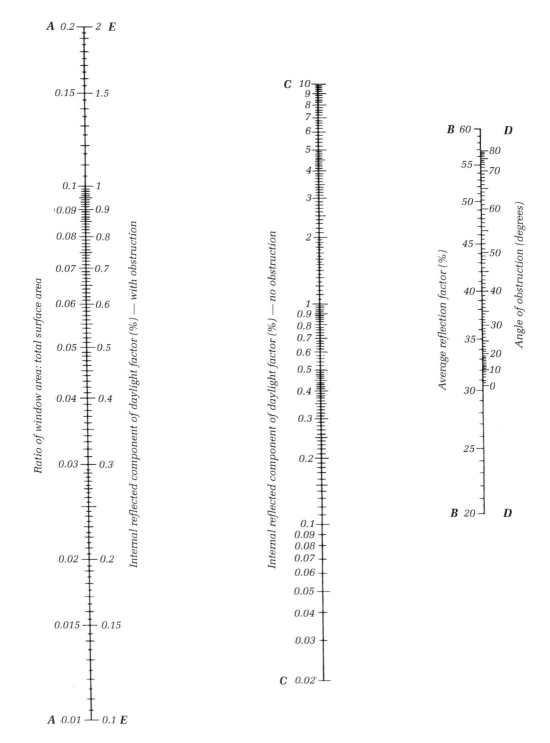

Figure 4.3   Average internally reflected component for side-lit rooms. (After *Architectural Physics: Lighting*, HMSO, 1963)

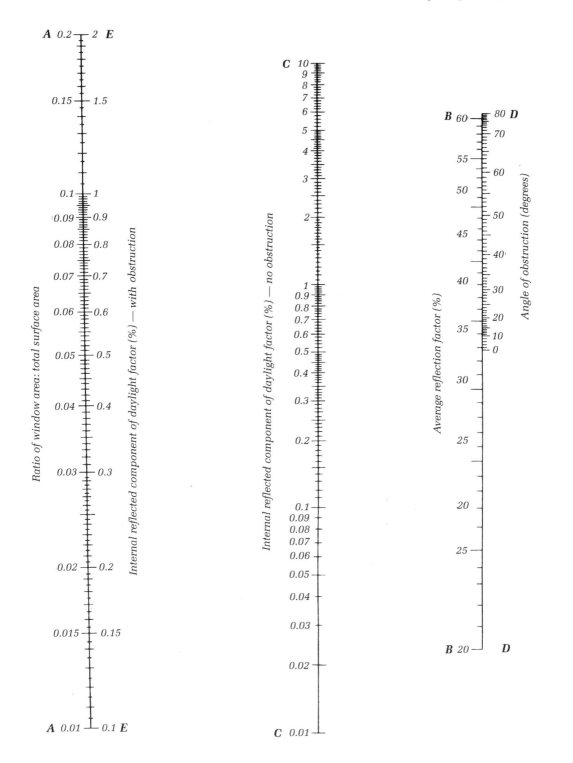

Figure 4.4 Minimum internally reflected component for side-lit rooms. (After *Architectural Physics: Lighting*, HMSO, 1963)

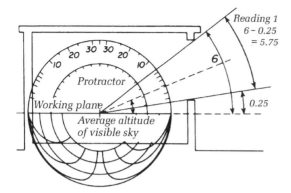

Figure 4.5   How to use a BRE sky component protractor.

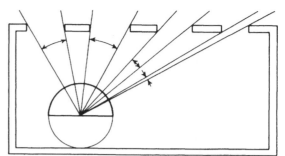

Figure 4.7   Plan correction for several windows using a BRE sky component protractor.

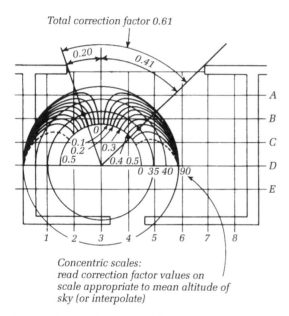

Concentric scales:
read correction factor values on
scale appropriate to mean altitude of
sky (or interpolate)

Figure 4.6   How to use a BRE sky component protractor on a plan. In this case the mean altitude of the sky is 30°.

To establish the sky component of the daylight factor multiply the value for long apertures found on the section by the plan correction factor.

In the example shown in Figs 4.5 and 4.6

$$5.75 \times 0.61 = 3.51$$

More than one window may contribute light:

- *For a window in the same wall as the main window*
  Add the plan correction factors for each window and multiply the section value by this sum (Fig. 4.7).
- *For a window in other walls or roof*
  Compute separate direct components and add.

Glazing bars may be allowed for by making a percentage reduction of light received.

Irregular external obstructions or non-rectangular windows may often be dealt with satisfactorily by considering them as equivalent to one or more rectangular windows. The rectangular windows assumed will be determined visually.

If window cleaning will be infrequent or the atmosphere is dust-laden, estimate and make a suitable percentage reduction of light received.

Direct components can be computed for planes other than horizontal. Consideration of the basic geometry in relation to the particular problem will usually reveal a method for solution.

In practice it will usually be found unnecessary to draw the lines delineating the edges of visible sky. Readings can be made by using a set square or straight edge. In school work, however, lines should be shown to demonstrate the method is correct.

Light from the landscape outside (the externally reflected component) can be computed if its brightness relative to the sky it obstructs is known. In practice it is found that obstructions have an average brightness of about one-tenth of the average sky brightness. A satisfactory allowance will be made if a conversion factor of 0.1 is used to reduce the appropriate sky component. A separate set of protractor readings with appropriate average altitude should be made.

It is usually convenient to carry out the calculations in tabular form (Fig. 4.8).

Presentation of data

Visualization of distribution

The direct components computed will often be marked on the plan against their appropriate points. Interpretation and understanding of the distribution can usually be assisted by graphical methods.

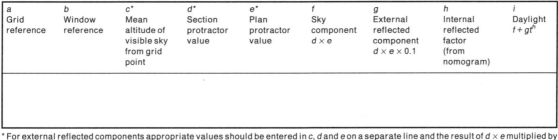

a Grid reference	b Window reference	c* Mean altitude of visible sky from grid point	d* Section protractor value	e* Plan protractor value	f Sky component $d \times e$	g External reflected component $d \times e \times 0.1$	h Internal reflected factor (from nomogram)	i Daylight $f + gt^h$

* For external reflected components appropriate values should be entered in c, d and e on a separate line and the result of $d \times e$ multiplied by 0.1 and entered in column g.

Figure 4.8   Daylight calculation: a typical layout.

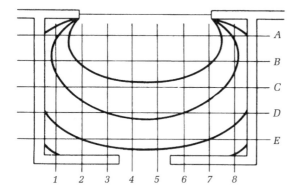

Figure 4.9   Contour plan showing daylight distribution.

Contour plans of the distribution   Figure 4.9 shows a typical contour plan of daylighting distribution which will be self-explanatory. More complex distribution patterns would result from additional windows or rooflights, but no special difficulty would arise from this.

Graphs of intensity   A visual indication of variation in intensity can be given by drawing on the section a graph of the varying intensity across the room (Fig. 4.10). The horizontal scale will represent the room. The extent of the vertical scale can be determined by convenience in including the curve within the section; its subdivision between the upper and lower limits of daylight forming the extremes of the scale will be logarithmic to make the graph variation correspond more closely with the sensitivity of the human eye (logarithmic scales can be projected from Figs 4.3 and 4.4).

4.4 Internally reflected component

The Building Research Establishment has developed a series of nomograms which enable the internally reflected component of the daylight factor to be

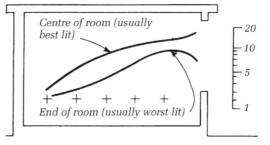

Figure 4.10   Section showing daylight distribution.

determined quickly. Separate nomograms are required for side-lit and top-lit rooms.

Figures 4.3 and 4.4 show nomograms to give the average and minimum values of the internally reflected component for side-lit rooms. The scales are self-explanatory.

The average reflection factor can be determined by summing the products of the areas of each different surface and their reflection factors and dividing this total by the total area. (The reflection factor of windows may be taken as 0.15.)

Use of the nomograms involves joining appropriate points on scales A and B then reading the unobstructed internally reflected component on scale C. If there are external obstructions they can be taken into account by joining the angle of the obstruction on scale D to the unobstructed value already found on scale C and projecting to cut scale E at the required value.

The decision to use average or minimum nomograms depends on the purpose for which the results are required and the form and window position in the room.

4.5 Waldram diagram

The Waldram diagram was the main method of daylighting analysis before the development of the BRE protractors and it still has applications where it is difficult to organize the visible sky into rectangular units.

Half of the hemisphere of sky is mapped on to a rectangular chart (Fig. 4.11) so the area of any portion of the diagram has the same relationship to the whole area as the light contribution for the part of the sky represented has to half the light from the whole sky.

To make use of the diagram, the outlines of the sky seen through window openings or above complicated roof lines are plotted on the diagram, the area enclosed measured and expressed as a percentage of twice the total area of the diagram, thereby giving the sky component. Vertical lines remain vertical in the diagram but horizontal lines such as window heads and sills curve down to vanishing points. The laborious task of plotting these curves by means of a series of altitudes and azimuths of points along the horizontal member can be simplified by the use of a droop line diagram (Fig. 4.12), which shows the curves representing horizontal lines running at right angles to the line of sight. Measuring the areas can be simplified by the use of a sheet divided into rectangles, the area of each representing 0.1% of the sky component (Fig. 4.13).

Using Waldram diagrams to estimate the sky component of a window at a selected point

1 On the plan draw a line from the point being considered normal to the window wall.
2 Measure azimuth angles of vertical edges of the sky seen between jambs.
3 Measure the altitude of the upper and lower limits of the sky seen through the window on the normal line (if the line does not cut the window, project the head and sill heights to the normal).
4 Place a piece of tracing paper over the droop line (Fig. 4.12) and draw vertical lines at appropriate azimuth angles (the normal is 0°). Select appropriate altitudes for head and sill at 0° azimuth, and using the droop lines as guides, trace curves representing the head and sill. The area enclosed between the vertical lines and curves represents the proportion of the light from the sky which will fall on the point being considered. (Where irregular obstructions exist, the outline may be transferred to the basic diagram, and altitudes and azimuths plotted.)
5 Transfer the tracing paper to Fig. 4.13 and count the number of rectangles enclosed within the

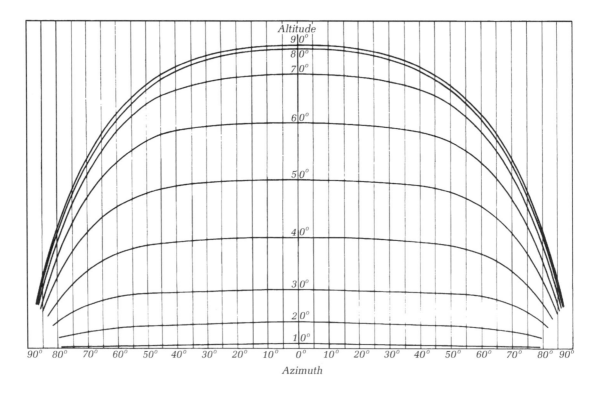

Figure 4.11 Waldram diagram for CIE sky and allowing for light transmission through clear window glass. [*See page 71.*]

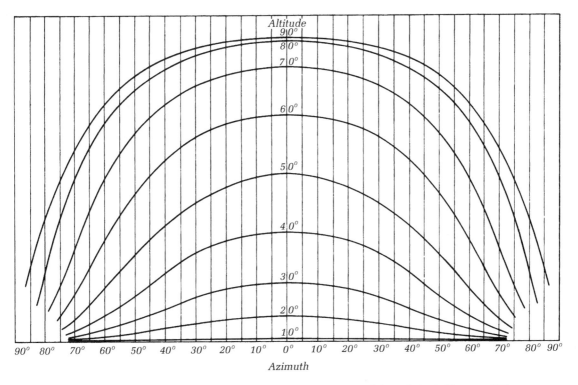

Figure 4.12   Droop line diagram for use with Fig. 4.11 (for obstructions parallel to window). [*See page 71.*]

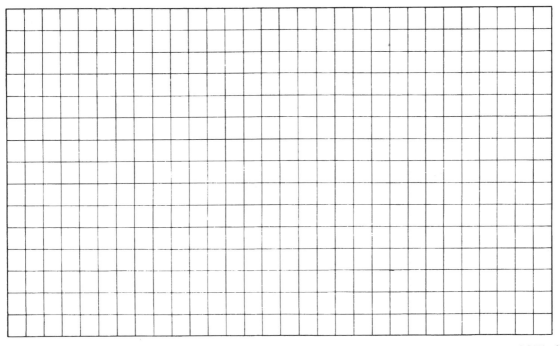

Figure 4.13   Calculation sheet for use with Waldram diagram. Each rectangle represents 0.1% of sky component or 0.01% of externally reflected component. [*See page 71.*]

window outline. Each rectangle represents 0.1% sky component at 0.01% externally reflected component.

4.6 Block spacing

When laying out groups of buildings, particularly in town planning situations where sites in different ownership are involved, it is necessary to have criteria which will control the spacing between the height of blocks so that, in later detailed design, it is possible to achieve reasonable daylighting. This is particularly important in town planning applications where the daylighting of future buildings, the details of which are unknown, must be safeguarded. In the past this was done by fixing street widths and building heights; or at the rear, by building heights, taking an angle from the first floor sills on one side of an open area and prohibiting buildings opposite from rising above this. While buildings remained limited in height (in Great Britain 80 ft with two storeys in the roof), there was no great problem apart from the uniformity of streets which resulted and the dreariness of light wells. The desire for more sophisticated town forms and higher buildings led to more flexible methods of ensuring adequate daylighting, but they do not take the form of statutory requirements. These recommendations will ensure that, in any building layout, adequate light will fall on the face of all the buildings.

There are separate recommendations for residential and non-residential buildings. For non-residential cases all sides of the buildings, except end or flank walls less than 15 m long, must have a sky component of at least 0.97% at all points 2 m above ground level. For calculating the sky component only sky visible within an area 45° on either side of a line normal to the wall and above 20° altitude but not above 40° can be included. These limits ensure that daylight can penetrate to adequate depths within the building. Similar tests can be applied along a boundary (or road centreline) to preserve the daylighting for an adjacent site to be developed. In this case it is ensured that an appropriate amount of daylight falls, from the inside of the site being developed on to all points 2 m above the boundary lines.

Table 4.2 shows the range of criteria.

The Department of the Environment (DoE) issues a set of indicators which can be used to check whether the criteria are met. There are sets of special indicators for each of the cases to be considered. Figure 4.14 shows a typical indicator.

The indicators form a quick and simple means of making a check on daylighting for planning and block layout. Although any situation which satisfies the indicators is acceptable, there can be circumstances in which precise analysis becomes difficult. The Waldram diagram (Fig. 4.11) can be used to give a precise analysis of whether a particular building configuration satisfies the criteria given in Table 4.2. A diagram is required that makes no allowance for glazing. Figure 4.15 is specially prepared for this purpose. It is used in exactly the same way as described above. Note that the droop lines are superimposed on the diagram itself. Figure 4.16 is a calculation sheet with the limits of altitude and azimuth for the various applications indicated. Only squares within the outlines defining the altitude and azimuth limits may be counted for satisfaction of the daylighting criteria. Figures 4.15 and 4.16 may be used for normal analysis of daylight penetration for any situation where there is no glazing. In this case the whole of the calculation sheet may be employed.

4.7 Model analysis

It is difficult for graphical methods of daylighting analysis to cope with very complex geometrical situations, particularly if many repetitions are needed. Mathematical and computer methods can overcome the problems of repetition but complex forms, varying colours and complicated glazing bars and mullions, still present difficulties if precise results are required. Model studies are of considerable value in these circumstances. Provided the scale model is accurately made, it will cause light to behave in exactly the same way as the actual buildings. And there is the important additional benefit that visual inspection can also be made to assess the effects of colour and appearance. There seems little doubt that, in many cases, where close estimates of daylighting levels are required, visual appraisal of the actual appearance will also be important.

To permit easy visual inspection, models of interiors should be reasonably large scale. For small rooms 1:10 would be suitable.

A suitable lightmeter is needed to assess the daylight factor in a model. Readings are taken inside the model and compared with readings from the whole unobstructed sky to give the percentage of the total available light actually falling on the point investigated.

The natural sky may be used as a source of light, but a position must be found with an unobstructed view of the sky and, to establish the daylight factor, a uniformly cloudy sky must be used. Both these conditions are difficult to meet, and model studies

Table 4.2   Daylighting criteria

Type of development and application	Percentage sky component (2 m above ground level)	Limits within which sky component must be achieved		
		Horizontal (azimuth)[b]	Vertical (altitude)	
Residential				
Block spacing[a]	0.84	45°	≥10°	≤30°
Boundary	4.3	65°	≥19° 26′	≤49° 6′
Non-residential				
Block spacing	0.97	45°	≥20°	≤40°
Boundary	2.9	65°	≥36° 3′	≤59° 13′

[a] Façades of residential buildings facing south (or in any direction east or west of south, including due west) are tested for sunlight not daylight.
[b] Angle on both sides of normal line to wall.

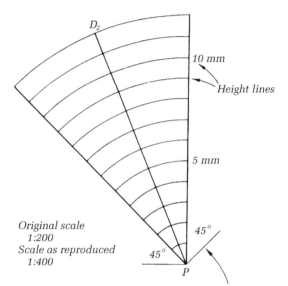

Original scale
1:200
Scale as reproduced
1:400

Guideline for positioning the indicator in relation to the building façade (only light falling beyond an angle of 45° with the façade may be counted)

Figure 4.14   Residential block spacing indicator (building to building).

are usually carried out in artificial skies. The most common pattern is the mirror type sky shown in Fig. 4.17. Few design offices are likely to possess artificial skies but many building research organizations and schools of architecture have them and make them available for use by designers.

4.8 Direct sunlight

Although it is at present not possible to produce any definite data of human requirements for sun penetration into buildings, it is clear that some degree of sun penetration is a strongly felt need on the part of many people. This is recognized in the building codes of a number of countries. Table 4.3 quotes the recommendations for Britain given in BSCP Chapter 1 (b) Sunlight. More recently, however, the recommendation made in *Sunlight and Daylight* by the DoE is that all sides of residential buildings having any southern orientation (i.e. any orientation from just south of east, through south and including due west) should have at least one hour of sunshine on 1 March (above an altitude of 10°). This is a much less stringent recommendation than in the code of practice, but its intention is to provide a realistic standard capable of achievement which takes into account the external as well as the internal environment (sunshine bearing less than 22.5° from the side of the building is not excluded since it contributes to pleasant external conditions). It is to be hoped that the DoE intention will be achieved and that residential developments will conform to its new suggestions.

Remember that solar radiation has problems as well as advantages and the penetration of the sun's rays into a building can give rise to acute discomfort because of glare. Direct sunshine can give illumination levels on working planes in buildings of 60,000 lx or more, which may be compared with 300–500 lx, perhaps a good level of artificial lighting or daylighting away from windows. The eye is able to accommodate itself over a very wide range of illumination levels. Moon quotes a range of adaptation of 100,000 to 0.1 lx at 1,000,000 to 1, but the eye can only be adapted to one level at a time and patches of bright sunshine can create a bright and cheerful environment or may cause intolerable glare. Careful consideration of form, fenestration and orientation in relation to sunshine is vital in the design of buildings and may have to be supplemented when necessary by

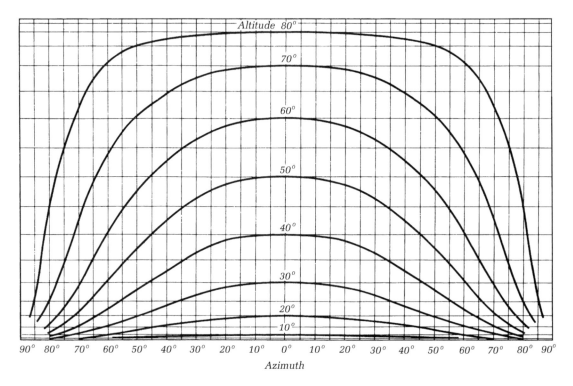

Figure 4.15   Waldram diagram for CIE sky with light falling on the horizontal plane (droop lines are for plotting horizontal obstructions). [*See page 71.*]

screens, louvres or blinds to control the degree of sun penetration.

Sunshine entry to rooms is not taken into account until the horizontal angle between the sun's rays and the plane of the window is more than 22.5° and more than 5° above the horizon.

Solar heat gain

Overheating due to solar gain has not, in the past, been a serious problem in Britain. Buildings had massive walls with small openings which minimized the direct effect of the sun's rays on the interior. Since heights were limited, vegetation and adjacent buildings provided additional screening. Many modern buildings are quite different in their reaction to solar heat. Large proportions of the walls are glazed, the construction and cladding are lightweight and quickly warm up, and tall buildings are not shaded by trees and other buildings. These changes, coupled with the property of glass which enables the high temperature radiation of the sun to pass through while preventing the passage of the lower temperature radiation from within the building, mean that acute overheating is now

encountered in buildings. The situation is made worse by increased external noise levels, which make people reluctant to dissipate heat through open windows. Ventilation problems in tall buildings may have dictated the use of mechanical ventilation and the windows may not even be openable.

The solar radiation striking the earth's atmosphere brings about $1.5\,\mathrm{kW\,m^{-2}\,h^{-1}}$ (measured normal to the sun's rays). Part of this is reflected, part scattered and absorbed by the atmosphere. Some of this scattered radiation reaches the ground as diffuse sky radiation. The radiation which penetrates directly to impinge upon buildings will rarely strike them squarely and the amount of heat per square metre of the building's surface is therefore substantially less than that originally present in the sun's rays. CIBS Guide A2, *Weather and solar data 1986*, gives a maximum value, depending on the term of orientation, of $635\,\mathrm{W/m^2}$ for solar radiation falling upon walls. Solar heat falling on walls will take time, of the order of some hours with most typical constructions, to pass through the wall. During this time external temperatures will have dropped and a substantial part of the heat gain is likely to be lost externally. The small windows and

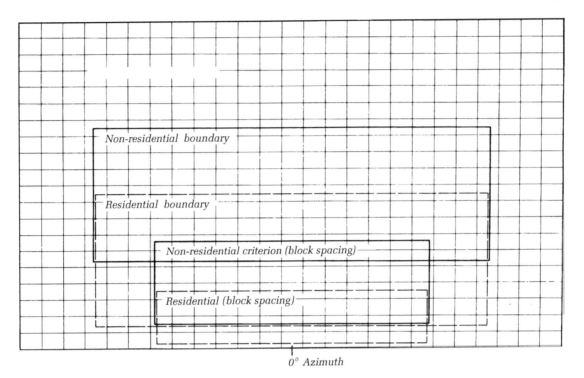

Figure 4.16 Calculation sheet for use with Fig. 4.11. Limits are marked for the satisfaction of non-residential boundary and block spacing criteria. Each square represents 0.1% of the sky component. For satisfaction of the daylighting criterion, only squares within appropriate rectangles may be counted. [*See page 71.*]

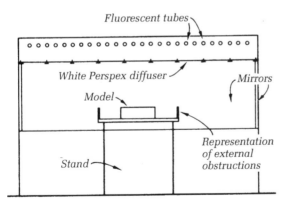

Figure 4.17 Mirror-type artificial sky. The interreflections between the mirrors make the Perspex diffuser appear to extend to infinity; they can be arranged to give a distribution of light similar to the CIE sky.

Table 4.3 Minimum sunlighting standards

Room	Preferable time of day for sun penetration	Minimum period for penetration[a]
Living-room	Afternoon	At least 1 h at some period of the day during not less than 10 months of the year from February to November
Bedroom	Morning	
Kitchen	Morning	

[a] Recommended minimum standards for sunlight penetration for dwellings.

massive walls of some types of tropical buildings show how buildings have been constructed to minimize the effects described.

In England an extension to a school has been constructed in which the heat gain from occupants, lighting and the sun made it possible to dispense with a normal heating installation. It is not possible to do without heating in many buildings but this example does demonstrate the importance of the sun's rays and the need to take them into account in design, both to conserve energy requirements for heating and to avoid summer overheating.

Correct orientation of windows is critical if maximum thermal advantage is to be taken of solar heat gains. The amount of heat which enters a south-facing window during the heating season is substantially greater than for any other orientation. If, as is usual in domestic buildings, the south-facing windows are curtained at night during the period of maximum heat loss, more heat will enter an unobstructed south-facing window during the winter than will be lost through it. This is not an argument for providing very large windows on southern elevations specifically to save energy, since mean radiant temperatures internally will be reduced and, although energy may be saved overall, larger heating installations will be needed for overcast periods. It is, however, a powerful argument for considering the orientation of buildings more carefully than at present and for planning internally so that where large windows are appropriate they may be concentrated on southern façades. It is important to remember that substantial obstructions will reduce the seasonal heat gain and that only elevations oriented very close to due south enjoy the full benefit.

Figure 4.18 shows the notional seasonal heat balance through 1 m^2 of south-facing single glazing with curtains drawn at night. The heat gains and losses are approximately equal. This does not mean that such windows have no effect on energy consumption since heat input to match the window losses will be required at times of little sunshine and when the sunshine is abundant much of it may be wasted.

Figure 4.19 shows a similar analysis for a single day of bright sunshine in January. The diagram shows how curtaining and reduced internal temperature at night can reduce energy consumption and how the window will make a significant contribution to heating during the day under those conditions. The heat gain will only be realized if the control of the heating system can provide the graduated rate of heat input needed.

Overheating in summer may be a risk in buildings with large south-facing windows. The admittance method may be used to establish the temperatures likely to be reached in summer. South-facing windows can have relatively easy control of solar heating. Special window glasses to reject solar heat are not appropriate since they will limit winter gains. Internal blinds, although they prevent the sun's rays from falling directly upon occupants of rooms, allow the sun's heat to pass through the glass, heat the blind and thereby the room itself. External blinds are particularly effective as are blinds in ventilated spaces between glazing. But the most convenient aspect of solar control for south-facing windows is that a relatively small projecting canopy over the window can give control of sun penetration in summer and allow full benefit from the rays in winter. Figure 4.25 can aid in the design of this type of canopy.

Geometry of the sun's movement

Control of sun penetration from the point of view of either illumination or warmth depends initially on the form of the building in relation to the sun's movement. A clear appreciation of the geometry of this movement is essential in the early stages of design. In the final stages, very detailed analysis may be required not only of the angles of the sun's rays but also of the heat gain resulting. The basic relationship between earth and sun is that the earth moves round the sun, making a complete orbit once a year. It is in fact the completion of the orbit which defines the year itself. As it moves along its orbit the earth revolves on its own axis, exposing its whole surface to the sun's rays during each 24 h and thus producing night and day. The axis is not normal to the plane of the orbit but at an angle of approximately 22.5° from the normal. Consequently, opposing hemispheres incline towards and away from the sun as the orbit proceeds and thereby benefit more

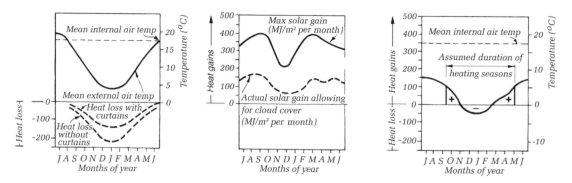

Figure 4.18   Seasonal heat balance for 1 m^2 of obstructed south-facing single glazing. Heat losses and gains are in MJ/m^2 per month.

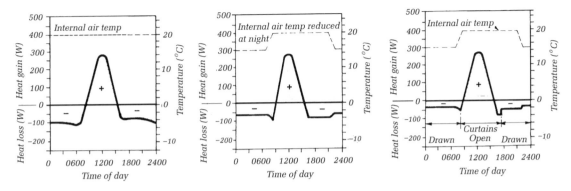

Figure 4.19 Heat balance through 1 m² of unobstructed south-facing single glazing during 24 h with clear sky in mid-January. Also shown are the effects of night-time temperature setback and the use of curtains at night.

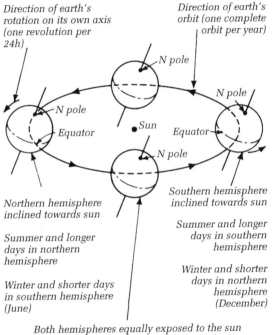

Figure 4.20 Principles of earth and sun relative positions.

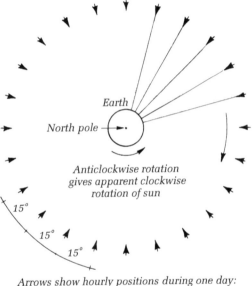

Figure 4.21 A plan of the earth looking down on the north pole.

or less from the sun's rays, thereby giving the winter and summer seasons. Figure 4.20 shows the principles of the relationships. It is not easy to imagine the motion in relation to the sun of a building sited at a point on the earth's surface and subject to the patterns of movement described. It is very much easier to consider the earth as stationary and the sun as moving and this also has the advantage of corresponding much more closely with the way in which the sun appears to behave when viewed from the earth's surface. This way of conceiving the relationship of earth and sun is

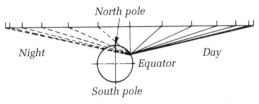

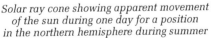

Figure 4.22 A view of the earth from above the equator.

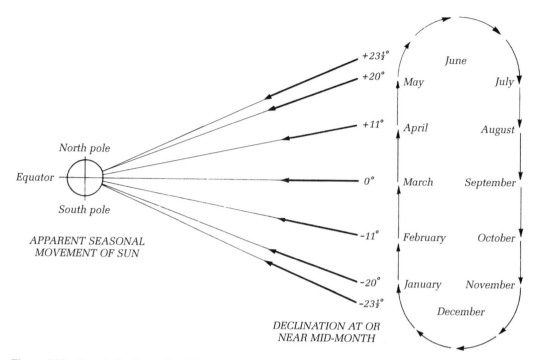

Figure 4.23 How declination varies during the year.

easily expressed geometrically. Figure 4.21 shows a plan of the earth looking down on the north pole with arrows indicating the changes in direction of the sun's rays at hourly intervals. Figure 4.22 shows a view of the earth from above the equator showing how the direction of the sun's rays resulting from the inclination of the earth at a particular time of year and the hourly movement of the sun throughout the day may both be represented by a solar ray cone. A series of such cones would be required in order to represent the apparent movements of the sun through a season because of the constantly changing relationship of the earth's inclined axis.

The angle, in relation to the equator, at which the sun's rays fall on the earth through the year is known as the declination. Figure 4.23 shows how the declination varies during the year. These diagrams show the principle of the apparent movement of the sun in relation to the earth. They do not, however, give a comprehensive picture of the apparent movement in relation to specific points on the earth's surface. Figure 4.24 uses lines on a hemisphere to show how the sun moves in relation to a point on the ground at 52°N latitude. The long lines running round the hemisphere represent the path of the sun at various times of year. The highest line represents the path of the sun in mid-June, and the lowest and shortest the path for mid-

December. The shorter cross-lines represent hours. The diagrams show a perfect symmetry of movement. In fact, there are minor alterations. These are important in navigation but of no significance in building, and the perfect geometry may be assumed for building design.

Figure 4.25 is a sun path diagram of 52°N which may be used to give information about the angles of

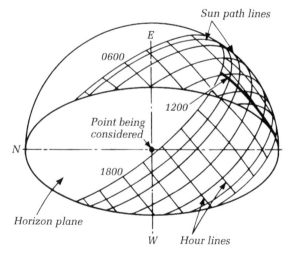

Figure 4.24 Apparent movement of the sun.

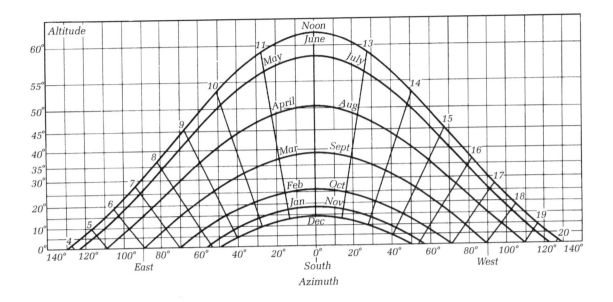

Figure 4.25   Sun path diagram for 52° N. All times are solar times with 1200 due south; correction must be made for BST. [*See page 71.*]

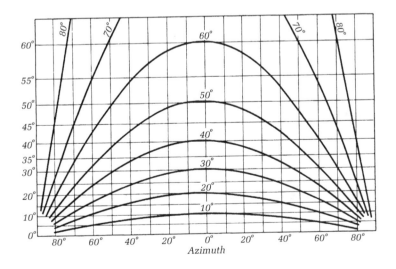

Figure 4.26   Droop line diagram for use with Fig. 4.25. [*See page 71.*]

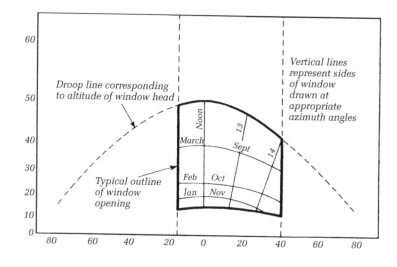

Figure 4.27 How to use sun path and droop line diagrams. [*See page 71.*]

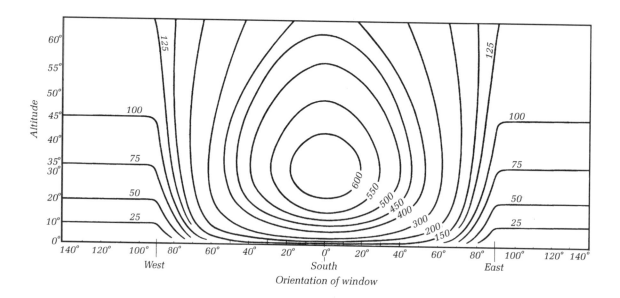

Figure 4.28 Solar gain (W/m^2) through 4 mm vertical glazing; for use with Fig. 4.25. [*See page 71.*]

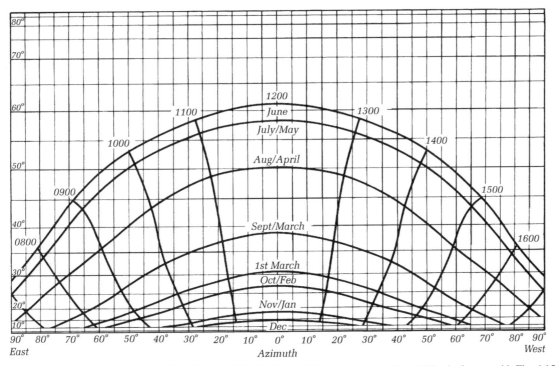

Figure 4.29 Sun path diagram for 52° N (London, Birmingham and Southern England) on 1 March; for use with Fig. 4.15. [*See page 71.*]

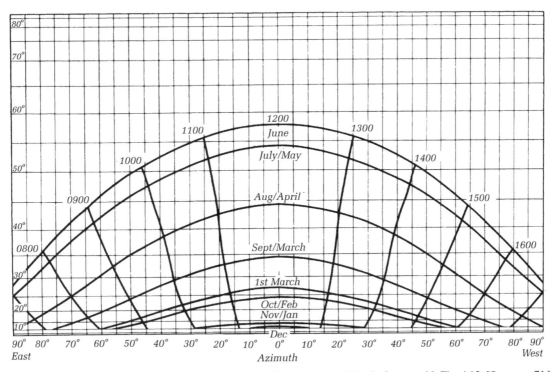

Figure 4.30 Sun path diagram for 54° N (Manchester and Newcastle) on 1 March; for use with Fig. 4.15. [*See page 71.*]

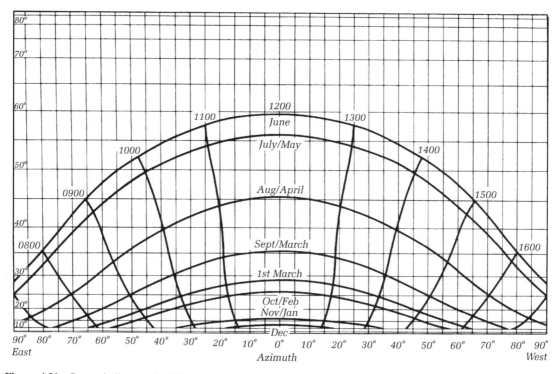

Figure 4.31   Sun path diagram for 56°N (Edinburgh and Glasgow) on 1 March; for use with Fig. 4.15. [*See page 71.*]

the sun. Azimuth angles are measured from 0° at due south.

The insolation of a particular point throughout the season may be established by projecting an outline of the window opening on to the diagram in a similar way to that used with the Waldram diagram. Since it is not convenient to have the droop lines marked on the sun path diagram itself, a separate diagram is provided (Fig. 4.26). A sheet of tracing paper is used to draw the window outline, which is then transferred to the sun path diagram. The normal line to the window is positioned on the azimuth scale at a point corresponding to its own angle from south. Periods of insolation can be identified within the outline of the opening. Figure 4.27 shows the principles involved. In addition to window opening it is possible to project the outline of buildings round open spaces to establish insolation there.

The chart is graduated horizontally in degrees measured from south (azimuth) and vertically in degrees of altitude. The curved lines represent the movement of the sun during one day at the middle of each month. The cross-lines represent hours in solar time or GMT. The slight variations between solar time and GMT are not significant for buildings. The diagram can be used with reasonable accuracy

anywhere in England and Wales. (Noon altitude error is less than 1° at all latitudes between Southampton and Nottingham and less than 2° at York.)

Use of the diagram

1   Azimuth and altitude of the sun for any time of day and year may be read from the chart. These angles may be transferred to drawings to carry out graphical analysis of insolation.
2   The outline of the sky seen through a window from a particular point may be projected on to the sun path diagram and the periods during which the sun can fall on the point under consideration observed within the outline. External obstructions should be taken into account and excluded from the outline of the visible sky. In the case of exterior situations the skyline of the surroundings can be projected on to the diagram in a similar way.

Recently another method of insolation analysis has been published in this country. It is based on gnomonic projection and allows the use of normal elevations of windows rather than distorted diagrams. It is of particular interest where appearance of the windows from inside or the extent of the view must be considered at the same time as sun penetration. To

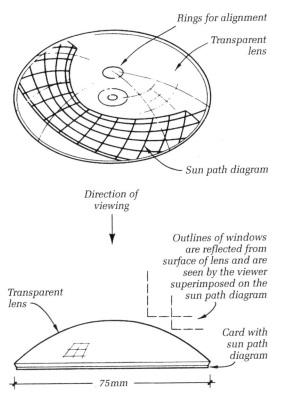

Figure 4.32 The TNO meter.

use the method an extensive range of diagrams is necessary. Sets have been published in the *Architects Journal Environmental Handbook* and in *Windows and Environment*, published by McCorquodale, 1969.

Figure 4.25 may also be used to estimate the heat gain through windows at any time of day or year. A tracing is made of the sun, and hour lines and the base of the diagram, with the 0° azimuth marked. The tracing is then transferred to Fig. 4.28 and positioned with 0° azimuth on the tracing corresponding to the orientation of the window. The heat gain per square metre may now be read off against any time of day or year. The horizontal lines on Fig. 4.28 represent the diffuse radiation from the sky.

Sunlight criteria

The methods described above can clearly be applied to the sunlighting criterion of the DoE. To simplify checking the DoE criterion, Figs 4.29, 4.30 and 4.31 show sun paths for 52, 54 and 56°N capable of covering England, Wales and Scotland with reasonable precision and constructed so that the outline plotted on Fig. 4.15 for checking the daylight criteria may be applied to the appropriate sun path diagram. Periods of insolation will be indicated by unobstructed sun path lines.

The sun path line for 1 March is specifically indicated, making it simple to check the DoE criterion. The 0° azimuth from Fig. 4.15 must be positioned over the appropriate window orientation on Figs 4.29, 4.30 or 4.31.

Field measurements of insolation

Study of insolation in existing buildings used to be laborious, either involving a survey and graphical analysis or the use of a Robin Hill Camera. The Dutch Public Health Laboratories in Delft have produced a very simple device called the TNO meter which enables very rapid determination to be made.

The device, shown in Fig. 4.32, consists of a circular sun path diagram under a transparent lens. When noon is oriented due south and the meter viewed from directly overhead, the bright reflection of windows or sky will be seen reflected from the surface of the lens, apparently directly over the sun path diagram. It is possible, therefore, to establish immediately the duration of insolation at that spot throughout the year. There are two sizes of meter; the smallest, about 75 mm in diameter, can easily be carried in a pocket.

Note:
Daylighting and solar worksheets

For greater accuracy in calculation the daylighting and solar worksheets may be enlarged. It is critical, however, that the scale of enlargement is the same for each sheet.

5 Heat

5.1 Thermal comfort

Heat presents a complex situation since a wide variety of factors, including most of those already mentioned, must be considered in balance with one another. The appropriate balance is governed by the human body's requirements for thermal comfort. Although it is not appropriate here to attempt to cover the physiology of the body's need for and reaction to heat,* general comfort is essential to any appreciation of the aspects of the natural environment which are concerned. Heat is constantly produced by bodily processes and must be dissipated to keep the body temperature at its correct level. The body normally loses heat by radiation, convection and evaporation, and if human beings are to be thermally comfortable, not only must the appropriate quantity of heat be lost, but a proper balance must be maintained between the various modes of loss. The rate of heat loss in each of these aspects is governed by the surrounding environmental conditions.

The net heat loss by radiation is controlled by the mean radiant conditions; the rate of loss by convection is dictated by the air temperature and rate of air movement; and evaporation losses by breathing and sweating depend on air temperature, relative humidity and air movement. The acceptable value of each of these features is not fixed but can vary in conjunction with one or more of the others. It is also possible for the body to vary its own balance of losses (e.g. increased sweating). There are, however, limits to each of the factors beyond which a satisfactory balance for comfort is not possible.

* The fundamental work in this field is described in *Basic Principles of Ventilation and Heating*, Thomas Bedford, H. K. Lewis, 1974.

There have been several attempts to produce a unified means of representing thermal comfort. The katathermometer, now used as a sensitive omnidirectional anemometer, was originally used by Sir Leonard Hill as a measure of warmth. Unfortunately it is more sensitive to air movement than to the human body and is consequently not a reliable measure of human reaction to warmth. A series of experiments in the laboratories of the American Society of Heating and Ventilating Engineers led to the development of a more sophisticated device called the scale of effective temperature, which takes into account air temperature, air movement and relative humidity. No account is taken of radiation. In England an instrument called a *eupatheoscope* reacting to thermal conditions in the same way as the human body, was developed by Dufton. The instrument is influenced by air temperature, air movement and radiation but takes no account of relative humidity. The readings of this instrument are in terms of equivalent temperature. This scale can also be established by measuring air temperature, air movement and radiation separately and combining them by a nomogram. Table 5.1 shows for standard air movement conditions a table of air temperatures and mean radiant temperatures to give equivalent temperatures of 18.5 °C and 25 °C. The heavy lines on the table indicate conditions beyond which one of the factors will cause discomfort which cannot be balanced by variations in the others. Extensions to the effective temperature and equivalent temperature scale have been made to cover mean radiant temperature and relative humidity, respectively; the resulting scales are described as *corrected effective temperature and equivalent warmth*. Here is a brief chronological summary of comfort indices.

- *Dry bulb temperature*
  Long established.
- *Wind chill – sensible cold*
  Concepts developed in the early nineteenth century to measure not temperature, but rapidity of heat loss. One instrument developed was the Heberden thermometer.
- *Wet bulb temperature*
  Nineteenth century.
- *General Board of Health 1857 Report*
  Called for walls warmer than air, air temperature higher at floor than at head.
- *Globe thermometer temperature, 1877 Aitken (and Vernon 1930)*
  Combined effect of radiation and air movement. Also resultant dry temperature, Missenard 1935.
- *Katathermometer cooling power, Hill 1914*
  The cooling time of the katathermometer gave an indication of the cooling power of the environment and consequently of comfort. The thermometer did not cool in the same balance as the human body and the thermometer is now used as an omnidirectional anemometer.
- *Effective temperature, Houghton and Yagloulou 1923*
  A single scale of temperature which combined air temperature, relative humidity and air movement. A basic scale dealt with unclothed subjects and a normal scale with fully clothed subjects.
- *Corrected effective temperature, Vernon and Warner 1933*
  Developed effective temperature by taking radiation into account using the globe thermometer in place of a dry bulb.
- *Equivalent temperature, Dufton 1929*
  Based on the eupatheoscope. Cooled in the same balance as the human body in relation to air temperature, air movement and mean radiant temperature. Did not take humidity into account.
- *Resultant temperature, Misserard 1933*
  Effective temperature with radiant component.
- *Equivalent warmth, Bedford 1936*
  Based on observations made in actual (not laboratory) conditions and analysed statistically.
- *Equatorial comfort index, Webb, BRS 1961*
  Similar method to Bedford, based on data related to Singapore.
- *Calidity, Webb, BRS 1967*
  Term developed to apply to the thermal comfort aspect of the environment.
- *Predicted mean vote (PMV) and percentage people dissatisfied (PPD), Fanger 1970*
  Sophisticated method of analysis based upon extensive laboratory and field studies which enables the percentage of dissatisfied occupants of a room with particular thermal characteristics to be predicted.
- Some studies have appeared to demonstrate a 70% acceptance of comfort by people in rooms with the following values of the indices.

Globe thermometer	16.5–19.5 °C
Effective temperature	14–17 °C
Equivalent temperature	14.5–19 °C
Air temperature (dry bulb)	15.5–19.5 °C
	(England)

These comfort scales do not, however, solve the problem of defining human comfort, since they mask in one unified value directional variations of radiation, temperature gradients in the air or other factors which could cause unsatisfactory conditions and which must be designed separately. And recent research has cast doubt on the validity of specific fixed temperatures as a close guide to thermal comfort. There is evidence that people adapt their clothing and immediate environment to suit prevailing conditions and, within reasonable limits, are most comfortable when thermal conditions remain fairly stable, irrespective of actual values.

Inextricably associated with thermal comfort and with ventilation is another set of comfort factors described as 'freshness'. An environment which is thermally satisfactory may yet give rise to complaints of 'stuffiness'. A variety of phenomena have been investigated in order to discover the basis of this,

Table 5.1 How to achieve equivalent temperatures of 18.5 and 21 °C

Air temperature (°C)	Mean radiant temperature (°C) to give these equivalent temperatures[a]		
	18.5 °C	21 °C	
10	32	37.5	} 1
12	28	34	
15.5	25	31	} 2
18.5	21	27	
21	18.5	24	
24	14.5	20	} 3

[a] For any given equivalent temperature the conditions with higher MRT than AT are normally preferred:
1 12 °C minimum acceptable air temperature
2 Most satisfactory
3 Noticeably less comfortable when mean radiant temperature is less than air temperature

including ionization of the air and the influence of radiation on the nasal membranes. Bedford concludes, however, that important factors in freshness are the correct relationship between some of the factors involved in thermal comfort (e.g. mean radiant temperature higher than air temperature) and adequate skin stimulation of the occupants of heated interiors by variation of air velocity. The following requirements for a pleasant environment were proposed by Bedford.

- Rooms should be as cool as is compatible with comfort.
- The velocity of air movement should be about 10 m/min in winter (less than 6 m/min may cause stuffiness). Higher rates are desirable in summer.
- Air movement should be variable rather than uniform.
- The relative humidity should not exceed 70% and should preferably be substantially below this figure.
- The average temperature of internal surfaces should preferably be above or at least equal to the air temperature.
- The air temperature should not be appreciably higher at head level than near the floor and excessive radiant heat should not fall on the heads of occupants.

The dry bulb temperatures in Table 5.2 represent currently recommended standards of warmth. For most normal conditions relative humidity should be 30–65%.

The standards quoted for thermal comfort and freshness usually give a particular value for general application and make no provision for sex or other

Table 5.2 Recommended standards of warmth

Room	Dry bulb temperature (°C)
Domestic	
Living-rooms	20–21
Bedrooms	13–16
Kitchens	16
Offices	
General	20
Machine rooms	19
Shops	19
Classrooms	17
Factories (according to type of work)	
Sedentary	18
Light	16
Heavy	13
General spaces	
Entrance, stairs, etc.	16
Lavatories	18
Cloakrooms	16
Hospital wards	19

differences. There seems to be no basic general variation in comfort requirements of men and women but different conventions of dress usually mean that women call for higher temperatures than men. Old people require higher values than young people. Different countries have different standards. This may be influenced by the extremes of climate experienced and by the relative economics of providing and running heating installations in the countries; different clothing may help to sustain the differences. The North American and British temperature standards have been significantly different but with the rapid extension of the use of central heating in this country there is evidence that the standards here are tending towards the American levels. Apart from general differences in heating standards, individuals vary greatly in their reaction to thermal environment, and these differences are more marked than geographical ones. It is possible for different individuals to feel too hot and too cold in the same thermal conditions. There is no one set of conditions which will satisfy everyone even in one locality. The aim in design is therefore to satisfy a majority and to reduce to a minimum the inevitable proportion of dissatisfied occupants.

Humphreys of the Building Research Establishment has demonstrated in several studies (published as BRE Current Papers) that globe thermometer temperatures form a satisfactory measure of thermal environment in temperate regions. Figure 5.1 shows a globe thermometer.

Humphreys has also demonstrated that, although occupants of buildings cannot easily respond to fast changes of temperature, they can adapt themselves to a wide range of long-term conditions. Clothing in particular can be varied, and location of desks and seats adapted. Control of heating ventilation can also be adjusted to counterbalance long-term variations. A scale of 'clo' units has been developed to represent the effects of different clothing levels. Table 5.3 shows typical combinations of dress and their corresponding clo values together with the appropriate comfort temperatures for sedentary people.

The degree of activity also has a considerable effect upon the comfort temperature. Table 5.4 shows the metabolic rate for young adult males for various types of activity expressed, as is now conventional, in watts per square metre of body surface area. Figure 5.2 shows the relationship between activity, globe thermometer temperature and comfort. Active people have a significantly wider range of tolerance than sedentary people; on the other hand, comfort can be maintained through substantial temperature variation by varying clothing.

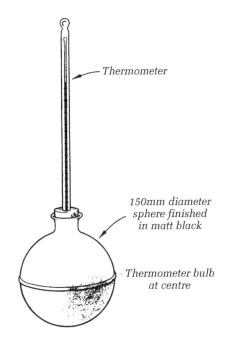

Figure 5.1   Globe thermometer.

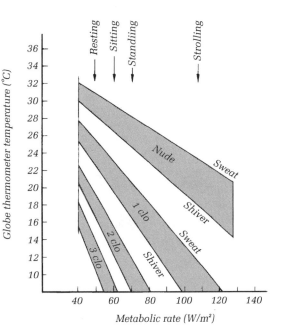

Figure 5.2   Relationship between activity, globe thermometer temperature and comfort.

Table 5.3   Combinations of dress and clo value

Clo value	Type of clothing	Comfort temperature for sedentary subject (°C)
0	Nude	28.5
0.5	Light slacks and T-shirt	25
1.0	Business suit or slacks and pullover	22
1.5	Heavy suit and waistcoat, woollen socks	18
2.0	Heavy suit and overcoat, woollen socks and hat	14.5

Table 5.4   Young adult males: metabolic rates per square metre of body surface

Activity	Power (W)
Sleeping and digesting	47
Lying quietly and digesting	53
Sitting	59
Standing	71
Strolling (2.5 km/h)	107
Walking (4.2 km/h)	154

5.2 Thermal properties of buildings

These considerations make it clear that the important factors in the natural environment affecting thermal comfort are air temperature, air movement, relative humidity and radiant conditions. Radiant conditions range from exposure to the rays of the sun to those of a clear cold night, when the radiation balance is reversed and the body is losing heat by radiation to the very cold sky. Experience tells us, without the need for elaborate analysis, that these conditions rarely combine in nature to give satisfactory thermal conditions, and buildings have an important function in providing thermal comfort. The interaction of building and thermal conditions is complex. Apart from the fact that several factors are involved, the fabric of the building itself reacts to thermal changes and does so over a period of time. This contrasts with the behaviour of the building in relation to wind or light. In these cases the presence of a wall gives, in effect, a total and immediate barrier. Heat, on the other hand, is not suddenly arrested; it is delayed but passes through over a period, and it is necessary to take into account the resistance of the materials to the passage of heat, the way in which they warm up or cool down and the time taken to do so. One of the major effects resulting from the delay in heat transfer to and from buildings is a smoothing out of the extremes of temperature. Figure 5.3 shows in principle the type of relationship which exists between internal and external temperatures.

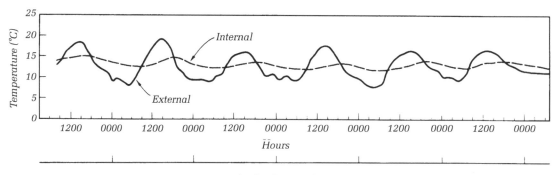

Figure 5.3 Effect of enclosure in modifying amplitude and timing of the diurnal temperature cycle. The effect will be most marked in massive well-insulated buildings. Lightweight constructions will have an internal cycle that is more similar to the external cycle.

The cooling effect on occupants of wind and radiation to the cold night sky is clearly reduced inside buildings and, during the winter, solar radiation will contribute to a general raising of internal temperatures. Without any heat input, the interiors of buildings in temperate zones in the winter therefore provide improved thermal conditions. Air movement is controlled, extremes of air temperature and radiation conditions are reduced and a more equable temperature is maintained, which may be significantly above the mean external temperature.

But in the summer, even in temperate zones, it is possible for the heat from the rays of the sun to raise inside temperatures to beyond comfort levels. Buildings with large windows may present a major problem, particularly if they are of lightweight construction, which implies low thermal capacity and fast warm-up. The situation may be made very much worse if external noise makes it difficult to open windows and obtain adequate ventilation. Towns often present environments of this type. It is quite possible to find summer temperatures so high that parts of buildings are uninhabitable and common to find summer discomfort.

It is possible to exploit the phenomena described and to design buildings which remain in good thermal equilibrium; at least one building has been designed to use solar heat, heat from occupants and lighting, and to dispense with any heating installation. But most buildings will require an installation that can provide additional heat when required and in some cases extract excess heat. It is clear, however, that the proper design of the building fabric itself is vital to the success of the thermal environment inside and will be a controlling factor in the economy of the initial costs and the running costs for the heating installation.

Figure 5.4 shows the balance of the human body with its thermal environment and Fig. 5.5 shows some of the factors of the building's thermal behaviour together with aspects of the natural environment that must remain in balance if thermal comfort is to be maintained inside. Tables 5.5 and 5.6 show the main factors which must be maintained in proper balance. Note that only the contribution of the heating and refrigeration installations and, to some degree, the ventilation can be varied in operation. The other factors are either not capable of modification, such as the heat gain from occupants, or must be settled by the basic design, such as the thermal properties of the fabric of the building and the form and fenestration in relation to solar heat gain.

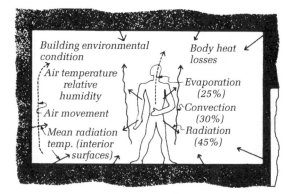

Figure 5.4 Heat balance of the human body in relation to its environment.

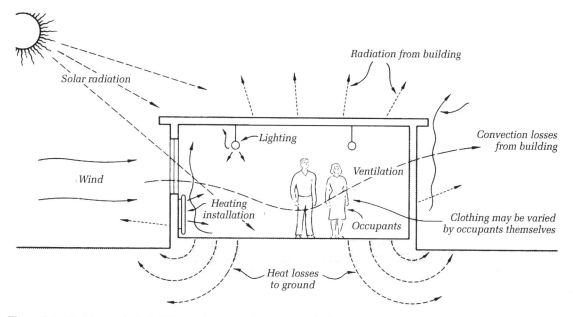

Figure 5.5 Heat balance of a building under external and internal influences.

Table 5.5 Factors that affect the heat balance of a building

Initial conditions	+ Heat gains	− Heat losses conditions[a]	= Satisfactory comfort
Thermal properties of the building Absolute humidity of the air (unless air-conditioning with humidity control is provided)	Solar radiation Heat from occupants, lighting, mechanical installations and equipment Heating installation	Radiation to sky Convection to air outside Ventilation losses to ground Refrigeration plant	Mean radiant temperature Air temperature Relative humidity

[a] A satisfactory rate of air movement must also be given by ventilation arrangements.

5.3 Significant thermal phenomena in buildings

Heat transfer

Heat will flow in a solid object, in liquid or gas, or between them until the temperature of each is equal. Transfer of heat can be by conduction, convection or radiation.

Conduction

Conduction is the direct transmission of heat through a material. The rate of conduction, i.e. conductivity, depends partly upon the density of the material. Metals have a high conductivity, wood has a low conductivity, and gases have even lower conductivity. This *conductivity* (λ) is the amount of heat that passes through 1 m^2 of the material of 1 m thickness for 1 K difference in temperature of the inner and outer surfaces.

Convection

Convection is the transmission of heat in fluids and gases by circulation. When a liquid or gas is heated it is displaced by the colder more dense liquid or gas round it, and it tends to rise. In doing so it will impart some of its heat to anything in its path. The greater the movement, the greater the speed of transfer. This is why air movement in insulating cavities must be avoided.

Radiation

Radiation is emitted from all surfaces and can transmit heat energy through space. The heat transfer at a particular surface will depend upon the balance of losses and gains, the net radiation exchange. The rate at which radiation is emitted from a surface depends upon the temperature and the nature of the surface. Dark matt surfaces both absorb and emit heat more

Table 5.6 Typical heat gains and typical heat requirements

Heat gains (wild heat)			Heat requirements	
Source	Extent of input, sensible heat gain (W)	Actual input governed by	Nature	Governed by
Occupants	100–350 per person	Number of occupants Sex and age of occupants Activity Period of occupation	Warming up	Thermal properties of building and contents Initial and final temperatures
Solar gain	0–530 W/m² of surface	Orientation Form and construction Time of day Time of year Situation of building	Ventilation	Rate of air change Outside and inside air temperatures
Lighting	1 W of lighting contributes 1 W to heating	Wattage and period of use of lighting	Losses through fabric	External radiant temperature External air temperature Wind velocity Moisture content of construction
Electrical apparatus	1 W of load contributes 1 W to heating	Wattage and use of apparatus	Heating cold materials brought into building	Not normally critical for heating
Gas apparatus	17,000 kJ/m³ of gas	Rates of consumption and periods of use of apparatus		
Water heating	Depends on nature, extent and use of system. Often a substantial proportion of the heat input is given out as space heating			
Other processes	Depends on circumstances			
Heating installation	Should be capable of maintaining balance between gain and requirements			

readily than smooth shiny surfaces. The wavelength of radiation has an important effect on its behaviour. Radiation from a surface at very high temperatures, such as the sun, has a very short wavelength and will pass freely through glass. Surfaces at lower temperatures, up to and including red heat, produce long-wave radiation which cannot pass through glass. These properties are important in buildings. The greenhouse effect of glass is used to take advantage of solar radiation. White surfaces reflect much of the short-wave radiation and can therefore protect buildings from some of the heat of the sun. Shiny surfaces are resistant both to emission and absorption and aluminium foil is often used to improve insulation, particularly in cavities.

Thermal transmittance

When heat is passed out of a building through the structure, all three methods of transference are used.

Heat is conducted through the solid parts of the wall or floor or roof, it is radiated across cavities and from the outside surface; it is also convected from the outside surface by wind passing across that surface.

The overall rate of transmission is known as the thermal transmittance. It is the heat in watts that will be transferred through $1\,m^2$ of the construction when there is a difference of 1 K between the temperature of the air on the inside and the temperature of the air on the outside. This is called the *U value* or *air-to-air* heat transmittance coefficient; its units are $W\,m^{-2}\,K^{-1}$. These coefficients are calculated from the conductivity k of each material and the surface resistance of each material.

Thermal resistance

Building materials present resistance to the flow of heat. The *resistivity*, r ($m\,K\,W^{-1}$) of the material is the inverse of the conductivity, i.e. $1/\lambda$, and the *resistance*,

R (m^2 K W^{-1}) of a given thickness of the material is the product of the resistivity and the thickness in metres.

The overall resistance of an element of construction is composed not only of the sum of the resistances of the materials but also of the resistances to heat flow of the external and internal surfaces and any cavities in the construction.

A surface resists the net transfer of heat according to its emissivity, its absorptivity and its reflectivity. At normal temperatures the emissivity and the absorptivity of a surface are the same and the surface reflects what heat it does not absorb, e.g. the emissivity of aluminium foil may be 0.05, so its absorptivity will be 5% of the heat falling in it (from a body at normal temperature), and it reflects 95% of such heat. The absorptivity of a surface for high temperature, i.e. solar radiation, may be quite different. This is important when insulating against the sun's heat. Surfaces can lose heat by convection, so the resistance of outside surfaces is governed by climate (temperature and speed of wind). The effect of these factors on the resistance of a surface is also influenced by the following considerations:

- Cooling wind across an external surface will reduce its resistance.
- The resistance of a corrugated surface can be about 20% less than a plain surface of the same material; this is because the corrugated surface has a larger area.
- Surfaces of low emissivity, e.g. bright metallic surfaces, will have a high resistance, but this may be nullified if convection takes place.
- When radiating to an area of very low temperature, e.g. to a clear sky in very cold, calm weather, the resistance of a surface can be decreased considerably.
- The resistance of a horizontal surface will depend upon whether the transfer of heat is upwards or downwards, as convection will assist in taking the heat away above and to keep it near the surface below.

The values of resistance of non-metallic surfaces have been computed and those normally used in Britain are set out, with much other useful data, in section A3 of the *CIBSE Guide*, from which Tables 5.7 and 5.8 are taken.

These surface resistances are in the same units as the resistances of specific thicknesses of building materials. It is possible, therefore, to add up the resistances for the materials making up a wall, floor or

Table 5.7   Internal surface resistance for building elements

Surface[a]	Resistance (m^2 K W^{-1})
Walls	0.123
Floors or ceilings	
For upward heat flow	0.104
For downward heat flow	0.148
Roofs, flat or sloping	0.104

[a] All high emissivity.

Table 5.8   External surface resistance for ordinary materials

Surface[a]	Resistance (m^2 K W^{-1}) for given exposure[b]		
	Sheltered	Normal	Severe
Walls	0.08	0.055	0.03
Roofs	0.07	0.045	0.02

[a] All high emissivity.
[b] Sheltered includes the first three storeys of buildings in the interior of towns.
   Normal includes the fourth to eighth storeys of buildings in towns and most suburban and country premises.
   Severe includes ninth and higher floors of buildings in towns, fifth and higher floors in suburban districts and buildings exposed on coasts and hills.
   Ordinary building materials covers the great majority of materials and colours in common use. For corrugated surfaces reduce values by 20%. For bright metallic surfaces see the *CIBSE Guide*.

roof and the external and internal surface resistances, to obtain a total resistance (Fig. 5.6). From this the thermal transmittance of the construction may be determined by calculating the reciprocal of the total resistance.

$$\text{Thermal transmittance,} \ U \ (\text{W m}^{-2}\,\text{K}^{-1}) = \frac{1}{\substack{\text{sum of resistances of} \\ \text{surfaces and layers} \\ \text{of material (m}^2\,\text{K W}^{-1})}}$$

Much of the basic data on thermal properties of building materials is given in terms of their thermal conductivity (λ) which is the amount of heat in watts that will pass through 1 m^2 of material 1 m thick for a 1 K temperature difference between the faces. The resistance of a given thickness may easily be established as follows:

$$\text{Resistance of a layer of given thickness (m}^2\,\text{K W}^{-1}) = \frac{\text{thickness (m)}}{\lambda \ (\text{W m}^{-1}\,\text{K}^{-1})}$$

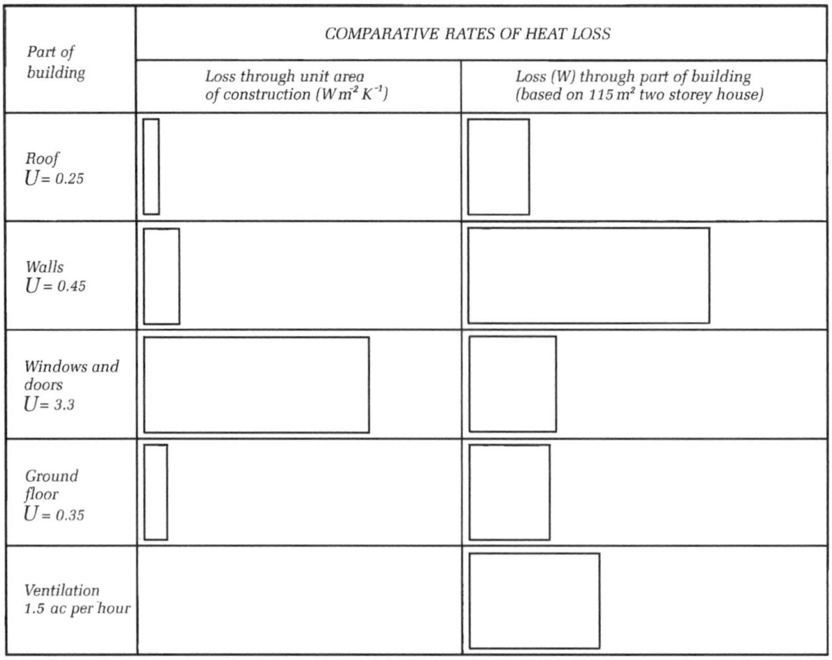

Part of building	COMPARATIVE RATES OF HEAT LOSS	
	Loss through unit area of construction ($W\,m^2\,K^{-1}$)	Loss (W) through part of building (based on $115\,m^2$ two storey house)
Roof $U = 0.25$		
Walls $U = 0.45$		
Windows and doors $U = 3.3$		
Ground floor $U = 0.35$		
Ventilation 1.5 ac per hour		

Figure 5.6 Comparative rates of heat loss through various building elements for unit area and for a typical dwelling. (U values are based on the elemental method of Approved Document L : 1995 of the Building Regulations : 1991).

Resistance of cavities and air spaces

Many forms of building construction create air spaces between layers of building materials and these air spaces contribute significantly to the insulation. With surfaces of traditional building materials, e.g. brick, stone, concrete, having high emissivity, the normal practice is to take the resistance value of a cavity as $0.18\,m^2\,K\,W^{-1}$, provided that it is at least 19 mm across and unventilated. It has been found that smaller cavities have lower resistance, but increasing the width of the cavity does not appreciably affect the resistance. If the cavity is ventilated, the resistance will be reduced depending on the speed of circulation of the air. If the cavity is horizontal and the heat flow is downward (e.g. through a floor), the resistance is increased.

If one surface of an unventilated cavity is of low emissivity, such as aluminium foil, the resistance is increased since radiation is substantially reduced. Multiple foil sheets providing a series of low emissivity surfaces and cavities give excellent insulation.

Roof spaces for pitched roofs also have an insulation value, although it is difficult to seal them and they must normally be considered as ventilated.

Table 5.9 gives values for the thermal resistance of cavities and air spaces.

Thermal capacity

Until recently little account was taken of the thermal capacity of buildings. It is now appreciated, however, that economy of energy use in winter and comfort in hot summer weather can be influenced critically by the capacity of building materials to absorb heat as well as by the resistance to the flow of heat (Table 5.10). Lightweight, low thermal capacity materials respond rapidly to control and warm up quickly. They therefore enable energy consumption to be kept low in buildings where heating is not continuous.

Apart from hospitals and prisons, this applies to most building types where the heating is turned off or reduced at night. Massive, high thermal capacity construction requires large preheating to reach working temperature and remains at a higher average temperature than lightweight construction; for the same thermal resistance it therefore loses more heat. In the summer, however, lightweight construction is likely to warm up very rapidly when the sun's rays

Table 5.9   Thermal resistance of air spaces

Space			Thermal resistance ($m^2\,K\,W^{-1}$)	
Type	Width (mm)	Surfaces[a]	Heat flow horizontal or upwards	Heat flow downwards
Sealed	6	obm	0.11	0.11
	6	ref 1	0.18	0.18
	19+	obm	0.18	0.21
	19+	ref 1	0.35	1.06
Reflective multiple-foil insulation and air space			0.62	1.76
Ventilated	19+	Loft space with asbestos-cement pitched roof	0.14	
		As above with reflective foil on upper surface of ceiling	0.25	
		Loft space with pitched roof, felt and building paper	0.18	
		Air space between tiles and felt or building paper	0.12	

[a] obm signifies ordinary building materials; ref 1 signifies one or both sides faced with reflective materials.

pass through windows. Heavyweight construction absorbs part of the solar heat and tends to give improved comfort in summer.

An optimum balance of thermal capacity is clearly needed in the design of buildings to take account of summer overheating, winter economy and the need to avoid a too rapid drop of temperature at night when the heating may be off. As yet, no optimum balance has been established. In the absence of such information the most reasonable objective for the designers appears to be to lean towards winter economy in the design of the fabric of the building while controlling summer overheating by careful consideration of fenestration and orientation.

5.4 Energy conservation

For many years during the 1960s and 1970s energy for heating buildings was relatively cheap and designers felt able to work without taking energy conservation into account. This situation, which was unusual in historical perspective, has now disappeared and both the increasing price and potential shortage of energy means that designers of buildings need to give careful account not only to Building Regulations which prescribe a minimum standard, but also to the other factors in design which will reduce energy consumption.

The main factors which govern energy consumption in buildings are described in the next few sections.

Table 5.10   Typical thermal response of common materials

Material	Thickness (mm) required for $U = 1\,W\,m^{-2}\,K^{-2}$	Temperature rise (°C) caused by application of $1\,kW/m^2$ for 1 min
Concrete	830	0.04
Brickwork	700	0.06
Timber	120	0.68
Lightweight concrete	250	0.24
Wood wool	83	1.4
Fibreboard	42	4.8
Expanded polystyrene	25	96.0

Selection of site

Site selection used to be informed by detailed local knowledge of the microclimate. This is now seldom the case and the very great increase in town sizes means that many exposed sites are developed. Architects are seldom able to have much influence on sites to be allocated for development and this point is one that should be borne in mind by town planners. On exposed sites increased wind speeds can increase ventilation rates and decrease surface resistances. A 20% increase in energy consumption may result. Elevated sites will be colder and more subject to rain and mist.

Apart from climatic conditions, manmade atmospheric and noise pollution may give rise to higher energy consumption on adjacent sites if windows cannot be opened and air-conditioning has to be used.

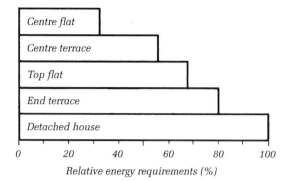

Figure 5.7 Relative energy requirements of a dwelling of 200 m³ volume in various groupings.

Layout and grouping of buildings

In general site layout should aim to position buildings to minimize wind effects but to allow maximum solar penetration. Grouping of individual buildings into composite blocks has a major effect on energy conservation. Figure 5.7 shows the extent of variation in energy requirement between detached and terraced houses and flats, all with the same extent of accommodation.

Shape and size

Within quite a wide range, shape has little effect of energy requirements. Figure 5.8 shows that the variation of heat loss is quite small for a variation of aspect ratio of 1:1 to 1:3 (i.e. from a square plan to a form with the long side three times as long as the short side). Functionally, however, such a variation is a major change.

More extreme variations of shape, such as transforming a compact building into a tower block, can have major effects on energy consumption. Not only is the surface area very greatly increased, the exposure of the upper part is made worse and in many cases the solar gain wind and stack effects will dictate that air-conditioning must be used.

Energy requirements vary almost directly with the volume of the building. Figure 5.8 demonstrates the relationship. It is apparent, therefore, that an efficient and economic basic design saves running costs as well as capital.

Planning

Open plans do not lend themselves to energy conservation since no partial use is possible and zone control is not as effective.

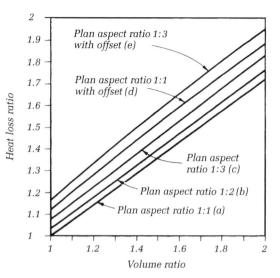

Figure 5.8 How the ratio of heat loss varies with volume for various plan forms. Based on a 100 m² two-storey dwelling with square plan form as unity.

Source of heat should be given internal locations, where possible, so that other parts of the building will benefit. In dwellings, if the boiler and flue can be given a central position, not only does the flue contribute heat to upper floors but also the warm flue works more efficiently and is less liable to condensation. Some 5% of total requirement may be saved in this way.

Figure 5.9 shows the principles involved.

Thermal performance of the fabric

The materials of construction should provide adequate insulation and appropriate thermal response. Minimum standards of insulation are prescribed by the Building Regulations. Some designers provide very much increased levels of insulation.

Construction with low thermal capacity internal layers (usually achieved by internal insulation) used with heating systems capable of controlled and variable output can give considerable economy during the heating season in buildings where heating is not required continuously. This includes most types of building other than prisons and hospitals. High thermal capacity linings absorb a great deal of heat to raise them to comfort temperatures then lose part of it when temperatures are reduced at night. Low thermal capacity linings require less heat to warm them and also lose less at night.

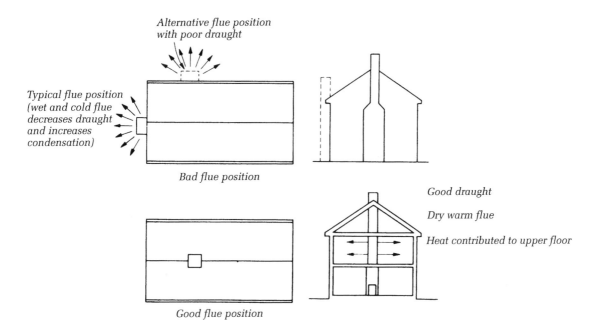

Figure 5.9 Good and bad flue positions.

Very low thermal capacity linings can result in low internal temperatures and some heat input may be needed in extreme cases to limit the temperature drop and reduce the risk of condensation. This does not spoil the energy-conserving properties since, as the internal temperature is low, the heat losses will also be low.

In summer conditions low thermal capacity linings may warm up very quickly and give rise to overheating. High thermal capacity in this case would improve comfort. As yet, no optimum balance has been struck between summer and winter conditions. Where it is practicable it seems more appropriate to control fenestration and orientation, and provide adequate ventilation and shading of window openings, than to modify the thermal capacity.

Fenestration and orientation

Unobstructed windows facing south which are curtained at night gain approximately as much heat from the sun during the heating season as they lose. On other orientations the gain is very much reduced and there is a net heat loss. It is therefore important for energy conservation that windows are orientated as far as possible towards south. Table 5.11 shows the significance of the energy balance variation with orientation.

Table 5.11 Seasonal heat balance for 1 m² single glazing curtained at night in a centrally heated house

Orientation	Solar heat gain through 1 m² unobstructed window (MJ)[a]	Net heat loss through 1 m² glazing over heating season (MJ)
South	680	0
East and west	410	270
North	250	430

[a] $U = 3.4\,\mathrm{W\,m^{-2}\,K^{-1}}$ (BRE value for windows curtained at night)

Average temperature difference = 10 °C (BRE value)

Length of heating season = 20×10^6 s (33 weeks)

Heat loss through 1 m² single glazing = $3.4 \times 10 \times 20 \times 10^6$
$$= 680\,\mathrm{MJ}$$

Not all the solar gain can be utilized for space heating but the thermal significance of window orientation is amply demonstrated.

Thermal installations

In addition to the normal requirements for comfort, heating installations should be selected to have a response and control system which will enable advantage to be taken of sunny periods when the heat output can be reduced and also to take advantage of any periods when heating is not needed. Both thermostatic and time controls should be provided.

Use of building

Reduced internal temperatures, which may require the occupants to wear more clothes, save considerable amounts of energy. The average temperature difference between inside and outside in the heating season is approximately 10 °C for buildings with night setback of heating. If the daytime temperature is reduced by 1 °C the energy savings must be 5–10%.

Multiple use of one building, where this is possible will give very substantial energy economy compared with the use of separate buildings.

Reduction of ventilation rates gives substantial energy economy. Care must be taken when controlling ventilation not to reduce the rate to a level where condensation becomes a problem.

5.5 Calculation of U values

An example of the calculation of the U value of a 275 mm brick cavity wall plastered internally is set out below: t = thickness, k = conductivity.

	Resistances m^2 K W^{-1}
External surface resistance	0.06
112 brick outer skin	0.13
($t/k = 0.11/0.84 = 0.13$)	
Unventilated 50 mm air space	0.18
50 mm insulation	1.6
112 mm brick inner skin	0.13
15 mm plaster	0.03
($t/k = 0.015/0.5 = 0.03$)	
Internal surface resistance	0.12
Total	2.25

$$\therefore U \text{ value} = \frac{1}{2.25}$$
$$= 0.44 \, \text{W m}^{-2} \text{K}^{-1}$$

U values for any desired construction may be calculated in this way. (Note that with pitched roofs the area to be used for calculation of heat loss is the plane of insulation, whether this is in the plane of the ceiling or in the plane of the roof covering.)

When considering the improvement that might be obtained by using an insulating lining, the appropriate U value can be re-expressed as resistance, the resistance of the insulation added; the reciprocal of this total gives the U value for the whole.

In constructions such as post-and-panel walling, where different parts may have different resistances, the proportion of the areas having different values must be calculated and multiplied by the appropriate resistances, and these must be totalled to get the true mean U value.

Where parts of the construction have a very much lower resistance than others, as with a metal framework exposed to both inside and outside, and having panels of good insulating material between, heat may pass through these metal members quickly and draw heat even from the edges of the panels. This is called the *cold bridge effect*. It is not normally a critical element in heat loss, but because of reduced surface and interstitial temperature, it may give rise to trouble from condensation. Figure 5.11(b) and (c) show the effect and typical temperature distributions.

Tables 5.12 to 5.17 and Figures 5.26 to 5.28 (pages 115–117) show typical U values.

Table 5.12   Extra layers of insulation: a rapid calculation

Quantity	Value
U value	$\dfrac{1}{\text{resistance}}$
Resistance	$\dfrac{\text{thickness (m)}}{\text{conductivity (W/m}^{-1} \text{ K}^{-1})}$

To establish the effect of additional layers calculate the resistance (see Appendix 1 for basic thermal properties).
Add value found to the resistance for the type of construction given in Table 5.13.
Divide 1 by the value found to establish new U value.

5.6 Calculation of surface temperatures

In problems both of thermal comfort and condensation, it is necessary to estimate surface temperatures. In steady-state conditions, assumed for normal heating calculations, the temperature difference Δt between the air and any part of the construction has the same relationship with the total temperature difference ΔT from air (internal) to air (external) as the thermal resistance of the section of construction giving rise to the difference it has with the total thermal resistance. In the case of interior surfaces:

$$\frac{\Delta t \text{ (room air to wall surface)}}{\Delta T \text{ (room air to external air)}} =$$

$$\frac{\text{internal surface resistance}}{\text{total thermal resistance of construction}}$$

When intermittent heating is used or other marked fluctuations from steady-state conditions are anticipated, then the resulting effects should be taken into account (e.g. heavy construction will be slow to warm up and surface temperatures lower than those estimated by the above method may be present for some time and give rise to discomfort and condensation).

Table 5.13 Typical wall constructions: representative *U* values

Type of wall		Details of construction[a]	*U* (W m^{-2} K^{-1})
Solid brick	1	110 mm brick 13 mm hard plaster	3.0
	2	220 mm brick 13 mm hard plaster	2.1
Cavity brick	3	Ext. leaf 110 mm brick Cavity 50 mm Int. leaf 110 mm brick 13 mm hard plaster	1.5
	4	Ext. leaf 110 mm brick Cavity 50 mm Int. leaf 100 mm insulating block	1.0
	5	As above but with inner leaf of 150 mm special insulating block	0.6
	6	Ext. leaf 110 mm brick Cavity 50 mm filled with insulation Int. leaf 100 mm brick 13 mm hard plaster	0.6
	7	As above but with 75 mm cavity fitted with insulation	0.4
Timber frame with brick skin	8[b]	Ext. leaf 110 mm brick 50 mm cavity 19 mm plywood 50 mm insulation 13 mm plasterboard	0.5
	9[b]	As above but with 10 mm insulation	0.3
Timber frame	10[b]	Ext. cladding of the tile hanging 19 mm plywood 100 mm insulation	0.35

[a] Ext. = exterior
  Int. = interior
[b] Vapour barriers must be considered with this type of construction.

Table 5.14 Properties of some building materials

Material	Conductivity (W m^{-1} K^{-1})	Resistance (m^2 K W^{-1})
Brickwork		
105 mm	0.84	0.125
220 mm	0.84	0.262
335 mm	0.84	0.399
Plaster		
15 mm hard	0.5	0.03
15 mm lightweight	0.16	0.09
10 mm plasterboard		0.06
Cavity		
Unventilated	–	0.18
With foil face		0.3
Behind tile hanging		0.12
Tile hanging	0.84	0.038
25 mm expanded polystyrene	0.033	0.76
13 mm expanded polystyrene	0.033	0.39
Fibreglass		
50 mm	0.035	1.43
75 mm	0.035	2.14
100 mm	0.035	2.86
Aerated concrete		
100 mm	0.22	0.45
150 mm	0.22	0.68
Softwood, 100 mm	0.13	0.77
Weatherboarding, 20 mm	0.14	0.14

Table 5.15 Typical *U* values for windows: normal exposure

Type	*U* (W m^{-2} K^{-1})[a]	
	Wood frame	Metal frame
Single-glazed	4.3 (3.2)	5.6 (4.0)
Double-glazed	2.5 (2.1)	3.2[b] (2.5)

[a] Figures in parentheses are mean values for heavily curtained windows at night.
[b] With thermal break in frame.

The effects of corners and cold bridges must be considered in relation to condensation (Fig. 5.11(a) and (b)).

5.7 Thermal bridges

Construction

As standards for insulation of walls and roofs become more rigorous the effects of local areas with higher rates of heat loss become critical. A concrete lintel (*U* = 2.0 W m^{-2} K^{-1}) bridging a cavity wall is a typical example. Set in a wall with two brick leaves (*U* = 1.5) the concrete provides a local increase in heat loss rate of about 30%. When it is set in a wall with 100 mm cavity insulation and a lightweight concrete block inner leaf (*U* = 0.25) the increase in heat loss rate is about 400%. What was previously a marginal difference in thermal performance between the lintel and the general wall area has become a major difference in percentage terms.

Cold bridges can take many forms. Figure 5.10 shows typical examples of cold bridging for domestic and similar types of construction, illustrated in plans (a) to (f) and in sections (g) to (m). Note that locally increased heat loss is not limited to situations where

Table 5.16 Typical *U* values for roof glazing: normal exposure

Type	U (W m^{-2} K^{-1})
Roof glazing	6.6
Horizontal daylight with skylight over (ventilated space)	3.8

Table 5.17   *U* values for intermediate floors

Construction	U (W m^{-2} K^{-1})	
	Heat flow down	Heat flow up
Timber: 20 mm on joints with 20 mm plasterboard ceiling	1.4	1.6
Concrete: 150 mm with 50 mm screed	2.2	2.7
Concrete: 150 mm with 50 mm screed and 20 mm wood floor	1.7	2.0
Concrete: 200 mm with 50 mm screed	1.6	1.9
Concrete: 200 mm with 50 mm screed and 20 mm wood floor	1.3	1.5
	1.3	1.5

(*Source*: After *IHVE Guide*, 1970)

high conductivity materials bridge areas of lower conductivity. Geometrical forms, such as corners, which present locally increased surface areas, can have the same effect.

The influence of a cold bridge is not confined to the area in immediate contact. Heat will flow from surrounding areas to the cold bridge itself. Figure 5.11(a) and (b) shows patterns of temperature distribution at cold bridges, demonstrating the extent of their influence on temperature.

Figure 5.11(c) gives the overall pattern of internal temperature distribution for a cavity wall in a bungalow. It shows the effects of the lintel, cavity closures, corners, eaves and ground. This case is an extreme one in terms of the extent of cold bridging. However, for this uninsulated cavity wall ($U = 1.5$) the overall heat loss rate, when compared with a wall containing no cold bridges, is only increased by 35 W from 210 W to 245 W, an increase of about 15%. Had the construction employed lightweight concrete blocks in the inner leaf, giving a *U* value of 1.0, the actual heat loss through the cold bridges would remain virtually unchanged but would represent a higher proportion of the total, some 23%. And for a highly insulated wall ($U = 0.5$), well within current Building Regulations standards, the actual loss of about 35 W due to cold bridges would represent 46% of the total loss through the wall.

Environmental effects

Cold bridges can lead not only to increased heat loss but also to other undesirable effects. If the surface temperature of the inner face of the wall falls below the dew point temperature of the air then condensation will take place and moulds will begin to grow. This is a situation that becomes more likely as ventilation rates are restricted to save energy. The appearance of moulds is, and should be, quite unacceptable to the majority of building occupants. The most immediate remedies available to the occupants to overcome this problem, apart from controlling the moisture sources, are to increase the internal air temperature so that the inner surface temperature rises above the dew point temperature of the air, to increase the ventilation rate so that the dew point temperature of the air is lower than that of the internal surface, or some combination of these.

Whichever of these remedies is adopted, the result will substantially increase heat loss. With an internal air temperature of 20 °C and 60% relative humidity, condensation would be very marked in the examples shown in Fig. 5.11(a) and (b). To overcome condensation by increased heating, the internal temperature would have to be raised by about 4 °C. If this were to be sustained over the heating season the energy consumption would be increased by about 40%.

The Building Regulations have been late in addressing the problem of cold bridges. In 1989 BRE published *Thermal Insulation: Avoiding Risks*. It identified the problem and proposed solutions. Approved Document L:1990 made reference to the document, but the 1995 edition includes very specific coverage of cold bridges round openings, including diagrams of acceptable solutions. It also provides numerical standards and a calculation procedure. Two standards are given. A comprehensive one which is acceptable in all circumstances and a minimum standard which is acceptable in terms of condensation risk but requires that a compensation is made for the excessive heat loss of the cold bridge. Details are given in the Building Regulations 1991, Approved Document L1 *Conservation of fuel and power*.

Thermal breaks

Many cold bridges can be avoided in detailed design. The most effective solution is to provide a thermal

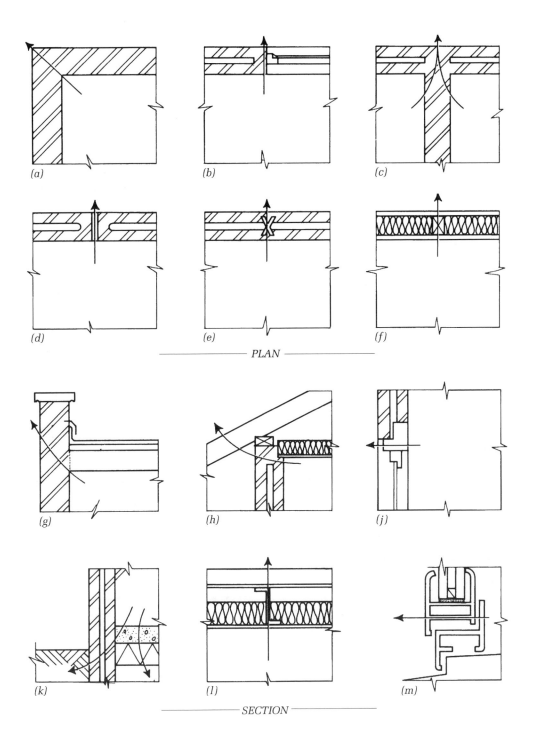

PLAN

SECTION

Figure 5.10   Cold bridges: as well as areas of lower insulation, increased surface area, such as (a) and (g), can create cold bridges.

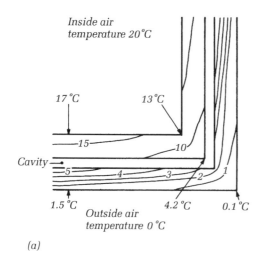

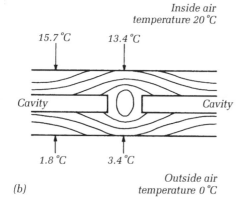

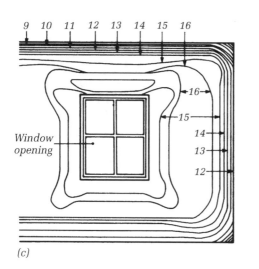

Figure 5.11 Temperature distribution at inner surface of wall.

break. An excellent example of this is at lintels across windows and door openings. Traditionally a single member supported both leaves of cavity walls and thus provided an easy passage for the flow of heat. In contrast, Fig. 5.12 shows the two leaves of a cavity wall over an opening supported separately. Window jambs and sills can be dealt with even more simply since, apart from more wall ties along unclosed jambs, no structural provisions are required.

Some of the problems presented by metal frames and supporting members can be solved by thermal breaks. Aluminium, widely used for window frames, is highly conductive to heat, so condensation often occurs. The incorporation of a lower conductivity element to break the continuity of the metal while maintaining the integrity of the frame can be very effective in overcoming the problem (Fig. 5.13). An ingenious thermal break has been used for metal roof purlins (Fig. 5.14). Galvanized steel channels form the top and bottom members, separated by a web of loose mineral fibre. The channels are linked tightly by stainless steel strips at intervals, giving the composite purlin structural strength and integrity. Stainless steel has a low thermal conductivity compared with other metals. The stainless steel straps are in tension, so they can have minimal cross-sectional area and there is very little heat transfer.

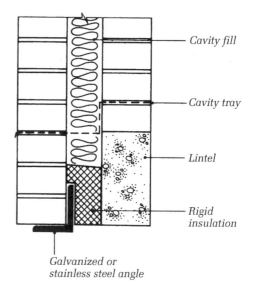

Figure 5.12 How to avoid cold bridging with separate lintels to masonry leaves at the head of the opening.

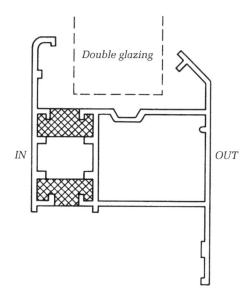

Figure 5.13   Thermal breaks in a window frame.

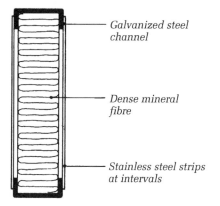

Galvanized steel channel

Dense mineral fibre

Stainless steel strips at intervals

Figure 5.14   The Korrugal thermal break purlin.

Corners

It is impossible to avoid corners and difficult to deal with the perimeters of solid ground floors. It might be thought that 100 mm cavity insulation would completely overcome the problem of increased heat loss at corners. Such insulation does certainly improve the situation but does not solve it completely. The drop in surface temperature in the corner compared with the general surface may be reduced from about 4 °C to 2.5 °C and the zone of reduced temperature made more narrow. The cold bridge phenomenon will still be present, however, and will place a possible constraint on energy conservation.

Calculation

In some simple cases it is possible to make reasonable estimates of the effects of cold bridges. Where an element of low thermal resistance makes a simple bridge across a wall of higher resistance a combined overall value for resistance can be estimated on the basis of electrical theory. This says that resistance in parallel can be represented by a single resistance (Fig. 5.15). The relationship is given by the following formula:

$$\frac{1}{R_1} = \frac{1}{R_2} + \frac{1}{R_3}$$

An example of this for a wall is given in Fig. 5.16. The examples given represent masonry construction. The procedure can equally be applied to metal or timber members bridging insulation materials.

The method in Fig. 5.16 assumes there is no lateral transfer of heat along the wall. In fact, there will be lateral heat flow but the method gives a useful approximation appropriate for building design.

Section A3 of the *CIBSE Guide* gives a method for estimating the U values of materials with a complex pattern of internal voids. Few architects would find the method easy.

Estimating heat loss at corners   Corners cannot be avoided. Fortunately very simple methods can be applied to estimate both heat loss and surface temperature. A close approximation to real heat losses can be made by assuming that the areas of walls adjacent to corners are larger. In calculation of wall area the plan dimension is increased by half the thickness of the wall forming the other (return) part of

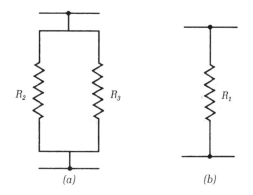

Figure 5.15   Electrical analogy for calculating thermal resistance of a construction by combining the thermal resistances of different parts in parallel.

The units used are as follows:

Temperature (T)	°C
Element thickness (t)	m
Thermal resistance (R)	$m^2\,K\,W^{-1}$
Thermal conductivity (λ)	$W\,m^{-1}\,K^{-1}$
U value (U)	$W\,m^{-2}\,K^{-1}$

Simple bridge

The basic thermal resistance per m^2 of the bridge, R_B, can be expressed as equations 1 and 2:

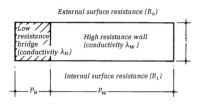

$$R_B = R_O + \frac{t}{\lambda_B} + R_i \qquad \text{[equation 1]}$$

Similarly for the wall

$$R_W = R_O + \frac{t}{\lambda_W} + R_i \qquad \text{[equation 2]}$$

If the bridge and the wall were of equal area the two resistances could be summed as shown in the electrical example. Since they are not equal the proportion of each type of construction must be taken into account. The combined resistance, R_C, can be determined by taking into account the relative proportions of the total area occupied by the bridge, P_B, or the wall, P_W ($P_B + P_W = 1$).

$$\frac{1}{R_C} = \frac{P_B}{R_B} + \frac{P_W}{R_W} \qquad \text{[equation 3]}$$

The overall U value is thus determined

$$U = \frac{1}{R_C} \qquad \text{[equation 4]}$$

Internal surface temperature
For considering condensation risk the surface temperature on the inner face of the cold bridge, which is a critical factor for condensation risk, can now readily be determined. With a temperature difference between inside the room and outside T_{io}, then:

$$\text{Internal surface temp} = \text{room temp} - \left(T_{io} \times \frac{R_i}{R_B}\right) \qquad \text{[equation 5]}$$

U value
Taking the values shown in the diagram below:

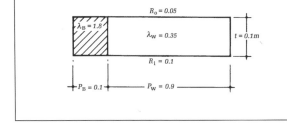

$$R_B = 0.05 + \frac{0.1}{1.8} + 0.1 \qquad \text{[equation 1]}$$
$$= 0.21 \; (U = 5)$$

$$R_W = 0.5 + \frac{0.1}{0.35} + 0.1 \qquad \text{[equation 2]}$$
$$= 0.45 \; (U = 2.2)$$

$$U = \frac{1}{R_C} = \frac{0.1}{0.21} + \frac{0.9}{0.45} \qquad \text{[equations 3, 4]}$$
$$= 2.48 \; W\,m^{-1}\,K^{-1}$$

Internal surface temperature
Assume a room temperature of 20 °C and an external temperature of 0 °C, thus a temperature difference inside to outside of $20 - 0 = 20$ °C. The internal surface temperature of the cold bridge, T, is given by

$$T = 20 - 20 \times \frac{0.1}{1.8} \qquad \text{[equation 5]}$$
$$= 18.9\ °C$$

Composite construction

The method is not confined to homogeneous walls. Composite construction can be taken into account, such as the wall shown below.

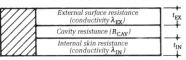

$$R_W = R_O + \frac{t_{EX}}{\lambda_{EX}} + R_{CAV} + \frac{t_{IN}}{\lambda_{IN}} + R_i \qquad \text{[equation 6]}$$

The cold bridge could also be composite in construction, and its thermal resistance calculated similarly.

Composite construction
Taking the values shown in the diagram below:

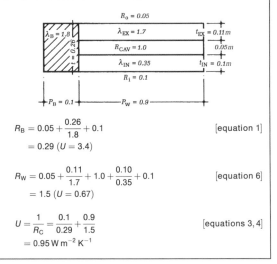

$$R_B = 0.05 + \frac{0.26}{1.8} + 0.1 \qquad \text{[equation 1]}$$
$$= 0.29 \; (U = 3.4)$$

$$R_W = 0.05 + \frac{0.11}{1.7} + 1.0 + \frac{0.10}{0.35} + 0.1 \qquad \text{[equation 6]}$$
$$= 1.5 \; (U = 0.67)$$

$$U = \frac{1}{R_C} = \frac{0.1}{0.29} + \frac{0.9}{1.5} \qquad \text{[equations 3, 4]}$$
$$= 0.95 \; W\,m^{-2}\,K^{-1}$$

Figure 5.16 Cold bridge calculations.

the corner. The internal surface temperature can be estimated using a German method that involves calculating the resistance of the construction in the normal way – see equations 2 and 6 in Fig. 5.16 – but taking the value of the external surface resistance as zero and the internal surface resistance as three times its usual value. The internal surface temperature can then be determined by using equation 5 in Fig. 5.16.

Complex geometrical forms such as eaves and parapets and particularly perimeters of solid ground floors defy simple calculation. In Belgium and Germany information for designers is provided in comprehensive collections of details for which the performance has been calculated. The Belgian information appears in Note d'Information Technique 153, *Problèmes d'humidité dans les batiments* from the Centre Scientifique et Technique de la Construction. It gives factors which enable temperatures to be predicted at various cold bridge locations. The German

information, which is particularly comprehensive, is given in *Wärmebrücken Katalog* by G. W. Mainka and H. Paschen, published by B. G. Teubner, Stuttgart (Fig. 5.17). It is a great pity that no comparable data is available in this country.

All the techniques and data that have been described are based on steady-state conditions. For heat loss calculations this is very reasonable, particularly for calculations of energy consumption over longer periods of time. But in the case of surface temperatures, intermittent heating may give rise to surface temperatures lower than those that would be determined by steady-state methods. Calculations of this effect are not practicable as design aids. It is important that data should be made available to designers. Until this happens, designers should make common-sense allowance where this problem is likely to occur. At present, Building Regulations do not take account of such dynamic effects.

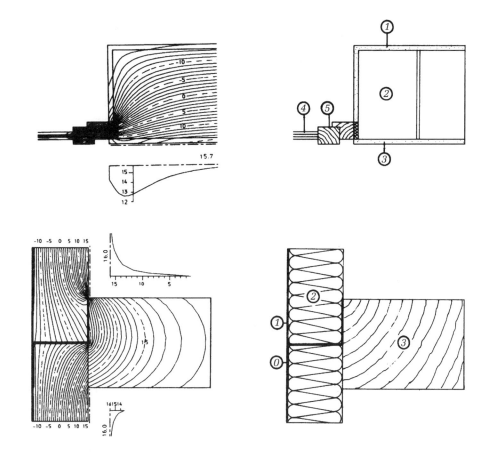

Figure 5.17 German guidance is more specific than Belgian guidance; it includes temperature contours.

Principles

Assume the conductivity of the metal is infinite, and there is a temperature difference between inside the room and outside, T_{io}. In the case of a uniform section metal bridge the temperature of the metal may be determined as follows:

Internal surface temp = room temperature

$$-\left(T_{io} \times \frac{R_i}{R_i + R_o}\right) \qquad \text{[equation 1]}$$

Often the metal will not be of uniform shape and will have flanges of different widths – W_i inside, W_o outside – as shown in the diagram below. In this case the internal temperature can be estimated as follows:

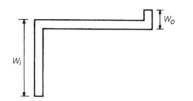

Internal surface temp = room temperature

$$-\frac{(R_i/W_i)}{(R_o/W_o) + (R_i/W_i)} \qquad \text{[equation 2]}$$

If a layer of insulation, resistance R_L, is applied to the internal surface to control condensation its inner surface temperature can be estimated by adding the resistance of the insulation to the internal surface resistance.

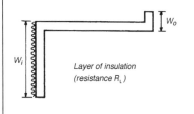

Layer of insulation
(resistance R_L)

Internal surface temp = room temperature

$$-\frac{(R_i + R_L)/W_i}{(R_o/W_o) + [(R_i + R_L)/W_i]} \qquad \text{[equation 3]}$$

Examples

Assume a room temperature of 20 °C and an external temperature of 0 °C, thus a temperature difference inside to outside (T_{io}) of 20 − 0 = 20 °C. The internal surface temperature of a uniform metal section, T, is given by

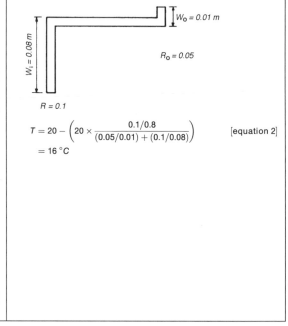

$$T = 20 - \left(20 \times \frac{0.1}{0.1 + 0.05}\right) \qquad \text{[equation 1]}$$

$$= 13.3\ ^\circ C$$

For a flanged metal section, flange dimensions W_o and W_i are given in the diagram. Temperatures are as in the previous example. The internal surface temperature, T, is now given by

$$T = 20 - \left(20 \times \frac{0.1/0.8}{(0.05/0.01) + (0.1/0.08)}\right) \qquad \text{[equation 2]}$$

$$= 16\ ^\circ C$$

Figure 5.18   Metal window frame calculations.

Metal window framing

A special case arises when metal framing members surround windows or inset panels. The effect of such bridges on overall heat loss is not likely to be significant. However, condensation can present a problem. A useful approximation can be made, particularly in the case of aluminium, which has a very high thermal conductivity, by assuming infinite conductivity in the metal frame compared with the surrounding building materials. This is illustrated in Fig. 5.18.

5.8 Heat loss calculations

When the thermal transmittance (U value) of the various elements of construction and the ventilation rates can be established (pages 172–3) it is possible to estimate the overall heat losses from whole buildings. There are two important applications of this type of calculation. The first is the determination of the maximum hourly rate of heat loss. This figure is needed in order to select appropriate heaters for rooms and to size boilers and other parts of the heating installations. (Figs 7.47 and 7.48 on pages 172–3 enable sizes for boiler rooms and flues to be estimated approximately on the basis of maximum hourly rate of heat loss.) The second is the estimation of heat requirements over a whole heating season. This enables the cost of fuel to be estimated, which may be an important factor in the selection of a heating system.

Maximum hourly rate of heat losses

This value is composed of heat losses through the fabric of the building plus the heat required to warm the air which is ventilating the building. The fabric loss for any space in the building is found by taking the areas of different types of construction multiplied by the thermal transmittance (U value) multiplied by the temperature difference which it is desired to maintain between the two sides of the construction.

The normal design temperature difference assumed between the interior of habitable rooms and the exterior is 20°C (outside temperature 0°C, inside temperature 20°C). Where rooms inside are maintained at different temperatures the heat passing through internal partitions and floors must be included in the calculation.

The ventilation losses in a room are based on the volume of air passing through the room requiring to be warmed multiplied by the volumetric specific heat of air multiplied by the temperature difference. The

volume of air to be warmed is usually determined by multiplying the appropriate number of air changes per hour (Table 3.3) by the volume of the room. In some cases, as the table shows, a specific volume of air per person is used instead of a general rate of air change. Rates of air change are almost invariably quoted in terms of hours. For convenience in calculation this must be reduced to seconds by dividing by 3600. The amount of heat required to raise a cubic metre of air at 0°C by 1 K is 1210 J (volumetric specific heat capacity).

For practical calculations involving numbers of rooms, each with windows and different types of construction, it is essential for clarity to set out the calculation in tabular form.

Table 5.18 shows the principles of the calculation.

The calculation is accomplished by establishing the U value of the various elements of construction and for each room in succession, multiplying the area of each particular type of construction by the U value and by the temperature difference between the air in the room and the air on the other side of the construction. The individual results for each type of construction are summed to give the total heat loss for the room which can be used to select appropriate heating appliances. The total heat loss from all the rooms plus an appropriate allowance for firing can be used to select a central boiler. Separate circuits would be designed in terms of the sum of the heat requirements in rooms served by each circuit.

Table 5.18 Heat loss: theoretical principles

QF	=	$\sum$	A	×	U	×			TD
Heat flow through fabric (W)	=	sum of all the cases	area (m²)	×	thermal transmittance W/m² K	×			temperature difference (K)
QV	=	$\sum$	V	×	n/3600	×	SHC	×	TD
Heat loss through ventilation (W)	=	Sum of all cases	volume (m³)	×	(number of air changes/hour)/3600	×	specific heat capacity of air (1210 J/m³ K)	×	temperature difference (K)

Or where ventilation is related to numbers of occupants:

$QV = \sum \times$ air change rate per person (m³/s) × specific heat capacity of air (1210 J/m³ K) × temperature difference (K)

Total rate of heat loss (W) = QF + QV

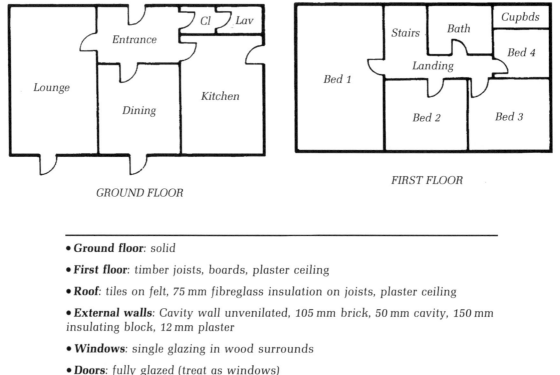

GROUND FLOOR

FIRST FLOOR

- **Ground floor**: solid

- **First floor**: timber joists, boards, plaster ceiling

- **Roof**: tiles on felt, 75 mm fibreglass insulation on joists, plaster ceiling

- **External walls**: Cavity wall unvenilated, 105 mm brick, 50 mm cavity, 150 mm insulating block, 12 mm plaster

- **Windows**: single glazing in wood surrounds

- **Doors**: fully glazed (treat as windows)

- **Rate of air change per hour** $1\frac{1}{2}$ changes

- **Hot water**: the heating installation will not provide hot water

- **Desired temperatures**
  Ground floor   20 °C
  First floor   16 °C

- **External temperature for design**: 0 °C

- **Exposure**: Normal

Figure 5.19   Plan of building used in heat loss calculation.

Some form of tabulation of the calculation is desirable to assist working and to record the results.

Table 5.19 shows a useful layout and gives an example calculation for the layout in Fig. 5.19.

Application of results

The individual room totals can be used to size the room heaters and the grand total to size the boilers. Allowances are added to the values calculated to allow for rapid heating up for firing losses and for losses from distribution pipework. In selecting a boiler an appropriate allowance must be made for the efficiency of the system (often taken as 70% for gas or oil installations) together with losses through exposed pipes and an allowance for water heating if this is needed.

Where a quick but approximate value is needed, for example, in the early stages of architectural design in order to estimate a boiler room size, the calculation can be simplified by taking into account only the external skin of the building and the total ventilation loss. (Fig. 5.21, on pages 100–101, shows a suitable form.)

Table 5.19 Heat loss: example calculation

| Construction | U | Ground floor[a] | | | | | | | | | | | |
| | | Kitchen | | | Dining | | | Lounge | | | Entrance | | |
		Area (m²)	ΔT (K)	Rate of heat loss (W)	Area (m²)	ΔT (K)	Rate of heat loss (W)	Area (m²)	ΔT (K)	Rate of heat loss (W)	Area (m²)	ΔT (K)	Rate of heat loss (W)
Ground floor (thermoplastic tile)	1.00	17.5	20	350	15.6	20	312	24.0	20	480	9	20	180
Intermediate floor (heat flow up)	1.4	17.5	4	98	15.6	4	87	24.0	4	134	9	4	50
Roof	0.4	–			–			–			–		
Walls[b]	1.0	18.4	20	368	5.9	20	118	29.8	20	596	5.9	20	118
Windows	5.7	3.0	20	342	3.0	20	342	5.0	20	570	3.0	20	342
Ventilation	$1300^c \times \dfrac{1.5}{3600}$ [d]	Vol. 43 m³	20	466	Vol. 38 m³	20	412	Vol. 59 m³	20	639	Vol. 22 m³	20	238
Room total				1624			1270			2419			928
										Total carried forward		6241	

[a] Total wall area = 138.4 m², total window area = 27.8 m². Therefore overall U value = 1.8, which is acceptable.
[b] Where more than one wall material occurs in a room, additional lines should be used.
[c] Volumetric specific heat of air (J/m³ K).
[d] 1.5/3600 represents 1.5 air changes per hour as a rate per second; this is necessary if watts are to be used.

| | | Ground floor (continued) | | | First floor[a] | | | | | | | | |
| | | Cloaks | | | Bed 3 | | | Bed 2 | | | Bed 1 | | |
Construction	U	Area (m²)	ΔT (K)	Rate of heat loss (W)	Area (m²)	ΔT (K)	Rate of heat loss (W)	Area (m²)	ΔT (K)	Rate of heat loss (W)	Area (m²)	ΔT (K)	Rate of heat loss (W)
Ground floor (thermoplastic tile)	1.00	5.4	20	108	–			–			–		
Intermediate floor (heat flow up)	1.4	5.4	4	30	–			–			–		
Roof	0.4	–			11	16	70	11.0	16	70	24	16	154
Walls	1.0	10.3	20	206	13.2	16	211	6.3	20	126	28.5	16	456
Windows	5.7	1.0	20	114	2	16	182	2	16	182	4.0	16	365
Ventilation	$1300 \times \dfrac{1.5}{3600}$	Vol. 13 m³	20	141	Vol. 25 m³	16	217	Vol. 25 m³	16	217	Vol. 55 m³	16	477
Room total				599			680			595			1452

Total 3326
Brought forward 6241
Carried forward 9567

[a] Heat gains from floor below have been ignored.

First floor (*continued*)

Construction	U	Landing Area (m²)	ΔT (K)	Rate of heat loss (W)	Bed 4 Area (m²)	ΔT (K)	Rate of heat loss (W)	Bath Area (m²)	ΔT (K)	Rate of heat loss (W)	Area (m²)	ΔT (K)	Rate of heat loss (W)
Ground floor (thermoplastic tile)													
Intermediate floor (heat flow up)													
Roof	0.4	13.7	16	88	8.4	16	54	5.4	16	35			
Walls	1.0	3.0	16	48	11.6	16	186	5.5	16	88			
Windows	5.7	1.6	16	146	2	16	182	1.2	16	109			
Ventilation	$1300 \times \dfrac{1.5}{3600}$	Vol. 32 m³	16	277	Vol. 19 m³	16	165	Vol. 12 m³	16	104			
Room total				559			587			336			

Total 1482
Brought forward 9567
Total (maximum rate of heat loss) 11049

∴ Net rate of heat input required = 11 kW

5.9 Seasonal energy requirements

Nearly 30 years ago the Building Research Establishment proposed a simple method for estimating the seasonal energy requirements for houses. The method was based on extensive field observations. The energy crisis has given a renewed significance to this type of estimation and the procedure has been substantially improved and brought up to date in terms of current construction and heating standards. The mean internal temperatures employed are the result of extensive field investigations and predictions using the method have been validated against large numbers of buildings whose actual energy consumption was measured.

The original procedure has now been superseded by a family of more detailed methods described as BREDEM 1 to 9. The original method described below gives a good introduction to the approach and is still useful for simple applications.

The method involves the following stages:

A   Calculation of seasonal heat losses
B   Calculation of seasonable heat gains
C   Calculation of seasonal energy requirement: (losses − gains) × efficiency factor

Figure 5.20 shows worksheets for carrying out the calculation. The necessary data is given in Tables 5.20 to 5.23. Calculate the overall heat loss rate using Fig. 5.20 and the following values:

- U value for curtained windows (from Table 5.15)
- Temperature difference to be:
  Mean internal − mean exterior
  (Table 5.20)   (see below)
- Ventilation heat loss rate (from Table 5.22).

Table 5.20   Heating regimes and insulation standards

System	Insulation standard	t_i (°C)[a]
Full central heating	Regime 1	
	A	17.5
	B	18.0
	C	19.5
	Regime 2	
	A	16.5
	B	17.0
	C	18.5
Partial central heating	Regime 1	
	A	16.0
	B	16.5
	C	18.0
	Regime 2	
	A	15.0
	B	15.5
	C	17.0
Not centrally heated	Regime 1	
	A	14.5
	B	15.0
	C	16.5
	Regime 2	
	A	13.5
	B	14.0
	C	15.5

[a] Add 1 °C for heavyweight and subtract 1 °C for lighweight construction.

Table 5.21   Estimated 'house efficiency factors'

Fuel	Appliance	House efficiency factor
Electricity	Central heating	
	Storage radiators	1.1
	Warm air	1.1
	Underfloor	1.1
	Local heating, all types	1.0
Gas	Central heating[a]	
	Independent boiler	1.5
	Back boiler	1.5
	Warm air	1.4
	Local heating, all types	1.8
Solid fuel	Central heating, all types	1.7
	Local heating	
	Open grate	2.9
	Closed stove	1.7
Oil	Central heating[a]	
	Independent boiler	1.5
	Warm air	1.4
	Local heating, all types	1.1

[a] Values may be increased by 5% when heating is provided intermittently.

Mean internal temperatures

The assumed heating regimes are

1 Heating provided from 0600 to 2300
2 Heating provided from 0600 to 0900 and 1700 to 2300

The three insulation standards are

A No additional insulation
B Loft insulation only
C Loft and cavity-fill insulation

Thermal response of the building

The thermal capacity of the structure (often called its 'weight'), rather than its insulation standard, determines the response of the dwelling to any input or loss of heat. The 'weight' of the structure can be assumed to fall into one of three approximate groupings, namely heavy, medium or lightweight. Typical examples of these are as follows:

Heavy

Brick–cavity–brick external walls, single-layer brick internal walls and concrete floors (ground and intermediate).

Medium

Brick–cavity–concrete block external walls, lightweight concrete block internal walls, suspended timber intermediate floors and concrete ground floor.

Light

Timber frame, brick–cavity–plasterboard external walls, plasterboard internal walls and suspended timber floors.

Mean external temperatures (at sea level)

For most of England and Wales the mean external temperature for the heating season may be taken as 6.5 °C. For the south-west of England and Wales 7.5 °C may be used, and for the north-east of England and Scotland 5.5 °C. (For every 100 m above sea level subtract 0.6 °C.)

Table 5.22 Ventilation heat loss rate

Description of dwelling	Ventilation rate (air changes per hour)	Ventilation heat loss rate (W/m³ K)
Well-sealed dwelling in sheltered position	0.5	0.17
Average dwelling in sheltered position	1.0	0.34
Leaky dwelling in sheltered position	1.5	0.51
Well-sealed dwelling in exposed position	1.0	0.34
Average dwelling in exposed position	1.5	0.51
Leaky dwelling in exposed position	2.0	0.68

Table 5.23 Useful solar gain

Heating schedule	Orientation	Light	Medium	Heavy
Gains through glazing				
0600 to 2300	N	0.18	0.21	0.24
	S	0.57	0.69	0.79
	E/W	0.34	0.41	0.47
0600 to 0900	N	0.12	0.18	0.21
and	S	0.4	0.62	0.72
1700 to 2300	E/W	0.24	0.37	0.43
Gains through fabric				
0600 to 2300	N	0.005	0.006	0.007
	S	0.017	0.021	0.024
	E/W	0.011	0.012	0.15
0600 to 0900	N	0.004	0.005	0.006
and	S	0.012	0.018	0.021
1700 to 2300	E/W	0.008	0.011	0.013

[a] Values are in gigajoules per season per square metre of single glazing or opaque fabric, averaged over the period October to April inclusive at Kew. Reduce values by 10% for double glazing.

Thermal weight of dwelling column header spans Light, Medium, Heavy.

A   Heat Losses – overall heat loss rate

Fabric loss

Fabric construction	Area (m^2)	U value (W m^{-2} K^{-1})	Temperature difference (K)	Rate of fabric heat loss (W)

Ventilation loss

Ventilation location (having differing ventilation or temperature difference)	Volume (m^3)	Ventilation heat loss rate per m^3 (W m^{-3} K^{-1})	Temperature difference (K)	Rate of ventilation heat loss (W)

Total ventilation heat loss rate (W)   [　　　　]

Total fabric heat loss rate (W)   [　　　　]

Total heat loss rate (W)   [　　　　]

Overal heat loss rate (kW)   [　　　　]

Figure 5.20   Worksheet for heat balance.

B Heat gains

Source	Unit gain (GJ)				Gains (GJ)
Solar gain through windows	Orientation	Area (m²)	×	gain (GJ/m² per season) =	
	South		×	0.68 =	
	East & West		×	0.43 =	
	North		×	0.25 =	
People	At home during day			1.2 GJ per person	
	Out during day			0.6 GJ per person	
Electricity	2.6–6.2 GJ depending on standard of accommodation and use				
Cooking	Assume: gas 3.2 GJ electricity 2.6 GJ				
Water heating	Assume: 0.5 GJ per person				

Total seasonal gains (GJ)

C Seasonal energy requirements

Overall heat loss rage (kW)	1	× 18.3	=	Seasonal heat loss (GJ)	
		×	=		3
Seasonal loss (GJ)	3	− Seasonal gains (GJ) =		Net seasonal requirements (GJ)	
		−	=		4
Net seasonal requirement (GJ)	4	× Efficiency factor	=	Gross seasonal requirement (GJ)	
		×	=		

The output required from a heating installation will be the total heat loss rate × the house efficiency factor × an allowance for intermittency (usually 1.5 or 2).

5.10 Peak summer temperatures

Admittance method

In recent years many buildings have suffered from overheating in the summer. Large windows, lightweight construction and façades exposed to the unobstructed sky have become increasingly common features of building design and can all contribute to overheating, as can the limitation of window opening often imposed by external noise levels in urban areas. Common sense in design can avoid overheating in many circumstances but it is necessary to be able to make estimates of the degree of overheating so that appropriate window areas, materials and levels of ventilation may be employed.

There is no legislation governing this aspect of building design but some government departments are calling for control of summer overheating in new buildings for which they are responsible.

Prediction methods remain limited but one developed by the Building Research Establishment after extensive analogue studies is relevant and useful. It is called the admittance method and can establish peak temperatures likely to be reached in rooms in the summer. The method is appropriate for rooms with only one external wall (i.e. surrounded by similar rooms on both sides, above, below and at the rear). This is not an uncommon situation in office buildings, where overheating has been a particular problem, and may give a useful indication in many other cases.

The method first establishes the mean internal temperature over a 24 h period, then establishes the maximum variation from the mean. These two values enable the peak temperature to be determined.

Though many of the concepts and values are the same as in heat loss calculations, some new terms are employed. The most important of these are explained here.

Admittance factor

A measure of the heat entering a surface due to a 24 h cycle of temperature fluctuation. It takes into account the thermal capacity of the construction element. Dense materials can absorb more heat and have higher admittances than lightweight materials.

Environmental temperature

A single value which takes into account both air temperature and radiant conditions. Its use is essential in the admittance procedure. It is also being used in

very accurate heat loss calculations, particularly where radiant heaters are employed.

$$\text{Environmental temperature (}^{\circ}\text{C)} = \frac{2}{3}\left(\begin{array}{c}\text{mean radiant}\\\text{temperature}\end{array}\right) + \frac{1}{3}\left(\begin{array}{c}\text{air}\\\text{temperature}\end{array}\right)$$

Solar gain factor

The proportion of heat in the sun's rays transmitted into a room, taking into account the angle of the incident radiation and losses incurred in the glazing and any associated blinds.

Alternating solar gain factor

A measure of the solar radiation gain taking place through glazing and blinds modified to take into account the effects of the thermal capacity of the structure.

Procedure

The following step-by-step procedure enables peak summer temperatures to be estimated quickly and reliably with a minimum of study of the method itself. Using the layouts on pages 106–11 follow these three stages:

1 Establish mean conditions.
2 Determine variation from mean.
3 Calculate the environmental temperature.

In each case fill in the data required for each box and carry out the arithmetic indicated as shown in the example.

Figure 5.23 shows a worked example while Fig. 5.24 provides blank worksheets which may be copied for use in calculations.

Tables 5.24–5.30 and Figs 5.22 and 5.25 provide the necessary data for calculation.

Note   The time of peak internal temperature is not always known. Peak solar gain usually predominates in fixing peak times, especially in highly glazed buildings. It may be necessary to check more than one possible time, e.g. time of peak solar gain and time of peak external temperature.

In lightweight buildings all factors can be assumed to act simultaneously. If the construction is heavyweight, the solar gain for 2 h before the time under consideration should be used.

Example

Consider an office in the southern façade of an office building.

- **Room dimensions**: 4 m wide, 4 m deep, 2.85 m high.
- **Area of single glazed window**: 8 m^2; this leaves 3.4 m^2 of solid external wall.
- **Floor construction**: 200 mm concrete with carpet plus plasterboard and ceiling battens on the underside.
- **Partitions**: lightweight concrete block with plaster finish.
- **External walls**: brick cavity with insulating block inner leaf and plaster finish, having U value of 1.00 W m^{-2} K^{-1}.
- **Lighting**: no artificial lighting used during the day.

What is the peak temperature in June at 1300 for two people doing light office work? (Note that Fig. 5.22 shows the highest temperatures are likely to occur in July.) Following through worksheets, gives a peak temperature of 35.2 °C and hence fundamental changes are needed. (Related information is given for reference in Figs 5.21 and 5.25.)

One of the most effective steps would be to reduce window area and/or to control solar gain through the window by special glass, blinds or shutters. If feasible, additional windows giving through ventilation could be effective. Increasing the mass of the structure would be useful as would removing the carpet and the plasterboard from the ceiling.

Note on total input variation (stage 3)

An additional value for variation in thermal properties of the fabric can be estimated where there are opaque areas of the fabric of lightweight construction (though otherwise this variation is normally significant). It can be calculated as shown here then added to item 13 at stage 3.

Procedure and example for gain through fabric

1. From Appendix 2 establish the time lag in hours; in this case take the wall as construction 15. Lag = 7.1 h.
2. From Table 5.25 establish the sol-air temperature at the appropriate number of hours lag. In this case the time of interest is 1300 in June. Subtracting the time lag of 7.1 h gives 0600. From Table 5.24 for a south-facing building of dark finish, the sol-air temperature at 0600 is 14.5 °C. The daily mean is 24.8 °C.

3. Subtract mean sol-air temperature from sol-air temperature with time lag (the result may be negative). 13.6−23.4 = −9.8 °C. 14.5−24.8 = −10.3.
4. From Appendix 2 establish the decrement factor. In this case the decrement factor is 0.56.
5. Multiply together the area of construction, the U value and the results of steps 3 and 4 to give the structural variation.
   Area of opaque wall is 3.4 m^2
   U value is 1.0 W/m^2 K
   $3.4 \times 1.0 \times (-10.3) \times 0.56 = -19$ W.
6. Add this figure to item 13, stage 3, if it is significant. This result, −19 W (0.019 kW), is negligible. In construction of low thermal capacity the time lag and the decrement factor are smaller, so the structural variation value could be important.
7. In stage 2, when finding the fabric transfer rate, the main sol-air temperatures should be used.

Admittance calculation

Notes on the method

The calculation falls into three distinct stages:

1. Determination of the 24 h mean internal environmental temperature.
2. Determination of the variation and swing from this mean due to the peak gain rates.
3. Adding the swing to the mean to give the peak environmental temperature.

Mean conditions

(a) The mean 24 h gains are calculated for all sources of heat input (rate of input × hours of operation ÷ 24). Table 5.25 gives the mean 24 h solar gain already calculated.

(b) The total rate of heat loss per kelvin is calculated for all external elements of the fabric and for ventilation. Note that up to 2 ac per hour the normal formula for heat loss rate $nV/3$ is used. For ventilation rates above 2 there is a special procedure to take account of the resulting increase in heat transfer.

(c) The mean gains (W) are divided by the rate of heat loss (W/K) to give the mean temperature difference which results from the balance of input and loss.

(d) The mean external temperature and mean internal temperature difference are added to give the mean internal environmental temperature.

Variation and swing between mean and peak

This stage is similar to the previous stage except for the following differences:

(a) The difference between the mean heat input rate and the peak rate for each source are used for the input value.

(b) In calculating the heat loss rate, admittance values rather than U values are used; this takes into account the effect of the fabric in absorbing heat during cyclic temperature changes. Internal partitions and floors, which can be ignored in stage 1 if the temperatures on each side are equal, must be taken into account in this stage.

In the case of glazing and ventilation the capacity of glazing and air to absorb heat is negligible and the same values are used as in stage 1.

(c) The temperature swing is given by dividing the variation in input rate (step (a)) by the total rate of loss and absorption.

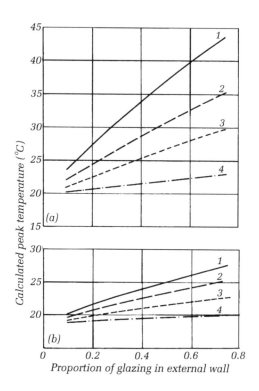

Figure 5.21 Example temperatures reached in typical building situations: (a) light, 2 air changes per hour; (b) heavy, 10 air changes per hour. Conditions: (1) unshaded; (2) internal blind; (3) blind between double glazing; (4) external blind. (After *IHVE Guide*, 1970).

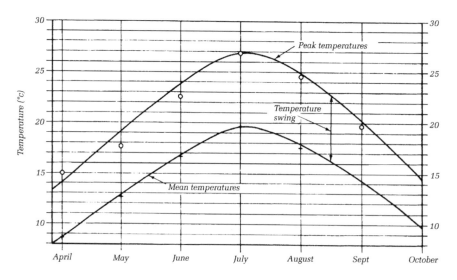

Figure 5.22 Mean and peak temperatures for typical sunny spells.

Admittance formulae

Symbols

A system of superscripts and subscripts has been developed to simplify the expression of the basic formulae of the admittance procedure:

- Superscript
mean	$'$
swing	$\sim$
peak	$''$

- Subscript
air	a
glazing	g
casual	c
ventilation	v
window	w
fabric	f
solar	s
total	t

The main symbols are as follows:

A = area (m^2)
I = solar gain (W/m^2)
Q = heat gain rate (W)
C = heat transfer rate (W/K)
U = U value (W/m^2 K)
Y = admittance (W/m^2 K)
t_{ei} = internal environmental temperature
t_{eo} = external environmental temperature
      (sol-air temp)
N = vent rate (air changes/hour)$\times$room volume (m^3)

Quantities and their units	Basic formulae
• Mean solar gain (W)	$Q'_s = \Sigma I' A_g$
• Mean casual gain (W)	$Q'_c = \dfrac{(q_{c_1} t_1) + (q_{c_2} t_2) + \dots}{24}$
• Total mean gain (W)	$Q'_t = Q'_s + Q'_c$
• Window transfer rate (W/K)	$C_w = \Sigma A_g U_g$
• Ventilation transfer rate (W/K)	$C_v = 0.33 N_v \quad \text{or} \quad C_v = \dfrac{1.58 \Sigma A N_v}{(4.8 \Sigma A + 0.33 N_v)}$
• Fabric transfer rate (W/K)	$C_f = \Sigma A_f U_f$
• Mean internal environmental temperature (K)	$t'_{ei} = \dfrac{Q'_t + t'_{ao}(\Sigma A_g U_g + C_v) + t'_{eo}(\Sigma A_f U_f)}{(\Sigma A_g U_g + C_v + \Sigma A_f U_f)}$
• Swing in effective heat gain due to peak solar ratiation (W)	$\tilde{Q}_s = (I_p - I') A_g S_a$
• Peak casual gain (W)	$Q''_c = q_{c_1} + q_{c_2} + q_{c_3} + \dots$
• Swing in casual heat gain (W)	$\tilde{Q}_c = Q''_c - Q_c \ (\text{peak} - \text{mean})$
• Swing in effective heat input due to swing in external air temperature (W)	$\tilde{Q}_a = (\Sigma A_g U_g + C_v)\tilde{t}_{ao}$
• Total swing in heat gain (W) [Q_f can normally be ignored = $\Sigma AU (t_{eo} - t'_{eo})$]	$\tilde{Q}_t = \tilde{Q}_s + \tilde{Q}_f + \tilde{Q}_c + \tilde{Q}_a$
• Area × admittance (W/K)	Subtotal = ΣAY Subtotal = ΣY^*_{cv}
• Swing in internal environmental temperature (K)	$t_{\bar{e}i} = \dfrac{\tilde{Q}_t}{(\Sigma AY + C_v)}$
• Peak environmental temperature (K)	$t''_{ei} = t'_{ei} + \tilde{t}_{ei}$

Stage 1 Mean conditions

		Select daily mean solar intensity	× solar gain factor for glazing	× window area (m²)		= mean solar gain (W)	
Gains	Solar	*161*	× *0.77*	× *8*		= *992*	1
	Other	sources	rate (W)	× duration (h)	÷ 24	= daily mean gain (W)	
		lighting (installed wattage)		×	÷ 24	=	
		occupants (no × rate /occupant)	*2 × 140*	× *8*	÷ 24	= *93*	
				×	÷ 24	=	
				×	÷ 24	=	
				×	÷ 24	=	
				×	÷ 24	=	
					total	= *93*	2
	Total	mean solar gain (1)		+ mean daily gain (2)		= total mean gain (W)	
		992		+ *93*		= *1085*	3

		window	area (m²)	× U (W m⁻² K⁻¹)		= transfer rate (W/K)	
Fabric, ventilation and window heat transfer rates	Window	*all*	*8*	× *57*		= *45.6*	
				×		=	
				×		=	
				total		= *45.6*	4
	Ventilation	ventilation rate air changes/h	× room volume (m³)	× 0.33		= mean ventilation heat transfer rate (W/K)	
		3	× *45.6*	× 0.33		= *45*	5
		for rates of ventilation over 2/hour complete					
		1 ÷ (5)	= (a)	0.21 ÷ total area of internal surfaces (m²)		= (b)	
		1 ÷ *45*	= *0.02*	0.21 ÷ *77.6*		= *0.003*	
		(a) + (b)	= (c)	1 ÷ (c)		= mean ventilation rate transfer rate (W/K)	
		0.22 + 0.0031	= *0.023*	1 ÷ *0.023*		= *43.5*	6
	Fabric	external element (wall, floor, roof)	area (m²)	× U (W m⁻² K⁻¹)		= heat transfer rate (W/K)	
		wall	*3.4*	× *10*		= *3.4*	
				×		=	
				×		=	
				×		=	
				total		=	7

Figure 5.23 Peak summertime temperatures worksheets.

Stage 2   Mean environmental temperature

	window transfer rate (4)	+ vent transfer rate (5) or (6) if	= subtotal I	× mean external air temp (K)	= subtotal II	
	45.6	+ 43.5	= 89.1	× 16.8	= 1447	
	fabric transfer rate (7)	× mean ext. air temperature (K)	= subtotal III	+ total mean gain (3)	+ subtotal II	= subtotal IV
	3.4	× 23.4	= 80	+ 1085	+ 1447	= 2662
	subtotal I	+ fabric transfer rate (7)	= subtotal V	subtotal IV	÷ subtotal V	= mean internal env temp (K)
	89.1	+ 3.4	= 92.5	2662	÷ 92.5	= 28.8	8

Stage 3   Variation of heat gains

	peak intensity solar radiation (W/m^2)	− daily mean solar int (W/m^2)	= effective peak input	× area of glass (m^2)	= subtotal	× alternating solar gain factor	= effective gain swing (W)	
Solar variation	540	− 161	= 379	× 8	= 3032	× 0.43	= 1304	9

casual gains at peak hours	rate (W)
lighting 2 occupants @ 140	280
total =	280	10

peak (10)	− mean (2)	= casual gain variation (W)	
280	− 93	= 187	11

area of glazing (m^2)	× U of glazing	= heat transmitted (W/K)	+ (5) or use (6) if more than 2/hour	= fabric plus ventilation losses	× external air temp swing (K)	= air temp input variation (W)	
8	× 5.7	= 45.6	+ 43.5	= 89.1	× 7	= 623.7	12

solar variation (9)	+ casual variation (11)	+ air variation (12)	= total variation (W)	
1304	+ 18.7	+ 623.7	= 2114.7	13

Stage 4 Temperature swing

internal surface	area (m²)	× admittance factor for construction (Appendix 2)	= area × admittance (W/K)	
window	8	× 5.6	= 44.8	
ext. walls	3.4	× 2.9	= 9.9	
partitions	34.2	× 2.6	= 89.0	
floor	16	× 3.1	= 49.6	
ceiling	16	× 5.8	= 92.8	
		×	=	
		total	= 286.1	14

area × admittance (14)	+ ventilation gains (5) or (6)	= subtotal	
286.1	+ 43.5	= 329.6	15
total swing in effective input (13)	÷ (area × admittance) + vent gains (15)	= swing in internal environmental temperature (K)	
2114.7	÷ 329.6	= 6.4	16

Stage 5 Peak environmental temperature

mean environmental temperature (8)	+ swing in internal environmental temperature (16)	= peak environmental temperature (K)
28.8	+ 6.4	= 35.2

Stage 1   Mean conditions

Gains	Solar	Select daily mean solar intensity	× solar gain factor for glazing	× window area (m²)		= mean solar gain (W)	
			×	×		=	1
	Other	sources	rate (W)	× duration (h)	÷ 24	= daily mean gain (W)	
		lighting (installed wattage)		×	÷ 24	=	
		occupants (no x rate /occupant)	-	×	÷ 24	=	
				×	÷ 24	=	
				×	÷ 24	=	
				×	÷ 24	=	
				×	÷ 24	=	
					total	=	2
	Total	mean solar gain (1)		+ mean daily gain (2)		= total mean gain (W)	
				+		=	3
Fabric, ventilation and window heat transfer rates	Window	window	area (m²)	× U (W m⁻² K⁻¹)		= transfer rate (W/K)	
				×		=	
				×		=	
				×		=	
					total	=	4
	Ventilation	ventilation rate air changes/h	× room volume (m³)	× 0.33		= mean ventilation heat transfer rate (W/K)	
			×	× 0.33		=	5
		for rates of ventilation over 2/hour complete						
		1 ÷ (5)	= (a)	0.21 ÷ total area of internal surfaces (m²)		= (b)	
		1 ÷	=	0.21 ÷		=	
		(a) + (b)	= (c)	1 ÷ (c)		= mean ventilation rate transfer rate (W/K)	
		+	=	1 ÷		=	6
	Fabric	external element (wall, floor, roof)	area (m²)	× U (W m⁻² K⁻¹)		= heat transfer rate (W/K)	
				×		=	
				×		=	
				×		=	
				×		=	
					total	=	7

Figure 5.24   Peak summertime temperatures worked example.

Stage 2   Mean environmental temperature

Mean environment temperature	window transfer rate (4)	+ vent transfer rate (5) or (6) if	= subtotal I	× mean external air temp (K)	= subtotal II	
		+	=	×	=	
	fabric transfer rate (7)	× mean ext. air temperature (K)	= subtotal III	+ total mean gain (3)	+ subtotal II	= subtotal IV
		×	=	+	+	=
	subtotal I	+ fabric transfer rate (7)	= subtotal V	subtotal IV	÷ subtotal V	= mean internal env temp (K)
		+	=		÷	=	8

Stage 3   Variation of heat gains

Solar variation	peak intensity solar radiation (W/m^2)	− daily mean solar int (W/m^2)	= effective peak input	× area of glass (m^2)	= subtotal	× alternating solar gain factor	= effective gain swing (W)
		−	=	×	=	×	= 9

Casual gain variation	Peak	casual gains at peak hours			rate (W)		
		total		=			10

	Variation	peak (10)	− mean (2)	= casual gain variation (W)	
			−	=	11

Air temp variation	area of glazing (m^2)	× U of glazing	= heat transmitted (W/K)	+ (5) or use (6) if more than 2/hour	= fabric plus ventilation losses	× external air temp swing (K)	= air temp input variation (W)
		×	=	+	=	×	= 12

Total input variation	solar variation (9)	+ casual variation (11)	+ air variation (12)	= total variation (W)	
		+	+	=	13

Stage 4   Temperature swing

		internal surface	area (m²)	× admittance factor for construction (Appendix 2)	= area × admittance (W/K)	
Internal environmental temperature swing	Area × admittance			×	=	
				×	=	
				×	=	
				×	=	
				×	=	
				×	=	
				total	=	**14**
	Swing	area × admittance (14)	+ ventilation gains (5) or (6)		= subtotal	
			+		=	**15**
		total swing in effective input (13)	÷ (area × admittance) + vent gains (15)		= swing in internal environ- mental temperature (K)	
			÷		=	**16**

Stage 5   Peak environmental temperature

	mean environmental temperature (8)	+ swing in internal environmental temperature (16)	= peak environmental temperature (K)
Peak env temp		+	=

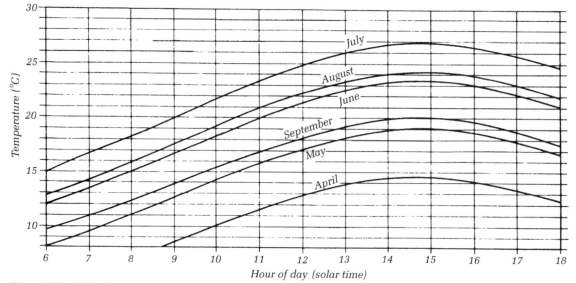

Figure 5.25   Typical hourly temperatures for sunny spells (based on studies of the sunniest fine day periods at Kew and Cardington).

Table 5.24   Solar intensities (W/m²) on vertical surfaces at 52° N. (Calculated using sun path diagrams and overlays for heat gain calculations compiled by Peter Petherbridge and published by HMSO)

Orientation	Month	Solar intensities (W/m²)													24-h mean
		Hour of day in solar time (south at noon)													
		6	7	8	9	10	11	12	13	14	15	16	17	18	
East	June	595	705	715	660	525	330	130	130	125	110	95	70	55	195
	May and July	550	680	710	660	520	330	130	130	120	105	90	70	50	187
	Apr and Aug	400	620	690	650	510	320	120	120	110	105	80	55	30	158
	Mar and Sept		410	555	555	465	295	95	95	85	70	55	25		112
	Feb and Oct			330	430	375	250								70
South-east	June	335	475	595	660	635	540	435	270	125	110	95	75	55	190
	May and July	330	485	630	665	650	580	450	330	120	115	90	70	50	197
	Apr and Aug	280	470	610	700	690	620	520	320	120	105	80	55	30	191
	Mar and Sept		350	575	680	715	670	560	390	195	70	55	30		179
	Feb and Oct			380	550	650	720	525	380	225	50	30			146
South	June	55	75	195	340	445	530	585	540	445	340	195	75	55	161
	May and July	50	70	230	395	500	580	610	580	500	395	230	70	50	177
	Apr and Aug	30	95	280	405	530	620	650	620	530	405	280	95	30	190
	Mar and Sept		110	285	450	585	695	715	695	585	450	285	110		206
	Feb and Oct			230	400	560	670	695	670	560	400	230			184
South-west	June	55	75	95	110	125	270	435	540	635	660	595	475	335	190
	May and July	50	70	90	115	120	330	450	580	650	665	630	485	330	197
	Apr and Aug	30	55	80	105	110	320	520	620	690	700	610	470	280	191
	Mar and Sept		30	55	70	195	390	560	670	715	680	575	350		179
	Feb and Oct			30	50	225	380	525	720	650	550	380			146
West	June	55	70	95	110	125	130	130	330	525	660	715	705	595	195
	May and July	50	70	90	105	120	130	130	330	520	660	710	680	550	187
	Apr and Aug	30	55	80	105	110	120	120	320	510	650	690	620	400	158
	Mar and Sept		25	55	70	85	95	95	295	465	555	555	410		112
	Feb and Oct								250	375	430	330			70

Table 5.25   Sol-air temperature during June for various wall orientations and horizontal surfaces (for admittance method)

Wall orientation and type		Hour of day, solar time [air temperature (°C)]												
		6 [12.0]	7 [13.2]	8 [15.0]	9 [17.0]	10 [18.5]	11 [20.0]	12 [21.3]	13 [22.4]	14 [23.3]	15 [23.3]	16 [23.0]	17 [22.0]	18 [21.0]
East	Dark	40.5	45.8	49.1	49.0	40.6	38.5	27.6	28.5	29.1	28.1	27.3	25.4	23.5
	Light	27.9	31.3	34.0	34.5	32.7	30.2	24.8	25.8	26.5	26.0	25.4	23.9	23.4
South-east	Dark	28.0	37.2	43.0	48.9	48.5	48.0	44.2	36.8	29.1	28.1	27.3	25.4	23.5
	Light	20.9	26.6	30.8	34.5	35.1	35.5	33.8	30.5	26.5	26.0	25.4	23.9	22.4
South	Dark	14.5	16.7	24.5	34.0	41.1	46.4	49.7	48.8	45.9	40.3	32.5	25.4	23.5
	Light	13.4	15.1	20.2	26.5	31.0	34.7	36.8	38.1	35.8	32.8	28.5	23.9	22.4
South-west	Dark	14.5	16.7	19.3	22.0	24.3	34.3	41.3	50.3	53.3	54.9	51.5	25.4	23.5
	Light	13.4	15.1	17.4	19.9	21.7	28.0	32.3	37.9	40.3	40.9	38.7	23.9	22.4
West	Dark	14.5	16.7	19.3	22.0	24.3	26.2	28.8	41.1	48.7	55.1	57.1	55.5	49.6
	Light	13.4	15.1	17.4	19.9	21.7	23.4	25.9	32.7	37.4	40.9	42.0	40.6	36.9
Horizontal	Dark	18.3	25.2	30.6	38.0	44.0	46.5	50.1	48.9	48.8	44.3	38.6	34.0	28.3
	Light	13.8	18.4	22.2	26.5	31.0	33.7	35.5	36.1	35.8	32.8	30.2	27.2	22.8

Table 5.26   Solar gain and alternating solar gain factors for various sun controls

System	Solar gain factor	Alternating solar gain factor	
		Heavy	Light
Single glazing			
Clear sheet glass (4 mm or 6 mm)	0.77	0.43	0.54
Clear plate (6 mm)	0.74	0.41	0.51
Heat absorbing (6 mm Antisun)	0.51	0.35	0.41
Heat absorbing (6 mm Calorex)	0.38	0.34	0.36
Lacquer coated glass (grey)[a]	0.55	0.37	0.42
Double glazing (outer + inner panes)			
Clear sheet glass (4 mm + 4 mm or 6 mm + 6 mm)	0.67	0.40	0.49
Clear plate (6 mm + 6 mm)	0.61	0.39	0.47
Heat absorbing (6 mm Antisun) + clear (6 mm)	–	0.30	0.35
Heat absorbing (6 mm Antisun) + clear (6 mm plate)	0.37	0.27	0.30
Heat absorbing (6 mm Calorex) + clear (6 mm plate)	0.23	0.19	0.20
Heat reflecting (6 mm Stopray) + clear (6 mm plate)	0.25	0.14	0.17
Internal sun controls + 6 mm glass			
Dark green woven plastic blind[a]	0.64	0.57	0.62
White venetian blind[a]	0.46	0.43	0.46
White cotton curtain[a]	0.41	0.25	0.33
Cream holland linen blind[a]	0.29	0.26	0.29
Blinds between double glazing			
Panes of clear glass (6 mm)[a]	0.28	0.23	0.25
External sun controls + 6 mm glass			
Dark green woven plastic blind[a]	0.20	0.13	0.18
White louvred sunbreakers with blades at 45° (horizontal or vertical)[a]	0.12	0.06	0.09
Canvas roller blind[a]	0.11	0.07	0.09
Miniature louvred blind[a]	0.10	0.05	0.06

[a] Typical value; data refers to British glass unless otherwise stated.
(*Source*: After BRE Current Paper 47/68)

Table 5.27 Ventilation rates for naturally ventilated buildings on sunny days

Position of opening windows	Usage of windows		Effective mean ventilation rate	
	Day	Night	Air changes per hour	Ventilation allowance (W m^{-3} K^{-1})
One side only	closed	closed	1	0.3
	open	closed	3	1.0
	open	open	10	3.3
More than one side	closed	closed	2	0.6
	open	closed	10	3.3
	open	open	30	10.0

(*Source*: After *IHVE Guide*, 1970)

Table 5.28 Building classification by weight

Building classification	Construction	Average surface factor, F
Heavyweight	Solid internal walls and partitions, solid floors and solid ceilings	0.5
Lightweight	Lightweight demountable partitions with suspended ceilings Floors: either solid with carpet or wood block finish, or suspended type	0.8

(*Source*: After *IHVE Guide*, 1970)

Table 5.29 Young males: heat output for various degrees of activity

Activity	Heat output (W)
Seated, at rest	115
Light work, office	140
Seated, eating	145
Walking	160
Light bench work	235
Moderate work, or dancing	265
Heavy work	440
Exceptional effort	1 500

(*Source*: After *IHVE Guide*, 1970)

5.11 Thermal legislation

Thermal requirements for walls and roofs were first introduced into the Building Regulations to combat condensation and were confined to dwellings. The need to conserve energy became apparent in the 1970s and since then the insulation standards for walls and roofs have been progressively increased and standards for windows, ground floors and other elements have been introduced. Simple standards for the control and insulation of heating and hot-water systems were introduced in 1985 and the requirements extended to buildings generally but, for buildings themselves, the explicit provisions of the regulations were, until 1995, confined to improving standards of insulation. Many other factors are involved in design for energy conservation. They include, among others, volume, form, thermal capacity, fenestration and orientation, air infiltration, ventilation and time and area control. It is clear that the guidance of the Building Regulations is directed towards a basic minimum standard rather than a rigorous objective for good design. On the other hand, it is clear that, until it became an element in the Building Regulations, the design of the great majority of buildings took little, if any, account of energy conservation. It is the regulations which govern the standards for energy conservation in the great majority of new buildings. The 1995 edition of Approved Document L marks a major step forward in the scope and sophistication of thermal regulations.

5.12 Building Regulations 1991 : Part L

Limitation of heat loss through the building fabric

The Building Regulations themselves are expressions of general principle and the Secretary of State for the Environment is responsible for determining what design features meet their requirements. Approved Document (AD) L1 sets out guidance for designers about ways of demonstrating compliance with the regulations which cover a comprehensive range of building situations.

The explicit provisions of Approved Document L1 are confined to standards of insulation. All other aspects being equal, improved standards of insulation in new buildings will economize on energy. However, many other factors are involved in design for energy conservation. They include, among others, volume, form, thermal capacity, fenestration and orientation, air infiltration, ventilation and time and area control. It is possible to take several of them into account by using calculation procedure 2 in AD L1, but the main thrust of the guidance in AD L1 leads to basic minimum standard rather than a rigorous objective for good design. On the other hand, the evidence is very clear that, until it became an aspect of the Building Regulations, the design of the majority of buildings took very little account of energy conservation. It is the Building Regulations which govern the standards for energy conservation in the great majority of new buildings. There is no doubt that the Building Regulations have made a major contribution to better standards of thermal performance in buildings. Table 5.30 and Fig. 5.26 show how the Building Regulations' insulation values have progressively increased.

In principle the Building Regulations allow the designer complete freedom to prepare designs and demonstrate, to the satisfaction of the Secretary of State, that they achieve adequate standards of thermal performance. In reality it is difficult to see how this could be done without demonstrating that the building either achieves the standards required by the regulations or exceeds them.

Approved Document L1 : 1995 covers insulation, fenestration, control for heating and hot-water systems and, for buildings other than dwellings, some standards for lighting efficiency. Dwellings and buildings other than dwellings are treated separately but the procedures are similar. There are three ways of demonstrating compliance with the regulations:

- *Elemental method*
  Standards of insulation are provided for both types of building.
- *Target U value (dwellings)*
  Calculation method (other than dwellings)
  Allows flexibility, within defined limits for *U* values and window areas, provided the overall standard does not fall below the standard of a building of similar shape and size that meets the requirements of the elemental method.
- *Energy rating method (dwellings)*
  Energy use method (other than dwellings)
  This method is intended to allow greater freedom of

design and to take into account solar and internal heat gains. The performance of the building has to be calculated.

Approved Document L1 gives a standard assessment procedure (SAP) for dwellings. The result of the calculation is a numerical value which is compared with numbers given in the document for various floor areas. This value is also used to determine the insulation standards appropriate to the elemental method.

Buildings other than dwellings require a calculation using the *CIBSE Energy Code*: 1981 Part 2a. To satisfy the regulations it must show that the proposed building requires no more energy than a building of the same shape and size conforming to the elemental method.

A series of tables and worksheets is provided to enable the standard assessment procedure to be followed. However, it is a complex procedure and manual calculation is very laborious. The practical way to make the calculation is by computer. The Building Research Establishment has the responsibility to test and approve programs to deal with SAP and only output from approved programs is acceptable for Building Regulations purposes.

In principle, Part L1 of the regulations allows complete freedom for the designer to prepare designs and to demonstrate to the satisfaction of the Secretary of State that they achieve an adequate standard. In

Table 5.30 Building Regulations: how *U* values have changed over time[a]

Element	1965	1976	1986	1990	1995
Roofs	1.47[b]	0.6	0.35	0.25	0.2–0.25
Exposed walls	1.56[b]	1.0	0.6	0.45	0.45

[a] In 1965 and 1976 the thermal regulations applied only to dwellings. In 1986 and 1990 all types of buildings over 30 m² floor area were covered. The domestic values are given for comparison.
[b] Converted from imperial values.

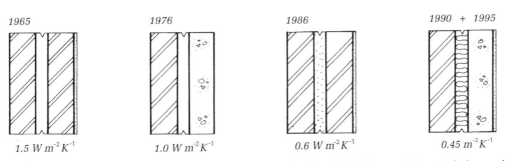

1965	1976	1986	1990 + 1995
$1.5\ W\,m^{-2}\,K^{-1}$	$1.0\ W\,m^{-2}\,K^{-1}$	$0.6\ W\,m^{-2}\,K^{-1}$	$0.45\ m^{-2}\,K^{-1}$

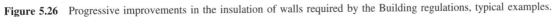

Figure 5.26 Progressive improvements in the insulation of walls required by the Building regulations, typical examples.

Table 5.31   Elemental method: standard *U* values

Element	*U* value from elemental method	
	Dwellings with SAP <60	Other buildings
Roofs	0.2	0.25
Exposed walls and roofs over 70° pitch	0.45	0.45
Semi-exposed walls and floors	0.6	0.6
Exposed floors and ground floors	0.35	0.45
Windows and rooflights	3.0	3.3
Large doors	–	0.7

(*Source*: Building Regulations, Approved Document L1 : 1995)

practice it would be impracticable to do this without demonstrating that the building either achieves the standards specified in the document or it exceeds them.

Fortunately the document is framed to allow the designer substantial freedom. A range of possible procedures for demonstrating compliance with the regulations is presented, extending from simple compliance with prescribed *U* values and window areas to calculations which take into account heat gain and thermal capacity.

Note on *U* values

Approved Document L calls for the calculation of *U* values to two decimal places. Walls and roofs are calculated in the ways described in Chapter 5. The Approved Document provides a special formulae for ground floor *U* values based on the exposed perimeter/area ratio. The basic formula is for both suspended and solid ground floors without insulation. An additional formula is provided for insulated ground floors. Figure 5.27 is based on the uninsulated formula and Fig. 5.28 gives more detail for floors with fully exposed perimeters.

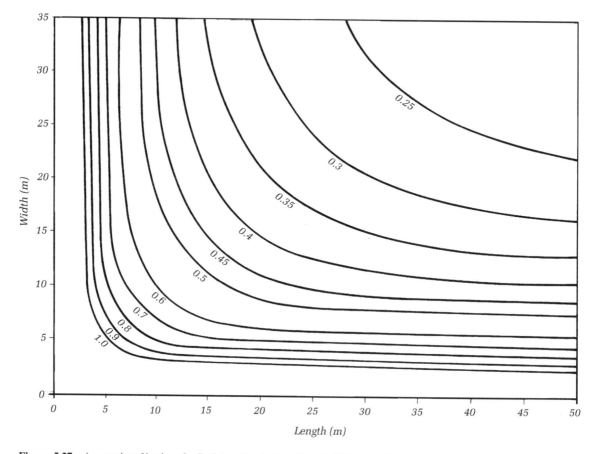

Figure 5.27   Appropriate *U* values for Building Regulations, Part L, Uninsulated Ground Floors.

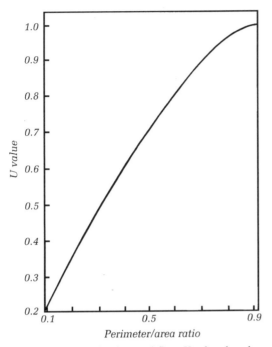

Figure 5.28 Uninsulated ground floor U values based on perimeter/area ratio

Technical risks

The higher levels of insulation arising from the new standards coupled with the tendency for ventilation to be restricted when compared with the past can give rise to problems. Most of these problems are associated with moisture. Thermal bridges across insulated constructions can cause cold surfaces which give rise to local condensation and mould growth. The temperature reduction across wall and roof constructions due to the presence of insulating materials can cause interstitial condensation. Provisions for insulation may allow external moisture to penetrate walls and roofs. Groundfloor insulation may trap water arising from leakages or from cleaning. Factors other than moisture can include overheating of electrical wiring where it is covered by insulation. Greater temperature variation on external surfaces as a result of insulation can give rise to failure and there are many other aspects of good practice in design which are important when higher levels of insulation are used. It is not appropriate to cover these points by Building Regulations. However, the Building Research Establishment has produced a document called *Thermal Insulation: Avoiding Risks*, which has been prepared to cover problems related to Part L of the regulations. It is an excellent document and should be used by all designers.

Thermal bridges

Approved Document L1 identifies the problem of cold bridging at openings. It provides a procedure for assessing the degree of risk, including the case where thin metal layers such as metal lintels are present. If the procedure shows there is a condensation risk at the cold bridge the design must be modified. If no condensation risk is identified but the rate of heat loss at the cold bridge is excessive, this could either be corrected by modifying the design or by compensating for the heat loss in other ways that are specified. Figure 5.29 demonstrates this procedure.

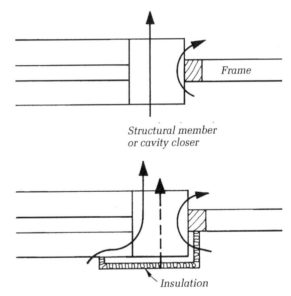

Figure 5.29 Estimation of U values for thermal bridges at wall or roof.

Example 1

Calculate the U value for proposed construction and, if necessary, the minimum thickness of expanded polyurethane board to give a final U value of 0.45 W/m² K.

Details of the construction are shown in Table 5.33, together with the U value. The U value is not acceptable.

Table 5.34 shows the calculation of the additional thickness.

Table 5.32 How to calculate a *U* value

Elements of construction		Thickness, t (m)	Conductivity, λ (W/m² K)	Resistance, $R = t/\lambda$ (m² K/W)
Layer	Description			
	External SR			0.06
1	Tiles	0.012	0.84	0.01
2	Cavity	0.025	–	0.12
3	Concrete block	0.10	0.51	0.20
4	Cavity	0.05	–	0.18
5	Concrete block	0.10	0.51	0.20
6	Plaster	0.013	0.16	0.08
.	.	.	.	.
.	.	.	.	.
.	.	.	.	.
	Internal SR			0.12

Total resistance = 0.97
U = 1/total resistance (W/m² K) = 1.03[a]

[a] To comply with the Building Regulations calculations should be to two decimal places.

Table 5.33 Additional insulation: calculation worksheet[a]

Enter U (W/m² K)		Calculate resistance = 1/U (m² K/W)	
Desired	0.45	2.22	a
Existing	1.03	0.97	b
Calculate desired − existing		1.25	$c = a - b$
Enter conductivity of insulation to be used (m K W⁻¹)		0.025	d
Calculate required thickness (m)		0.031	$c \times d$

[a] This method may be used not only with walls and roofs but also with solid and suspended ground floors.

Table 5.34 Stage 1: calculating the *U* value with an additional layer[a]

Original construction	U (W m⁻² K⁻¹)	0.45	a
	Resistance = 1/U (m² K W⁻¹)	**2.22**	$b = 1/a$
New layer	Thickness (m)	0.019	c
	Conductivity (W m⁻¹ K⁻¹)	0.14	d
	Resistance (m² K W⁻¹) = $\dfrac{\text{thickness}}{\text{conductivity}}$	**0.14**	$e = c/d$
New construction	Total resistance (m² K W⁻¹)	2.36	$f = b + e$
	New U = 1/resistance (W m⁻² K⁻¹)	**0.42**	1/f

[a] Layers may also be subtracted.

Table 5.35 Stage 2: calculating the additional insulation required[a]

Enter U (W m⁻² K⁻¹)		Calculate resistance = 1/U (m² K W⁻¹)	
Desired	0.35	2.86	a
Existing	0.42	2.36	b
Calculated desired − existing		0.5	$c = a - b$
Enter conductivity of insulation to be used (m K W⁻¹)		0.035	d
Calculate required thickness (m)		0.017	$c \times d$

[a] This method may be used not only with walls and roofs but also with solid and suspended ground floors.

Example 2

To take advantage of the simple alternative method where it is desired to achieve a *U* value of 0.35 for a concrete suspended ground floor of existing *U* value 0.45. Expanded polystyrene insulation is to be used of conductivity 0.35 W/m K with a 19 m chipboard finish, conductivity 0.14 W/m K.

The simplest method is to determine the effect of the known thickness of chipboard (stage 1) then to establish the insulation thickness required (stage 2).

Table 5.34 shows the effect of the chipboard (stage 1) and Table 5.35 shows the additional insulation (stage 2).

5.13 Controls for heating and hot-water supply

For both dwellings and other buildings the details of the control requirements for heating and hot-water supply are not specified explicitly since a wide variety of systems and controls are possible. The general requirements are that controls for heating must be provided appropriate to the building and type of installation and covering zoning, temperature, time and boiler controls. Similarly hot-water supply controls would cover heat exchanger size, temperature, time and boiler controls together with insulation and safety. Compliance with appropriate British Standards would satisfy the regulations.

6 Sound

6.1 General introduction

Very great increases of noise resulting from increased motor and air traffic have occurred at the same time as developments in building materials and techniques giving lightweight components erected with dry and sometimes not well-sealed joints. These factors, either separately or in conjunction, frequently cause discomfort and dissatisfaction in modern buildings. Noise transmitted from one room to another or from one part of the building to other parts can often cause annoyance. In addition to noise penetration and transmission it is necessary in many rooms to consider the way in which sound behaves within the space itself in order to ensure that speech or music can be heard intelligibly.

Some appreciation of the nature of sound and the units of measurement and terminology is essential to more detailed consideration of the practical applications of sound insulation and room acoustics. Sound itself is the sensation produced by a certain range of rapid fluctuations of air pressure affecting the ear mechanism. Vibrations from a source of noise excite a similar movement in the molecules of air, which produces a series of pulses of increased pressure moving outwards from the original source. These vibrations can be transmitted not only through air, but through any elastic medium, including the materials making up the fabric of buildings, and from one medium to another.

The ear is able to react to frequencies of vibration from 20 to 20,000 Hz (hertz or cycles per second), although with increasing age the ability to distinguish the higher frequencies diminishes. In normal usage the frequency is called pitch; low frequency sounds correspond to low pitch and high frequency sounds correspond to high pitch. It is rare that sources of sound produce a pure tone, i.e. vibrations at only one frequency. The sounds that we hear normally consist of vibrations at many frequencies.

From a free-standing source sound moves outwards in all directions forming a spherical wavefront. The energy is therefore progressively spread as the sound waves move outward from the source, and its intensity diminishes in proportion to the square of the distance from the source (Fig. 6.1). When sound waves impinge on surfaces much of the energy is reflected. The reflection is similar to a ray of light falling on a mirror; the angle at which the sound leaves the surface is the same as the angle of arrival (Fig. 6.2). Convex surfaces will give dispersed reflections and concave surfaces give concentrated reflections. Although sound is propagated in straight lines, obstructions to the waves do not cast sharp acoustic shadows. Particularly with

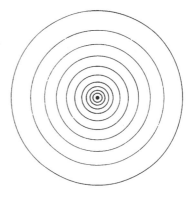

Figure 6.1 Sound propagates outwards from a source. The intensity of the sound diminishes with distance from the source. Multiple sources, although individually the same, will reinforce each other and the diminution effect will be considerably less apparent.

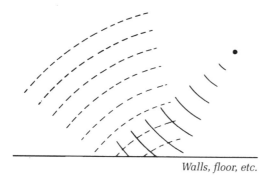

Figure 6.2   Sound waves impinge on surfaces.

Walls, floor, etc.

the lower frequency, longer wavelength sounds, diffraction round the object takes place (Fig. 6.3).

Measurements of sound are normally made in terms of sound pressure, which is the increase in air pressure created by the sound. Sound energy is the power per unit area. The ear is capable of distinguishing sounds over a very wide range of sound pressures. In order to simplify the expression of values, to give better correspondence with the subjective appreciation of sound and to simplify insulation calculations, a logarithmic scale gives the ratio of sound pressures to a base level chosen at the threshold of hearing. This is described as a decibel scale (symbol dB). Values on the scale range from zero (threshold of hearing) to 130 (painful sound). An increase of 6 dB at any point on the scale represents a doubling of pressure, whereas a decrease of 6 dB represents a halving. Doubling of sound energy is represented by 3 dB. Decibels cannot be summed by simple addition. A calculation is needed to establish the overall value resulting from the simultaneous occurrence of two or more sounds of known decibel value.

In addition to defining sound levels the decibel scale is used to specify the ratios of sound reduction achieved by various elements of building construction. In the same way the overall insulation value of a wall containing more than one type of construction cannot be obtained by taking a weighted average of decibel values in proportion to their areas. In fact, the overall value is likely to be little better than the value of the least effective part of the wall, unless the area of poor insulation is very small. CIRIA Report 127:993 *Sound control for homes* contains a more detailed treatment of the theoretical aspects of sound and the range of units employed than is appropriate here and contains a functional scale which enables combined noise levels to be established, a chart to enable the insulation value of non-uniform partitions to be established and a table of the insulation values of partitions with doors.

For the majority of purposes decibel values representing the total sound energy or pressure over the whole frequency range are used. It is recognized, however, that sounds at different frequencies, although having the same pressure of energy, can appear to have different loudness when judged subjectively. Several scales have been developed which divide the sound into octave bands of different frequencies, thereby enabling the importance of sounds at particular frequencies to be indicated. Noise criteria (NC), speech interference level (SIL) and dBA are examples of scales of this sort. The dBA scale is considered to correlate well with subjective judgments of loudness. Most data on typical noises, recommended standards and insulation performance of various constructions are given in normal decibel values averaged over 100–3200 Hz, and this is the scale that has been generally adopted here.

Some of the sound falling on surfaces is absorbed. The ratio of sound absorbed to the sound falling on the surface is described as the absorption coefficient. The coefficient varies at different frequencies. A value of 1 represents total absorption such as would be given by an open window.

Sound transmission through walls and floors has no direct connection with absorption. Transmission occurs as a result of the constructional members being set in vibration by incident sound and giving rise to vibrations in the air on the far side. The efficiency of walls and floors in preventing sound transmission depends upon their mass. The heavier the construction, the less easily it is set into vibration and the higher its insulation value. Transmission straight through a

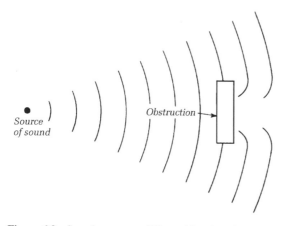

Source of sound

Obstruction

Figure 6.3   Sound waves are diffracted by obstacles.

partition or floor is known as direct transmission. It is clearly possible, however, for sound to be transmitted along structural members which link but do not divide rooms. Figure 6.4 shows some possible paths of this type of transmission, which is described as indirect or flanking. Each particular type of construction will reduce the amount of sound transmitted by a fixed proportion, irrespective of the actual noise level. The insulation values can therefore be expressed in decibels. The reduction in noise levels is found by subtracting the insulation value in decibels from the noise level on the other side of the construction. In practice this value may be modified by indirect transmission. Any air passages such as keyholes, gaps round doors (Fig. 6.5) or gaps between wall and ceiling can allow sound to pass and seriously diminish the insulation value.

Sound insulation may be improved by *discontinuous construction*. To achieve this a wall is divided into two separate skins, so vibrations are not easily transmitted from one to the other. It is less easy to make floors discontinuous. A resilient quilt is usually employed, laid on a structural floor and itself supporting the 'floating' floor finish. Floating floors should be isolated from structural walls and should not be continuous from room to room. Figure 6.6(a), (b) and (e) shows floors of this type.

Impacts on floors or other constructional members can cause sound transmission not only into adjacent spaces but also into other parts of the building, particularly if it has a framed construction. This problem can be met in two ways. Floating floors (described above and in Fig. 6.6) will limit the transmission of vibrations into the main structure and soft finishes such as cork tiles or carpet will reduce the generation of vibrations. Note that, although floating floors provide improved insulation against both airborne and impact noise, the soft finishes on solid

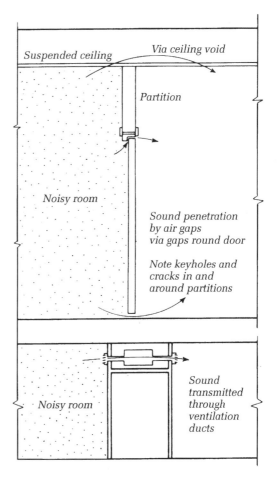

Figure 6.5 Possible path for sound transmission through gaps in the structure of a building.

floors do not improve insulation for airborne sounds nor would increased mass significantly reduce the impact noise transmission.

In addition to external noises and internal noises due to the occupants, considerable noise can be generated by services in buildings. In domestic buildings plumbing can be a source of considerable annoyance, as can boilers and pumps and power-operated appliances such as dishwashers, waste-grinders and washing-machines. In larger buildings lifts, boiler and refrigeration plants and other mechanical plant may give rise to noise.

6.2 Noise control

In the practical design of buildings consideration of sound is likely to have four aspects: minimizing sources of noise planning to keep noise sources as far

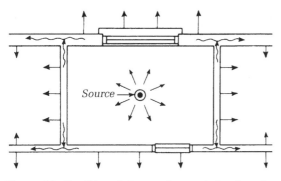

Figure 6.4 Possible path for sound transmission through the structure of a building.

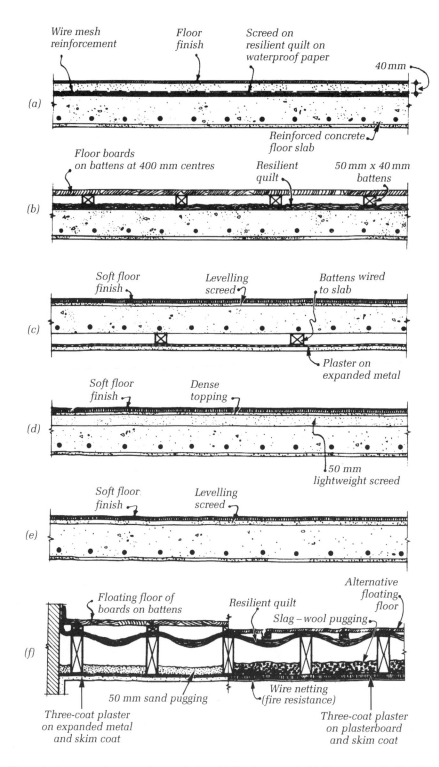

Figure 6.6 Floors designed to reduce sound transmission: (a) floating screed; (b) floating wood raft; (c) suspended ceiling and soft finish; (d) lightweight screed and soft finish; (e) heavy concrete floor and soft finish; (f) insulated wood joist floor.

as possible from quiet areas; structural precautions to reduce noise penetration and, in cases where it is appropriate, the internal acoustics of rooms.

Planning in relation to noise

It is clearly desirable to site buildings on sites as far as possible from sources of noise. In addition hard paving should be kept to a minimum and grass and planting used as much as possible. Hard paved re-entrant areas should be avoided. In some cases it may be thought desirable to interpose solid screens between noise sources and windows. For such screens to be effective it is important that they should be close either to the source or to the window. Their effect in intermediate positions may be negated by the diffraction of sound. High buildings do not usually give any useful protection from noise, for although upper floors receive less sound from areas immediately adjacent, they receive more sound from distant sources. It is important to recognize not only noise sources external to the site but also site activities which may themselves give rise to annoyance. Children's play areas, refuse collection, deliveries or garage areas may all give rise to annoyance.

Interior planning arrangements should attempt to concentrate quiet rooms on façades remote from external noise sources and should also attempt to group noisy and quiet areas separately and in isolation from each other.

Constructional precautions to reduce noise

The Building Regulations 1991 call for 'an adequate resistance to the transmission of airborne sound' for walls and for airborne and impact sound in the case of floors. Means of complying with the Regulations are set out in Approved Document E: 1992 *Airborne and impact sound*. The standards for airborne sound reduction are specified in terms of weight of walls and floors and in terms of soft coverings and floating layers for impact sound, together with special requirements for joints and the elimination of air passengers. The approved document gives a comprehensive schedule of acceptable and commonly used types of construction. Types of construction which are not covered by the approved document may be demonstrated to meet the Building Regulations requirements by testing an already built example. If this example meets the appropriate acoustic criteria then the construction may be repeated. It will be noted that, although measurable standards of sound reduction form the basis of the regulation, it is considered

impracticable to carry out acoustic testing, which is complicated and expensive, in every room of every building which is constructed. The problem is solved by allowing the replication, without further testing, of methods which have been demonstrated to be satisfactory. The advantages of this approach in terms of administration, simplicity and economy are apparent. However, to be successful the method requires careful control of materials and, in particular, of work standards.

The insulation standards which form the basis of the regulations are called the *weighted standardized level difference*, $D_{nT,w}$ (for airborne sound) and the *weighted standardized sound pressure level*, $L'_{nT,w}$ (for impact sound) (Table 6.1). The level difference for airborne sound is simply the difference between sound levels in decibels between the level on the source side of the construction and the receiving side. The level differences, D, are standardized to the value which would exist in a room with an 0.5 s reverberation time, D_{nT}. The standardized values at each one-third octave band from 12 to 2000 Hz are plotted in relation to a reference curve and the weighted standardized level difference, $D_{nT,w}$ is calculated over the whole range. *Sound Control for Homes* gives a more detailed general description of the method. BS 2750 deals in detail with the measurements involved whereas BS 5821 defines the method of calculating the weighted values.

The standards which must be achieved in testing vary with the number of examples tested. A higher standard is required if only one example is tested than is required for the mean of a number of tests.

Some of the typical types of construction which meet the Building Regulations requirements are shown in Table 6.2.

Surveys by BRE have found that, even when they were of constructions normally considered adequate to meet the Building Regulations standard, some 50% of party walls failed to achieve the standard described

Table 6.1 Standards for airborne and impact sound

Standard quality[a]	Single test	Mean of 4	Mean of 8
$D_{nT,w}$			
Walls	49	52	53
Floors	48	52	52
$L'_{nT,w}$	65	61	62

[a] Minimum value of airborne sound reduction weighted standardized level difference ($D_{nT,w}$), and maximum value of impact sound weighted standardized sound pressure level ($L'_{nT,w}$).

Table 6.2 Typical constructions

Wall construction		Minimum weight (kg/m²)
Solid masonry (with at least 12.5 mm plaster or plasterboard on each side)	Brickwork	375
	Concrete block	415
	Dense concrete (plaster optional)	415
Cavity masonry (including 12.5 mm plaster on each side)	Brickwork	415
	Concrete blockwork	415
	Lightweight concrete block	150
Masonry with lightweight panels (solid cone)	Brickwork	300
	Concrete block	300
	Lightweight concrete block	200
	Panels each of at least two layers of 12.5 mm plasterboard with 25 mm cavity	
Timber frame	Two separate supporting frames are required with two layers of 12.5 mm plasterboard on each and an absorbent quilt between the frames	

Floor construction	
Solid soft covering	365
Solid with floating layer on resilient carpet	220
Timber with floating layer on 25 mm resilient carpet density 60–80 kg/m³ and absorbent blanket or pugging	

above. As a result BRE is able to identify the important features associated with several types of construction if they are to perform satisfactorily.

Masonry walls

Mass is the most important factor in ensuring adequate performance. Dense concrete or brick should be used and bricks should be laid with frogs upward to increase the overall density and to avoid air passages. There is little advantage for sound reduction in cavity walls but, if cavity construction is employed, butterfly wall ties should be used rather than more rigid types.

Care should be taken to avoid any penetration of party walls and joist hangers should be used in preference to building in joists. Plaster finishes are superior to dry linings from the sound insulation viewpoint.

Framed timber construction

Satisfactory performance can be achieved by two separate stud walls lined with layers of plasterboard or plywood not less than 32 mm in total thickness. The

inner faces should be separated by not less than 200 mm and this space should be lined with one or two sound-absorbing quilts. Care is required to avoid air gaps in the construction.

In the BRE survey this type of construction performed more consistently than the other forms.

General

Partitions should not be built off floating floors. Pipes and conduits should not be allowed to cause direct contact between the floating floor and the subfloor. They can be on the subfloor but should be haunched up in mortar so the quilt passes over them and is continuously supported.

Lightweight screeds are not satisfactory for *impact* if too light. Minimum density should be 112 kg/m^3, and a dense topping is necessary to seal off all air passages.

Suspended ceilings are a benefit against airborne sound. It is important that the ceiling should be moderately heavy (not less than 25 kg/m^2), not too rigid, completely airtight and its points of suspension should be few and flexible.

Concrete floors are good for insulation because they brace and load the walls which they adjoin, so flanking paths are not as liable to transmit noise. With wood floors it is possible to add insulation, as shown in Fig. 6.6, to give better insulation, but sound will still travel round by the supporting walls if they are thin, e.g. 112 mm brick or less.

The ceiling of three-coat plaster on expanded metal lath is an important feature of several satisfactory constructions. It is important that the pugging, where it is used, should be directly on the ceiling.

Doors can provide easy passage of sound through otherwise well-insulated walls. Figure 6.5 shows how sound can penetrate through the cracks round doors and through keyholes. Modern doors are often of hollow construction and do not in themselves provide much insulation. A hollow hardboard door with normal cracks gives about 15 dB reduction. This may not be critical in most cases where doors open to corridors rather than directly into other rooms. Where insulation must be maintained, sealing of cracks by rubber strips and the use of heavier doors is desirable. If a sill is unacceptable then automatic draught excluders may be used to close the crack at the foot of the door. Sometimes to improve insulation, doors are used on both faces of a partition. If the air sealing is effective, shutting is likely to be difficult and a very much more effective system is to create a sound lock with two doors opening on to a small lobby finished with absorbent materials.

Suspended ceilings in office buildings where, for reasons of flexibility, partitions do not extend up to the structural floor, can provide an easy path for sound penetration between rooms. Not only can sound pass through the gaps in the construction but, if the ceiling is light and rigid it can itself transmit sound from one room to another. Remember that the usefulness of ceilings for sound absorption is in no way affected by sound transmission. To achieve satisfactory insulation using an airtight ceiling, as previously described, the space above should be as great as possible and in critical cases additional absorbent may be used above the ceiling.

External walls of traditional construction are normally heavy enough to give adequate sound insulation (275 mm cavity walls give 50 dB average reduction or more). Windows in external walls, however, present a critical problem. An open window will only give 5–10 dB reduction, depending on whether it is fully or only slightly open. A closed single-glazed window with 3 mm glass will give only about 20 dB. Sealing improves the effect up to about 25 dB and the use of heavy glass can improve things slightly more. To gain a significant further increase it is necessary to use double glazing. Unlike double glazing for thermal resistance, which has only a narrow space between the sheets of glass, double glazing for sound insulation is best spaced to give about 200 mm air space between the two sheets of glass. The performance is significantly improved if the reveals are lined with absorbent. Figure 6.7 shows a typical construction. One problem associated with this type of window is the prevention of condensation within the air space. It is not possible, in view of the very wide gap, to seal the two sheets of glass hermetically. The diagram shows one method for keeping the air space dry. A tube of silica gel is used. When it has absorbed all the water that it can, it is removed and heated, after which it can be replaced and will function again. Double glazing with 6 mm plate glass, and 200 mm air space lined with absorbent, can give an insulation value of 43 dB. For cleaning purposes it would be convenient if one leaf of the double window could open. It is difficult, however, at reasonable expense, to obtain a leaf that will open and also remain reasonably tightly sealed when closed.

From the values given above it is clear that, for most circumstances where external noise is a problem, windows will have to be sealed if satisfactory internal conditions are to be achieved. If this is to be the case air-conditioning will be required to give ventilation and avoid overheating.

Many of the new constructional methods coming

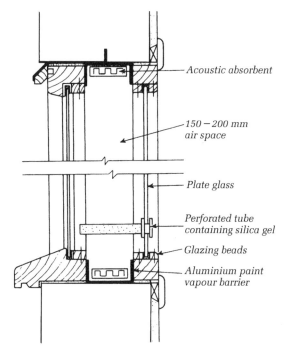

Acoustic absorbent

150 – 200 mm air space

Plate glass

Perforated tube containing silica gel

Glazing beads

Aluminium paint vapour barrier

Figure 6.7 Window designed to reduce sound transmission.

into use are based on lightweight panels with dry joints. With this type of construction it is particularly difficult to achieve satisfactory sound insulation. The problem is discussed in BRE Digest 96 *Sound insulation and new forms of construction*. The recommendation of the digest for party walls in lightweight construction is of two separate and separately supported skins with a wide (22 mm minimum) space between.

BRE Digest 143 *Sound insulation basic principles* describes the physical principles involved to complement the other digests. Digests 128 and 129 deal specifically with insulation against external noise.

External noise indices

Some of the major external sources of noise, such as motorways and air corridors, are variable in the intensity, duration and frequency of the noises they generate, and if standards are to be established they must take these factors into account. The problem is particularly important since householders can receive grants for precautions or compensation if they are exposed to some types of noise, and these payments have to be rationally based. A number of special indices have been developed. The traffic noise index (TNI) and the noise pollution level (LNP) take into account the factors described but are complex in application and the 10% level (L_{10}) is thought to be a reasonable and practicable unit. The value is based upon the noise level in dBA which is exceeded for 10% of the time between 0600 and 2400 measured 1 m away from the façade of the building.

The Noise Advisory Council recommends that for houses of normal construction, without special noise provisions, the L_{10} index should not exceed 70 dBA.

Aircraft noise presents a special problem. The loudness at the various frequencies differs from that of traffic noise and a special scale has been developed which weights the contributions of various frequencies so that the scale values correspond with subjective experience. The scale is called the perceived noise decibels (PNdB) scale. With aircraft in particular it is necessary to take into account the balance of noise and frequency and the noise and number index (NNI) has been developed. This index takes the loudness of the noise (L PNdB) and combines this with the number of aircraft by means of a formula.

Noise from building services

Recent years have seen a very great expansion of services installations in buildings. The plant itself may generate airborne noise and transmit vibration to the structure. Precautions against noise can be taken in the location of the plant, by the provision of antivibration mounting, and walls and floors capable of attenuating the airborne noise. More difficult problems arise, even in dwellings, as a result of the distribution of the services round the building. Noise problems of this sort can arise in many ways:

- Penetration of walls and floors can allow the passage of sound.
- Pipes and air trunkings can provide routes for the transmission of vibration. Pump and fan vibration, in particular, can be transmitted in this way.
- Excessive velocity of flow in pipes or air trunkings can give rise to noise, particularly at points where valves or dampers control flow.
- Waste soil and rainwater pipes flow in a partially filled mode with air entrained in the flow; this creates noise.
- Where pipes expand and contract as a result of temperature changes, the resulting friction with pipe supports can generate noise.
- In water supply pipework, particularly where it is under mains pressure, sudden arrest of the flow such as might result from the sudden closure of a sticking ball-valve or the loose jumper in a main

isolating valve, can cause pressure waves which reflect along sections of pipe. The resulting noise can be very loud.

Much can be done to reduce the noise by the design of the services themselves and the fixing of apparatus. Very noisy equipment should ideally be mounted on foundations independent of the building structure. Where this cannot be achieved, antivibration mountings may be fixed to an inertia block, sized to have a natural frequency of vibration different from that of the equipment. The whole can rest on a resilient pad to minimize noise transmission to the structure. Flexible sections of pipe can overcome transmission of plant vibrations to pipework and canvas couplings achieve the same result for ventilation systems. Velocities of flow can be controlled, valves and dampers carefully selected and pipe support, in particular, designed to allow thermal movement. Several types of ball-valves avoid the sudden closure which encourages water hammer.

Ducts should be sited as far as possible from rooms requiring quiet conditions. Substantial casings for ducts can minimize penetration into rooms. BRE have recommended a weight of 10–25 kg/m^2 for ducts in kitchens; heavier casings with careful attention to the elimination of air gaps would be appropriate for living-rooms and bedrooms.

Bathrooms, where flushing WCs will add to the noise levels, should be treated in a similar way to ducts. Partitions and doors should be substantial and air gaps sealed. Pipes and sanitary appliances should not be mounted on walls common with bedrooms.

There are no statutory requirements in this country for limitation of noise from plumbing and high levels of noise are not uncommon; 90–95 dB may be achieved. In some other countries there are statutory limits to the noise which is permissible.

6.3 Room acoustics

In domestic circumstances the internal acoustics of rooms are not usually critical, but in larger spaces such as lecture rooms, conference and music rooms or in assembly halls and auditoria the satisfactory use of space depends on satisfactory acoustic conditions. This is a problem of considerable complexity in auditoria and expert consultants will normally be employed to advise on the problem. There are several very detailed texts dealing with the subject. In the case of the smaller rooms it is possible to distinguish two principal aspects, apart from proper insulation, which govern satisfactory hearing. They are the planning and shape of the room to give good paths for sound from speakers or performers to the audience and the rate at which a sound dies away. This is called the reverberation time.

Direct sound

Ideally all the occupants of a space should be able to hear by means of a direct path of sound between the speakers or performers and each individual occupant. In conference venues, council chambers or board-rooms, where any occupant may be called upon to speak, this is achieved by layout of seating so that all the occupants can see one another clearly. A long, narrow board table will be much less satisfactory than a circular arrangement. In rooms where there is a large passive audience, it is possible to improve the direct paths of sound by raising the speaker and also by raking the seating so that each member of the audience is less obstructed by those in front. Sound reflected once from a surface (first reflection) can, if it is directed in the right direction, usefully augment the direct sound. Figure 6.8 shows a notional section through a small auditorium where the speaker has been raised, the seating raked, and parts of the ceiling arranged to reflect sound to specific audience areas. It is important that the extra distance which first reflections have to travel over the direct path should not be too great. If the additional distance is over 20 m the reflected sound will arrive at the listener noticeably later than the direct sound and give rise to an echo. BRE Digest 82 (second series) *Improving room acoustics*, recommends that the difference in length of the two paths should not exceed 10 m and preferably it should be less. Care is therefore needed in disposing reflecting surfaces to ensure intelligible reinforcement of sound without echo. In general reflecting surfaces should be flat, curved surfaces may give rise to sharp concentrations of sound in some places and lack of reinforcement elsewhere.

In the same way that properly disposed reflecting surfaces can assist in distributing sound, inappropriate surfaces will have an undesirable effect. Cross-beams projecting below a general ceiling level are very undesirable since they prevent part of the ceiling from reflecting sound to the audience. Reflecting surfaces do not necessarily have to be part of the main fabric of the building. In very large halls it is not unusual to hang a reflector over the speaker's position. The Building Research Establishment recommend as a minimum thickness and weight a surface of 25 mm plasterboard, painted on the side used for reflection.

Reverberation time

Sound emitted from a source will be reflected not once but several times within a space. Figure 6.9 shows the early stages of propagation. If the interreflections continue for too long the echo effect described above comes into effect for a multiplicity of reflected paths and intelligibility suffers. Standards to govern this effect are based on the time taken for the sound level in the room to decay by 60 dB. This is described as the reverberation time. The BRE recommend the reverberation times in seconds for various sizes and uses of rooms shown in Table 6.3.

These reverberation times apply over the whole range of frequencies except that a longer time at low frequencies may be desirable for music.

Reverberation time depends upon the volume of the room and the amount of sound absorption given by the surfaces. Sound absorbents perform differently at

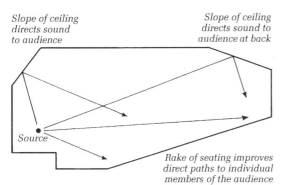

Slope of ceiling directs sound to audience

Slope of ceiling directs sound to audience at back

Source

Rake of seating improves direct paths to individual members of the audience

Figure 6.8 Shaping of an auditorium to improve the paths of sound from source to audience.

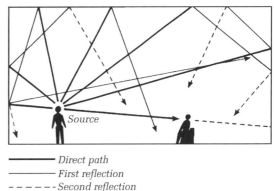

Source

———— *Direct path*
———— *First reflection*
- - - - - *Second reflection*

Figure 6.9 Early stages in the propagation of reflected sound.

Table 6.3   Recommended reverberation times

Use	Small rooms <750 m³	Medium rooms 750–7500 m³	Large rooms >7500 m³
For speech	0.75	0.75–1.0	1
Multipurpose, e.g. school halls	1	1.0–1.25	1.0–1.5
For music	1.5	1.5–2.0	≥2

different frequencies and it is necessary when estimating reverberation times to make separate calculations at different frequencies. For complex analysis a full octave band analysis may be undertaken but for most situations checking at 125, 500 and 2000 Hz is considered appropriate. Several formulae exist for establishing reverberation time, each with special relevance to particular conditions. The original formula and one which is widely used at present is that developed by Sabine:

$$\text{Reverberation time (s)} = \frac{0.16 \times \text{volume (m}^3)}{\text{total absorption (m}^3/\text{s})}$$

Total absorption (m³/s) = sum of each area (m²) × its absorption coefficient (m/s)

If satisfactory values are not found in a particular case, it will normally be necessary to vary the absorption by varying the surface finishes. Theoretically the volume could be varied but this is rarely a possibility in practice. A number of different types of absorbent may be required to achieve a proper balance since the performance of absorbents varies greatly with frequency. Fibreboards and similar soft materials commonly used for absorbents are very efficient at high frequencies but not at low frequencies. A thin panel concealing a space lined with absorbent is efficient for low frequencies. Some proprietary acoustic panels have a combination of materials which give a more balanced performance over the range of frequencies. When positioning absorbents, surfaces giving important reflections should clearly be avoided. In general back walls will be a first choice and ceilings a last choice of position.

Table 6.4   Sound absorption coefficients

Item	Unit	Absorption coefficient at specific frequency		
		125 Hz	500 Hz	2000 Hz
Air	m³	0	0	0.007
Audience (padded seats)	person	0.17	0.43	0.47
Seats (padded)	seat	0.08	0.16	0.19
Boarding or battens on solid wall	m²	0.3	0.1	0.1
Brickwork	m²	0.02	0.02	0.04
Woodblock, cork, lino, rubber floor	m²	0.02	0.05	0.1
Floor tiles (hard)	m²	0.03	0.03	0.05
Plaster	m²	0.02	0.02	0.04
Window (5 mm)	m²	0.2	0.1	0.05
Curtains (heavy)	m²	0.1	0.4	0.5
Fibreboard with space behind	m²	0.3	0.3	0.3
Ply panel over air space with absorbent	m²	0.4	0.15	0.1
Suspended plasterboard ceiling	m²	0.2	0.1	0.04

The absorption coefficients of some common surfaces are given in Table 6.4. See the MBS title *Materials* for a more comprehensive list. A reverberation time calculation is set out in Table 6.5.

Standing waves

In rooms with parallel walls 10 m or less apart it is possible for particular wavelengths of sound to appear intensified at the expense of others. This effect is emphasized if all the dimensions are the same (in effect a cube) or if the dimensions are related by simple ratios (e.g. 1:2 or 1:3). Carefully selected relationships of dimensions, opposing walls out of parallel, and absorbents to operate at the critical frequencies are means of avoiding difficulties from the phenomenon. A brief theoretical treatment of this problem is described in *Acoustics and Noise in Buildings*, by Parkin and Humphries.

Table 6.5   A reverberation time calculation[a]

Surface	Finish	Area (m²)	Low (125 Hz)		Medium (500 Hz)		High (2000 Hz)	
			Absorb coeff	Absorb units	Absorb coeff	Absorb units	Absorb coeff	Absorb units
First estimation, no absorbents[b]								
Ceiling	Plaster (or concrete)	60	0.03	1.8	0.02	1.2	0.04	2.4
Walls	Plaster	92	0.02	1.8	0.02	1.8	0.04	3.6
Window	Glass (4 mm)	20	0.2	4.0	0.1	2.0	0.05	1.0
Floor	Wood block	60	0.02	1.2	0.05	3.0	0.1	6.0
Occupants		10 people	0.17	1.7	0.43	4.3	0.47	4.7
Air		210 m³	–	–	–	–	0.007	1.5
Total absorption units				10.5		12.3		19.3
Reverberation time (s) $= \dfrac{0.16 \times 2.10}{\text{total absorption units}}$				3.2		2.7		1.7
Second estimation, using absorbents described[c]								
Ceiling	Suspended plasterboard	60	0.02	12.0	0.1	6.0	0.04	2.4
Walls	Plaster	62	0.02	1.2	0.02	1.2	0.04	2.5
	Ply panel with absorbent backing	30	0.3	9.0	0.15	4.5	0.1	3.0
Window	Glass	30	0.2	6.0	0.1	3.0	0.05	1.5
Floor	Wood blocks	30	0.02	0.6	0.05	1.5	0.1	3.0
	Carpet	30	0.1	3.0	0.3	9.0	0.5	15.0
Occupants		10 people	0.17	1.7	0.43	4.3	0.47	4.7
Air		210 m³	–	–	–	–	0.007	1.5
Total absorption units				33.5		29.5		34.6
Reverberation time $= \dfrac{0.16 \times 2.10}{\text{total absorption units}}$				1.0		1.1		0.97

[a] Case considered: small multipurpose room 6 m × 10 m × 3.5 m; desired reverberation time of 1 s at all frequencies; assumed occupancy of 10.

[b] This is not a satisfactory pattern. More absorption is required particularly at the lower frequencies. (A total of 33.6 absorption units are needed at each frequency to give the desired reverberation time.) A suspended plasterboard ceiling gives very much better absorption at low frequencies than plaster direct on concrete. Plywood panels backed with absorbent have characteristics similar to windows. It is therefore reasonable to attempt to correct the reverberation time by using a suspended ceiling, increasing the window size and introducing an area of absorbent-backed ply. Since a substantial reduction is required at all frequencies, an area of carpet – a very efficient absorbent – may be included with advantage. The result of these modifications is shown in panel B.

[c] The balance of absorbents selected appears to have produced a very satisfactory correction.

ENVIRONMENTAL SERVICES

7 Thermal installations

Technical development, changing fuel costs and availability, and environmental concerns have produced rapid changes in heating systems. However, many millions of buildings still have systems which would not be installed today. This chapter attempts to describe the new systems and some of the older systems still in widespread use. The government has established an Energy Efficiency Office which publishes a wide range of information on efficient use of energy in buildings.

Most buildings require an installation to maintain a satisfactory thermal balance. In some parts of the world this will involve only the extraction of heat, in others heating and cooling seasons are involved. In Britain an increasing number of buildings, particularly those with high levels of artificial lighting or high buildings with large proportions of glazing, must be provided with facilities for cooling. But the majority of buildings are thought to give adequate environmental control in summer by their form and fabric alone, and it is only in the winter that additional heat is required. To make a satisfactory provision for this it is necessary to decide on the source of energy which is to be used and the means of distributing the heat derived from that energy into the building. The main criteria involved in this decision are amenity and economy, although in most cases there will be constraints limiting the range of decision. These may include the preferences of the building owner; the non-availability or unsatisfactory supply record of a particular fuel; the Clean Air Act; impracticable or uneconomic flue requirements; limited site area precluding storage or many other factors arising from a particular situation. Consideration of amenity must include these factors:

- *Thermal comfort*
  Taking into account not only the appropriate type of

installation but also the degree of control required to maintain good conditions (also related to economy).
- *Appearance*
  - Internally: form and positions of radiators, exposed pipe runs, grilles, thermostats, etc.
  - Externally: form and position of boiler rooms, fuel stores, flue, etc.
- *Planning*
  - Generally: access for fuel delivery and positions of apparatus compatible with the general planning arrangements.
  - Individual interiors: satisfactory arrangement of heat emitters to allow freedom in disposing furniture and resolution of conflicting claim for space (e.g. radiators and windows may compete).
- *Maintenance*
  Both routine attention, and servicing and replacement should be possible without gross interference with the use of the building.

The economic factors to be considered are as follows:

- *Initial cost*
  - Cost of installation.
  - Cost of the accommodation (e.g. boiler room, flue ducts) and access (e.g. service road for fuel delivery) required for the installation.
- *Running cost*
  - Cost of fuel (which will certainly vary during the life of the building). Appropriate control of output may be important for fuel economy, particularly in buildings with rapid reaction to external changes and intermittently used buildings.
  - Repayment of loan charges or allowance for replacement.
  - Inspection, maintenance and insurance.

– Differences in decorating costs which result from different types of installation.

Some of these factors, such as the cost of fuel over the life of the installation, are speculative and it is difficult to arrive at a solution by averaging the costs over a period of years except in the case of temporary buildings whose life is defined. In such cases the most economical answer is often to reduce the initial cost as much as possible. A method of approach to the problem which may be helpful in reaching a decision is shown in Fig. 7.1. It shows the cumulative costs of two different installations having different capital and fuel costs expressed in relation to time. If the two lines did not meet the solution would be clear and simple. It will be more usual, if the systems being considered are reasonable choices, for the lines to cross after a period of years. If this period is short it will be clearly be more economical to employ the system giving lower cost in all but the first few years. Where the changeover is many years ahead the decision is more difficult and may involve speculative assumptions about future fuel costs. In cases like this, where the economic decision is not clear, it may be best to assume that economically the possibilities are the same and to base selection on considerations of amenity.

Although the above method will be useful in simple cases of comparison it has the disadvantage of giving equal importance to expenditure now and expenditure in the future. It is possible to use the principles of compound interest to evaluate the present investment that is the equivalent of a payment in the future. If all future payments are assessed in this way it is possible to arrive at a present value equivalent to all the expenditure involved and several schemes having

different balances of initial running cost can be evaluated and compared. The method is frequently used for business management and is called *discounted cash flow*.

An example of the use of this method is given in Table 7.1. One or more heating installations are to be compared. The table shows the costs of one installation, an oil-fired low pressure, hot-water, space heating system using radiators. The installation is to be considered as having a life of 15 years with no residual value at the end. The initial and running costs (including builders' work) are set out in the table, in the pattern in which they occur, together with the conversion factors enabling the expenditure each year to be converted to equivalent present value. A full description of these techniques and tables and graphs of the conversion factors required can be found in *Building Design Evaluation: Costs-in-Use*, by Peter Stone, published by Spon. The present value can then be calculated as shown. The various different schemes may then be assessed and a decision reached.

Use of this method involves a decision about the percentage of interest to be taken into account in selecting the present value conversion factors. This should be taken at the average rate of return of money which could be invested by the building owner. Higher rates of interest will reduce the present value of future expenditure whereas lower rates will increase its importance. When there is doubt as to the appropriate interest rate, more than one may be calculated and the results compared to see whether the effect is significant. Inflation will not normally be taken into account when assessing running costs; if it affects all prices equally, comparison is valid ignoring inflation.

In more complex examples than this a variety of other factors could have to be taken into account, including investment allowances and taxation rates in the case of business organizations.

In some cases in buildings, particularly those which are to be rented, it is important for the owner to be able to express the cost in terms of annual expenditure rather than present value. A conversion from present value to equivalent annual expenditure can be made also using tables based on compound interest.

The conversion factor giving annual equivalent values of an initial payment, taken over a period of 15 years at an interest rate of 7%, is 0.11. In the case under consideration, therefore, the annual value is £1886 × 0.11 = £207 per annum.

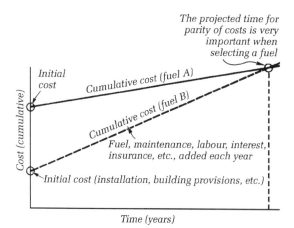

Figure 7.1 Graph comparing the cumulative costs of alternative heating systems.

7.1 Energy sources and heat transfer

The energy for heating is commonly derived from the

Table 7.1 Estimating equivalent present value: discounted cash flow method

Year	Installation	Fuel	Mortgage interest (less tax)	Annual maintenance	Periodic overhaul	Decoration	Total cash flow	Conversion factor for present value at 7% interest	Present value
Base	550	45	12			60	667	1	667
+1		90	25	5			120	0.93	110
+2		90	25	5			120	0.87	104
+3		90	25	5	20		140	0.82	114
+4		90	25	5		45	165	0.76	126
+5		90	25	5			120	0.71	85
+6		90	25	5	20		140	0.67	94
+7		90	25	5			120	0.62	74
+8		90	25	5		45	165	0.58	96
+9		90	25	5	20		140	0.54	76
+10		90	25	5			120	0.51	61
+11		90	25	5			120	0.48	57
+12		90	25	5	20	45	185	0.44	82
+13		90	25	5			120	0.42	50
+14		90	25	5			120	0.39	47
+15		90	25	5			120	0.36	43

Total £1,886

various types of solid fuels and oil, or from gas and electricity. It is also possible to employ the sun's energy or to extract heat from the air or from a river by the heat pump and to use it for heating. In Britain the balance of cost for some 20 years was heavily weighted in favour of the very cheap imported oil which progressively superseded the use of indigenous coal and limited other developments. Because of the imbalance between day and night demand, electricity has been offered at low prices during off-peak hours to encourage use during these periods. Both these situations have now changed. Oil prices have increased sharply and day and night consumption of electricity is becoming more balanced, making substantial inducements for off-peak use self-defeating. North Sea gas has, in recent years, provided a low cost fuel which has been very widely used.

Purely economic decisions about fuel selection are difficult, partly because of uncertainty about fuel prices and partly because problems of conservation, supply and ecology must now be borne in mind. It is clear, however, that in the future, whatever the problems, much closer attention will have to be given in the design of buildings to economy of fuel consumption.

It is important to remember that maintaining appropriate thermal conditions in buildings is thought to absorb almost 40% of total energy production and that in Britain the percentage is almost certainly higher. The balance of use between the main fuels is shown in Table 7.2, together with the approximate efficiency of the production and distribution processes (1973 data).

Each of the usual fuels must be converted to heat in an appropriate piece of apparatus. Traditionally fuel could be moved about buildings but the controlled distribution of heat was difficult, consequently the fuel was burnt in the space where heating was required and the heat emitted directly to the room. Control was achieved by varying the rate of combustion. Over the last 100 years effective methods of transmitting heat round buildings have been developed. They enable fuel to be converted to heat at a central point, thereby reducing the number of flues required to one, eliminating the need to distribute fuel and simplifying maintenance and control. In new buildings, including houses, this type of central installation is almost universal. The significant exception is where electricity is used as the fuel. The electricity is more

Table 7.2 Heating fuels: efficiency of production and distribution

Fuel	Energy used in domestic heating (%)	Efficiency in terms of primary energy required (%)
Electricity	20	27
Manufactured fuel	24	71
Coal	30	98
Natural gas	16	94
Oil	10	93

Table 7.3   Heat transfer mechanisms for various sources of energy

Fuel or energy source	Direct: heat emitted directly to environment	Central: heat transferred to a distributing medium for transmission to required situation	
		Apparatus	Usual distribution medium
Solid	Fires Stoves	Boiler Warm air is sometimes produced from solid fuel appliances for ducted distribution in houses, but is uncommon in larger buildings	Water
Oil	Oil heaters	Boiler Oil/air furnace	Water Air
Gas	Convectors Radiant heaters	Boiler Gas/air furnace	Water Air

Electricity	Convectors
	Thermal storage
	   Floors
	   Block heaters
	Radiant sources
	   High, medium and low
	     temperature
		Boiler: electrode immersion heater
		
		Duct air heater
			Water

Solar radiation	Governed by orientation and fenestration	Collector mechanisms	Various (often air or water)
General environment		Heat pump	Refrigerant to air or water

easily and economically distributed than heat would be and has no special requirements such as flues. Electric underfloor heating in particular is often installed in new buildings. Table 7.3 shows the fuels together with the main types of apparatus for heat transfer either by direct emission or by central distribution. Except for electrical appliances and the decorative use of solid fuel, direct heaters are rarely used except in low cost housing.

The selection of a particular type of fuel involves the provision of storage and delivery facilities. Even in the case of gas and electricity, which are not themselves stored, meters and switchgear have to be delivered and accommodated. In large buildings these items may be of considerable size, requiring access for large lorries. This is particularly the case where transformer stations are provided. Table 7.4 gives details of the calorific values, weights and volumes, storage provision and delivery and plant requirements for the main fuels and suggests further sources of more detailed information.

Figure 7.2 shows a typical oil fuel store. The sump to receive the oil should leakage occur, the fire valve to shut off the supply in case of fire in the boiler room, the sludge drain, and the fire protection of the store will be noted. As an additional precaution against overflowing it is possible to have an audible alarm

fitted to the vent. It is not necessary for the oil fuel store to have direct road access for delivery vehicles. A length of hose is carried by each vehicle and where this will not be adequate a fixed pipeline can be laid across the building site itself.

Figures 7.3 and 7.4 show domestic and large-scale storage for solid fuel. Delivery to the domestic store is usually by porter and sack, whereas in the case of the larger store with access for large lorries, tipping directly into the store is needed.

In some countries which have a substantial percentage of sunshine it is possible to have solar heating installations which can collect and store the heat and deliver it to the building as required. In Britain the dull periods without sunshine are too long for economic storage of the heat and, consequently, although the sun when shining can exercise a decisive influence on heating and give reduced fuel bills if the heating system is able to react and reduce output, solar heating installations have not been practicable. Some experimental installations giving hot water in summer have been built and plans published for long-term storage of solar heat underground so that heat collected in summer could be used in winter.

The principle employed in most solar installations is to expose glass-fronted collector boxes at an optimum angle and to pass air through or to pump water through

Table 7.4 Fuel data[a]

Fuel	Calorific value		Weight and volume (kg/m³)	Storage		Delivery	Plant
	MJ/kg	GJ/m³		Domestic	Non-Domestic		
Solid fuel							
Anthracite	35	29	840	Floor areas for fuel storage	Allow storage for 4–6 weeks	Domestic delivery in sacks; non-domestic delivery direct to fuel stores by lorry (access road required)	Central boiler room
Bituminous coal[b]	30	24	800	House with two types of fuel, 2 m²			Flue
Coke	28	11	400	House with one type of fuel, 1.2 m²			
Oil[c]							
Domestic (35 s)	45	37	830	Tank Minimum bulk delivery is 0.5 m³ Economic bulk delivery is 2 m³ (2.5 m³ tank is often used)	Tank for 3 weeks Heating of stored fuel may be necessary	Hose from road tanker to storage tanks. Storage remote from road may be served by site pipeline	Boiler room Flue (Note also oil/air furnaces)
Light (200 s)	42.6	40	940				
Town gas		0.02				Large meters may require lorry access	Boiler room Flue Meter
Natural gas		0.04					
Electricity	3.6 MJ per kWh					Transformers may require lorry access	Switch and distribution gear Transformer station in some cases

[a] Useful publications on fuels, fuel delivery and storage can be obtained from fuel suppliers or the Electricity Association.
[b] Store not more than 2.5 m deep and in batches of not more than 280 m³ to eliminate risk of spontaneous combustion.

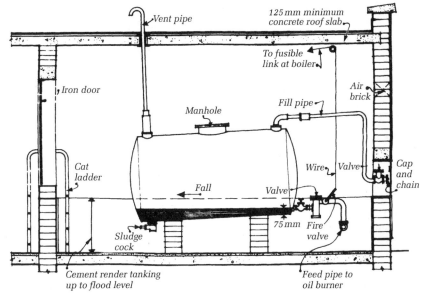

Figure 7.2 Typical store for oil fuel.

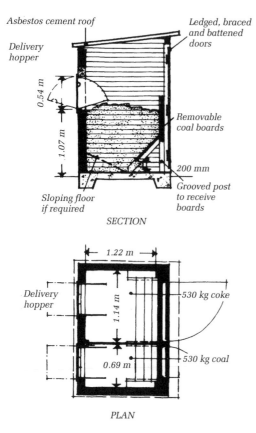

Asbestos cement roof

Delivery hopper

Ledged, braced and battened doors

Removable coal boards

200 mm

Grooved post to receive boards

Sloping floor if required

0.54 m

1.07 m

SECTION

1.22 m

Delivery hopper

1.14 m

530 kg coke

530 kg coal

0.69 m

PLAN

Figure 7.3 Typical store for domestic oil fuel.

pipes welded to black metal plates. The heat collected is stored, perhaps in a large vessel filled with gravel in the case of air, or in a storage vessel in the case of water. It is then available for use in the building when needed. Figure 7.5 shows a typical collector box. Present opinion favours the use of carefully designed orientation and fenestration to give a useful solar gain without the expense, complication and maintenance required of solar collectors. This is described as the passive use of solar energy. One interesting method which combines some simple mechanical devices with the capacity of the fabric of the building is the Trombe wall. Figure 7.6 shows the principles of operation.

Heat pump

A number of heat pump systems have been used for heating in this country, including one drawing heat from the Thames to warm the Festival Hall. This has been dismantled for many years. However, the costs of basic fuels have greatly increased and it is likely that the capability of the heat pump to deliver more energy

into a building than is required to operate the pump will become more important. It is particularly significant that heat pumps are most usually driven by electricity. Since the production and distribution of electricity operates at such a low efficiency of conversion from primary energy to electricity, the use of electricity directly for heating in the future seems likely to be inhibited. When used to drive a heat pump, however, the overall efficiency of operation compares favourably with other fuels.

The heat pump is a fundamental part of most air-conditioning systems for summertime cooling and essential to several methods of waste heat recovery. Figure 7.7 shows the cycle of operation of the heat pump. Its functioning depends on the use of a liquid with suitable vaporization properties; this liquid is known as the *refrigerant*, since most heat pumps are used for this purpose. The latent heat of this liquid, by circulation in a pipe system under low and high pressures, can be made to absorb heat at ambient temperature by vaporizing and to give out heat at more than ambient temperature by condensing. Ammonia is a cheap and efficient refrigerant often used for commercial applications. For domestic heat pumps, such as refrigerators and air-conditioning systems, less noxious refrigerants have been developed. One widely used refrigerant of this type is known as *Freon*. Concern for the ozone layer in the atmosphere is causing changes to refrigerants which cause less damage. If the heat pump is to be used for heating it is essential that the amount of heat transferred through the system represents a substantially greater fuel cost than the electric power used to drive the compressor. If this is not the case it will not be worth installing the expensive heat pump, and a simple boiler heating system would be more economical. The same consideration does not apply to the use of the heat pump for cooling since for most air-conditioning applications it is the only feasible system.

7.2 Direct heaters

Solid fuel

The traditional type of open fire is now rarely installed except for decorative purposes. They were troublesome, inefficient and could not burn the types of fuel required in Clean Air Areas. A range of specially designed fires and stoves is now available, however, which will burn smokeless fuels, stay alight overnight and operate with relatively little trouble. They can be fitted with back boilers to provide hot-water supply and perhaps one or two small radiators in

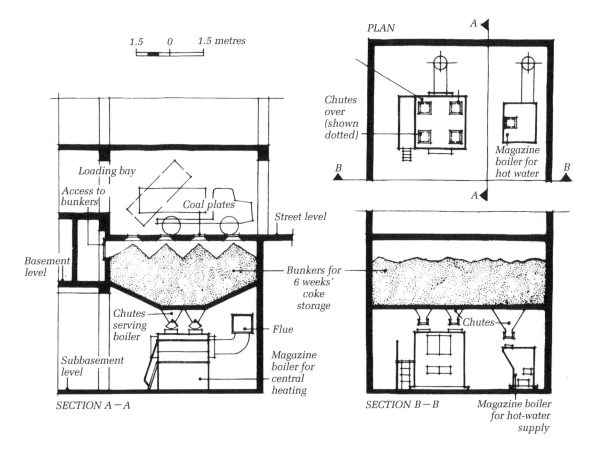

1.5   0   1.5 metres

Loading bay

Access to bunkers

Coal plates

Street level

Basement level

Bunkers for 6 weeks' coke storage

Chutes serving boiler

Flue

Subbasement level

Magazine boiler for central heating

SECTION A – A

PLAN

Chutes over (shown dotted)

Magazine boiler for hot water

Chutes

SECTION B – B

Magazine boiler for hot-water supply

Figure 7.4   Typical built-in storage for solid fuel in a large heating installation. (After *Fuel Stores and Solid Fuel*, Coal Utilization Council).

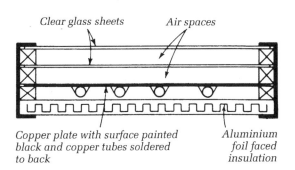

Clear glass sheets

Air spaces

Copper plate with surface painted black and copper tubes soldered to back

Aluminium foil faced insulation

Figure 7.5   Solar heat collector.

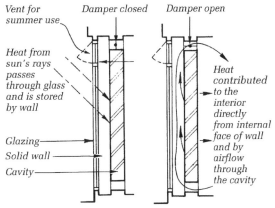

Vent for summer use

Damper closed

Damper open

Heat from sun's rays passes through glass and is stored by wall

Glazing

Solid wall

Cavity

Heat contributed to the interior directly from internal face of wall and by airflow through the cavity

Figure 7.6   Operating principles of the Trombe wall: (a) no integral heat requirement; (b) heat required internally.

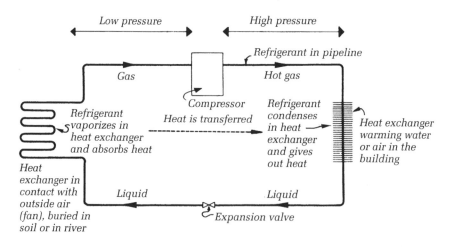

Figure 7.7 Operating principles of the heat pump. The action may be reversed so that heat is extracted from the building.

other rooms. This type of device is widely used in low cost housing. Figure 7.8 shows some typical installations.

Recent technical developments have greatly improved both the thermal and the smoke emission performance of solid-fuel appliances even at the domestic scale. Appliances are available which can be used with bituminous fuel even in smokeless zones. Downdraught combustion employed in room heaters can achieve acceptably low levels of smoke emission. Figure 7.9(a) shows an appliance of this sort.

New techniques have also been applied to boilers which enable bituminous coal to be employed and provide very high efficiency. They are based on the use of 4–16 mm bituminous coal with a self-cleaning underfeed stoker. Such a boiler can be used to provide central heating for houses with a heat requirement as low as 13 kW. Apart from filling the fuel hopper and emptying the ash tray at intervals, operation is fully automatic. In larger installations fuelling and ash removal can be automatic and automatic hot-air ignition can be provided.

Figure 7.9(b) gives a schematic explanation of the operation of the automatic stoker and fire retort.

Oil

The portable flueless type of oil heater shown in Fig. 7.10 is likely to be used in existing buildings not provided with central heating rather than as part of a system selected and provided with the building. Fixed oil heaters with flues and perhaps with oil supply piped from central storage are, however, often installed, particularly in large factory areas, where they provide

heating at low capital cost and with considerable flexibility of operation and rearrangement. The larger types of industrial heaters are fitted with fans to give better distribution and faster circulation of the warmed air.

Gas

Very small gas heaters can discharge the products of combustion into the spaces they warm, provided they are well ventilated. Heaters capable of giving full heating in rooms of even modest size will require a flue. The flue requirements for gas appliances are, however, relatively simple. The appliance does not depend for its operation on a draught from the flue. The products of combustion consist mainly of water vapour, and the flue does not require cleaning. It is possible to use very small flues and precast concrete flue blocks are available which will bond into brick and block partitions and walls. If the appliance can be fixed on an external wall the air for combustion and its products can be drawn and discharged through the wall immediately behind the appliance. Figure 7.10 shows such a balanced flue gas convector. Other types of gas heater provide both radiation and convected heat output for domestic use and high level radiant appliances of various types are available.

Condensing boilers

The main products of gas combustion are carbon dioxide and water vapour. In conventional gas boilers these products are discharged to the atmosphere via a flue at a temperature of some 200–250 °C. A

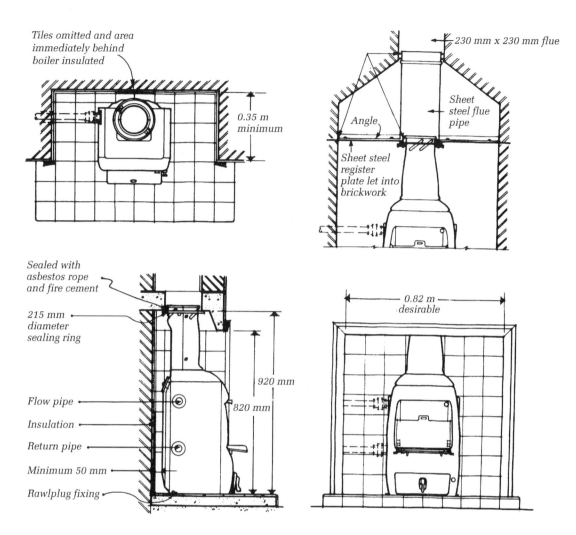

Tiles omitted and area immediately behind boiler insulated

0.35 m minimum

230 mm x 230 mm flue

Sheet steel flue pipe

Angle

Sheet steel register plate let into brickwork

Sealed with asbestos rope and fire cement

215 mm diameter sealing ring

Flow pipe

Insulation

Return pipe

Minimum 50 mm

Rawlplug fixing

920 mm

820 mm

0.82 m desirable

Figure 7.8 Solid fuel operable stove with back boiler gives space heating in one room as well as water heating. Some small radiators can also be served.

substantial amount of heat is lost. Care must also be taken to minimize condensation in the flue and prevent corrosion. In many boiler systems the water temperature in the boiler itself is sustained at a high level to avoid condensation and corrosion of the boiler itself.

Modern boiler materials, pumped water circulation and fan-assisted air circulation have made it possible to design boilers where the configuration of the boiler and temperature of the water circulation in the heat exchanger enables condensation of the vapour in the flue gases. This contributes the latent heat and also some of the sensible heat in the flue gases to the water flow.

To achieve continuous condensation conditions would require a low water flow temperature at all times, which would mean that radiators and other heat emitters would have to be substantially larger than is conventional in heating installations and would involve significant additional cost. However, normal boiler operation is intermittent, allowing variations in water temperature and only part of the full output is required for much of the heating season. This allows condensation to take place intermittently with normal sizing of heat emitters and gives a very substantial improvement in boiler performance. Seasonal efficiencies of over 90% can be achieved.

Condensing boilers must be fitted with a condensate

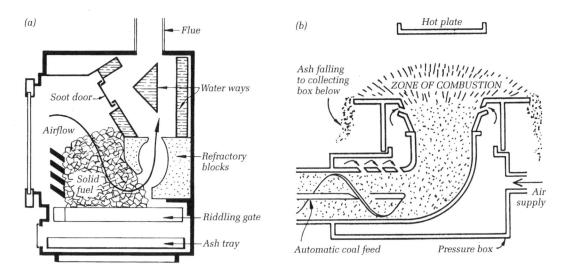

Figure 7.9   Self-cleaning underfeed stoker: (a) layout; (b) operation.

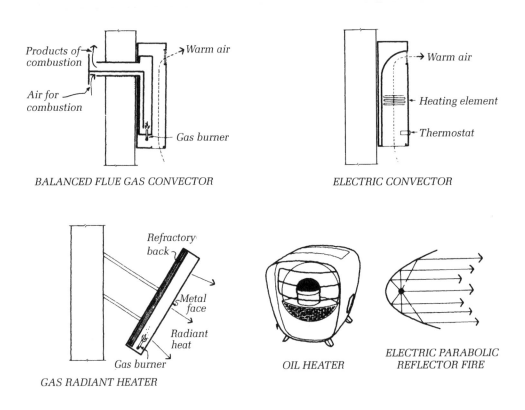

BALANCED FLUE GAS CONVECTOR

ELECTRIC CONVECTOR

GAS RADIANT HEATER

OIL HEATER

ELECTRIC PARABOLIC
REFLECTOR FIRE

Figure 7.10   Some local space heaters.

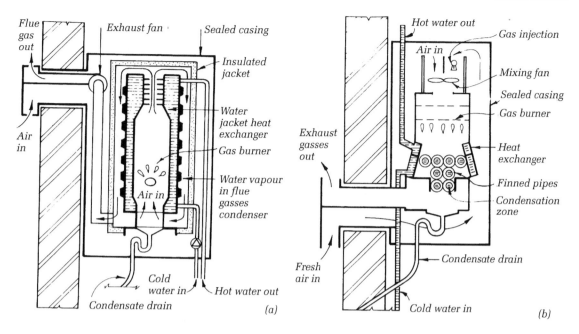

Figure 7.11   Domestic-scale condensing gas boilers.

drain. This can be discharged to the normal building drainage system. Concerns have been expressed that the condensate might be harmful to drains, but the method has been used successfully. The main problem in connecting the condensate pipe to the drains is to ensure that it cannot freeze in cold weather.

Figure 7.11 shows schematically the operation of domestic scale condensing boilers; Fig. 7.12 shows a larger type of boiler with a condensing facility.

Electric heaters

Radiant

There are three general types of electric radiant heater: high temperature, medium temperature and low temperature. High temperature radiant sources operate at red heat usually in the form of resistance wire spirals wound round a rod with a parabolic reflector to concentrate the direction of radiation. Figure 7.10 shows a section of such a heater. This type of heater is normally purchased for use in existing houses, although there are industrial versions employing a more robust heating element set in a reflecting trough; they are used for high level radiant heating in industrial buildings. Medium temperature radiant panels do not reach red heat but are too hot to be touched with safety. They are often fixed at high level, inclined towards the area being warmed. Electric underfloor heating is one type of low temperature

radiant source (see page 144). Another form employs an element mounted on building paper, which can be applied to ceilings and other surfaces.

Convectors

A natural draught electric convector is shown in Fig. 7.10. This type of heater is economical to install and can be thermostatically controlled. The heating

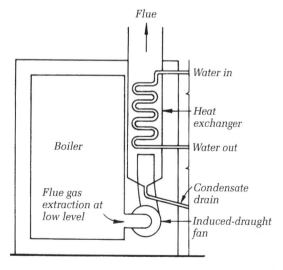

Figure 7.12   Larger-scale boiler with condensing facility.

element, although still at black heat, is considerably hotter than the heating element of a water-operated convector. The air that passes through the heater is therefore made relatively hot, which causes complaints from some people and can give rise to noticeable temperature gradients. Fan convectors, where the air is drawn through the heating element by an electric fan, overcome some of these disadvantages and provide very effective heat output. They are available in a very wide range of sizes, types and noise levels, including models for purchase by occupants and operation from a socket outlet, and models for permanent installations. Fixed heaters of this type are often installed at high level.

Electric radiators

Pressed steel radiators filled with oil and fitted with an immersion heater are available in a wide range of shapes and sizes. Their heat output is similar to that of a radiator mainly by convection but having a significant proportion of radiation. They react very much more slowly but are thought by some people to give a better standard of thermal comfort than convectors.

Tubular heaters

Tubes of 35–50 mm diameter containing a resistance wire coil with a loading of 20 or 30 W/m have a balance of radiant and convected output similar to radiators. They are most often installed in existing buildings and frequently fixed to skirtings below windows or below high level windows or skylights to counteract downdraughts. Multiple mountings support several tubes one above another since relatively long lengths are required for an output sufficient to keep a room at comfort temperature.

Thermal storage

Electricity supply undertakings have offered a substantially reduced tariff for current taken only during off-peak hours. Electric thermal storage heaters were designed to take advantage of this. A heating element is enclosed in a block of refractory material of high thermal capacity surrounded by an insulated casing. The block is heated during the off-peak period but gives off its heat slowly, thus providing warmth during the following day. This is not a very controllable system since the output cannot be varied to suit the needs for heat during the day and the building may tend to be very warm at first, slowly cooling as the day goes on. This is not a very suitable characteristic for domestic warming and the thermal storage heaters may have to be supplemented in cold weather. A solution to the problem is provided by

heaters which have a fan that can be switched on when desired to draw air through the block, thereby giving continuous background heating supplemented when desired by warm air input. All the heaters of this sort are heavy and care must be exercised in positioning them. On timber upper floors they should be placed against the walls on which the joists bear, never in midspan. The use of block thermal storage will normally be confined to existing buildings. Their successful operation depends to some degree on the heat storage of the building itself. High thermal capacity, particularly in internal partitions and walls, helps towards successful heating. In buildings with low thermal capacity it is important, if this type of heating is to be used, that the rate of ventilation should be controlled and the standard of insulation should be high. In new premises the same effects can be achieved by underfloor heating which will be very much neater in appearance. Figure 7.13 shows a section through a block-type electric thermal storage space heater with a fan for accelerated and controlled output.

Electric underfloor heating

In new buildings with concrete floors, or in existing buildings where concrete floors are to be installed, it is possible to lay a thicker screed than normal (50–75 mm thick according to the thermal capacity required) in which a grid of cables is embedded. The screed is then heated during the off-peak period and will give off its heat during the following day. As in the case of the

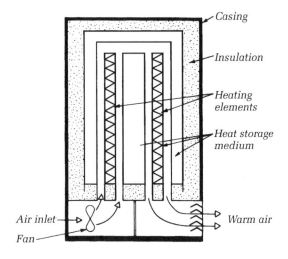

Figure 7.13 Block-type electric thermal storage space heater with fan. Background warmth is provided while the fan is not in operation. Accelerated and controlled heat output is given by the fan, which can be time-switched and thermostatically controlled.

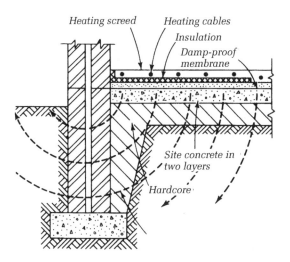

Heating screed   *Heating cables*
   Insulation
   Damp-proof
   membrane

Site concrete in
two layers

Hardcore

Figure 7.14 Heat loss from a solid ground floor and the appropriate position of edge insulation for electric underfloor heating.

block heaters the success of the system depends on storage of heat also taking place in the structure. Special attention must be given to the construction and insulation of the floors. Considerable heat loss would take place round the perimeter of solid ground floors, so edge insulation should always be provided. Figure 7.14 shows how the heat loss takes place and one method of providing insulation. If the ground were wet, heat would be lost rapidly and the insulation should be carried completely across the floor. Where overall carpeting is used above underfloor heating, the temperature of the screed will have to be higher in order to contribute the same amount of heat to the room. This will increase the heat loss downwards, so overall insulation should be considered, particularly in the case of rooms of modest size having a relatively large perimeter in relation to their area. The overall insulation prevents the thermal storage of the ground beneath the floor from contributing to the heating and consequently the thickness of the screed will normally have to be increased to compensate for this. The heat loss downwards from upper floors represents a useful gain to the space below but makes the control of heating difficult. It is therefore usually more satisfactory to provide overall insulation. In the case of flats with individually controlled heating overall insulation is essential to prevent large-scale heat loss into adjacent unheated dwellings.

For upper floors the insulation may be laid on top of the slab but it is also possible to apply it to the underside, either directly to the slab or suspended from

it. This enables the storage capacity of the slab itself to be used and will enable a thinner screed to be employed. Edge insulation is theoretically desirable where heated slabs are in contact with external walls but difficult to achieve in practice. The screeds for upper floors are sometimes laid on fibreglass quilts or other material for sound insulation (floating floors). The insulation will normally be too highly compressed for effective heat insulation.

The laying of screeds for electric underfloor heating is within the scope of normal good building practice. The mix must be as dry as possible and the floor should be laid in bays to minimize shrinkage effects and dried out very carefully (see Chapter 1 of the MBS title *Finishes*). The layout of the bays for screeding should be planned in conjunction with the layout of the heating cables to avoid crossing of construction joints. A rough guide to screed thickness is as follows:

- *50 mm thickness*
  - Groundfloor slabs with edge insulation only and midday boost.
  - Upper floors with insulation under the slab and midday boost.
- *62 mm thickness*
  - Groundfloor slabs with edge insulation but no midday boost.
  - Upper floors with insulation under the screed and midday boost.
- *75 mm thickness*
  - Ground and upper floors with insulation under screed and no midday boost.

The great majority of normal floor finishes, including cork and carpet, can be used in conjunction with underfloor heating. Detailed treatment of heat losses through the fabric of the building has already been given. When installing underfloor heating care must be taken about other wires and pipes which may have to be run across the heated areas. The problem may be acute on upper floors where it is conventional to lay the conduit for lighting in the screed above. Particular care in the wiring layout will be needed and it may be desirable to use mineral-insulated cable for the lighting wiring.

The design of electric underfloor heating must take into account not only adequate thermal capacity but also the limitation of the surface temperature to an acceptable value for comfort; 25–27 °C for standing or sitting occupants is often used as a design standard. This limits the heat output possible from the floor and, in the case of light rather than heavily insulated buildings or those with large areas of glazing or high ventilation rates, it may not be possible to obtain sufficient heat output. The additional requirement, which will only be called upon when the outside temperature reaches its design minimum must be provided by other means. Block heaters are a possibility and will be convenient because the control system for the floor can also be used for them.

Electric underfloor heating requires a grid of cables laid usually at between 75 mm and 225 mm centres. Two types of wiring systems are available. Conduit systems employ asbestos or fibreglass-insulated wiring, which can be renewed, or plastic-insulated cables, which are permanently embedded in the screed. Opinions are divided over which is preferable. With suitable instruments the position of faults in cables embedded in screed can be determined to within a few centimetres, thereby minimizing the inconvenience and expense of repair.

The main control of this type of heating is by a time switch, which limits the period of electrical input to the off-peak periods. The periods and the tariffs vary in different parts of the country and more than one tariff may be available in one area, usually related to the length of time during which the off-peak current is taken and particularly to whether a midday boost is included. Thermostatic control is also needed. For small buildings a simple room thermostat may suffice; larger installations may need controls that take account of the weather and which perhaps divide the building into zones.

In general this form of heating provides an extremely neat, clean and trouble-free installation giving a high standard of comfort. Close control is, however, not easy since it can only be achieved during the previous night. If the building overheats during the day then windows will be opened to maintain comfort, if more heat is needed it must be provided from other sources. These conditions are often turned to advantage by using the underfloor heating as a background, which at some times of year will be all that is required, and providing fast-reacting supplementary heating.

7.3 Central heating

Heat transfer media

In central heating installations fuel is converted to heat in a central plant and the heat is then distributed round the building to heat-emitting devices by a heat transfer medium. Fluid media for heat distribution have obvious advantages in terms of the ease with which they can be distributed through pipes and ducts. It is clearly necessary, however, to select materials for heat

distribution which will convey the maximum amount of heat for the minimum amount of material, and for the least cost in terms of material to be purchased. Specific heat gives an indication of the heat-carrying capacity of possible materials. In a building it will be space rather than weight which governs the choice, and the heat transfer efficiency in these terms is given by the volumetric specific heat. Table 7.5 shows the specific heat and volumetric specific heat of several substances which might be considered for the purpose.

Water is the most efficient medium from both points of view, and since this is also the cheapest material on the list, except for air, it is an obvious choice for the purpose, and is in fact the material most widely used. There are some circumstances when water cannot be used, most particularly in the case of pipes which may be subject to freezing (heating coils in roads or greenhouses are examples, although electric cables are now more likely to be used) or where the medium will be operating in any case below freezing (heat pumps, air-conditioning cooling plant, etc.). Other materials must then be used, and brine and various oils are often employed. The same problem does not arise with temperatures over boiling point, since high pressure systems can be used which raise the boiling point of water, or steam itself may be employed. High pressure hot water and steam, although they require more complex plant, have advantages in that a given quantity of water can carry more heat, and consequently smaller pipes can be employed. Since the specific heat of water is 4.2 kJ kg^{-1} K^{-1}, it follows that a given quantity of water will absorb and convey more heat as its temperature of flow is increased. The average flow temperature of low pressure hot-water heating installations is usually about 77 °C and cannot be very much increased before boiling takes place. If, however, the boiling point of water is increased by placing the whole installation under pressure, the flow

Table 7.5 Heat transfer efficiency of various materials

Material	Specific heat capacity (kJ kg^{-1} K^{-1})	Volumetric specific heat capacity (kJ m^{-3} K^{-1})
Water	4.2	415
Mercury	0.13	188
Alcohol	2.93	262
Petroleum	2.10	174
Iron	0.46	355
Glycol	3.24	355
Air[a]	1.01	1.2

[a] Air is compressible but this does not affect the general principle demonstrated.

temperature can be increased; 175–230°C is quite usual for this type of installation. So, for a given quantity of heat, smaller pipes can be employed and considerable economy achieved in the transfer of heat over long distances. On the other hand, the boiler, heat-emitting appliances and pipework have to be able to resist the high pressures involved ($1000 \, kN/m^2$ perhaps) special pumps and valves are required, stringent safety precautions are necessary and special consideration must be given to pipe expansion. These factors incur substantial installation costs, and high pressure hot-water systems have in the past been limited to large buildings or district heating systems. The system can be used to supply high buildings, or points above the level of the boiler plant and feed cistern, but there are limits to the possible height since the pressure must not be reduced to the point at which steam will be produced. The flow temperature can be modulated to any value below the maximum, often desirable in space heating installations. To maintain economy of main distribution coupled with local modulation it is possible to use high pressure in the mains and provide low pressure hot water for heating by means of calorifiers.

The use of steam appears to give further advantages for heat transfer since the latent heat of vaporization (2.25 MJ/kg) is also available for heat transfer and the steam will pass rapidly along comparatively small pipes to condense in the heat emitters, giving up its latent heat in the process, providing its own motive power and in a very tall building not imposing a great pressure at its foot. For space heating, however, steam is not often employed for several reasons. The steam-raising plant is more complex and requires more attention; the heat emitters must be protected from human contact since scalding would result; modulation of the heat output is more difficult since the steam temperature is not so easily varied as the water temperature. In addition the condensation of the steam in the pipes and emitters often creates noise. Where steam is available, as is the case in many industrial and hospital buildings, it is often employed for the main distribution of heat for space heating serving local calorifiers which supply conventional low pressure systems. Steam has some advantages for district heating schemes and metering the condensate provides a fairly easy method of estimating the heat used, which is often difficult with other systems.

Air, as can be seen from Table 7.6, is not a very efficient medium for heat transfer. It is, however, convenient if mechanical ventilation is to be provided in any case, and it has many advantages for cooling in an air-conditioned building. The cool air has a satisfactory thermal relationship to the warmer structure, and also avoids the condensation difficulties which might arise if panels or radiators were used to extract heat from the building.* The quantity of air which must be circulated for heat transfer will be several times that required for ventilation. This leads to ducts which are large enough to present a problem in planning the building and substantial installation costs. Some present developments in air-conditioning systems are based on attempts to provide adequate heating or cooling while limiting the amount of air to be moved to that required for ventilation.

Table 7.6 shows comparative pipe and duct sizes for a range of heat outputs for low pressure hot water, high pressure hot water, steam and air.

Distribution system

Basic distribution systems consist of a boiler, furnace or mechanisms for the production of heat from fuel, a system of pipes or ducts holding a heating medium,

---

* The ducts carrying air will, however, require careful insulation to avoid wasteful heat gain and also to prevent condensation. A vapour seal (in many cases paint is used) is necessary to prevent condensation on the inside of the insulation. (Chapter 2 describes the principles involved.)

Table 7.6   Comparative sizes of pipes and ducts used for space heating[a]

	Pipe diameter (mm): flow (return)			Area of air duct (m²)		
	Low pressure hot water	High pressure hot water	Steam[b]	Low velocity input	High velocity input	Extract
10	18 (18)	12 (12)	18 (12)	0.05	0.013	0.037
100	62 (62)	25 (25)	37 (18)	0.5	0.13	0.37
1000	150 (150)	75 (75)	100 (50)	5.0	1.3	3.7

[a] The sizes are based on typical temperatures, velocities and pressure losses. They give an indication of comparative size and might be used to access approximate space requirements at early design stages.
[b] The flow pipe is the steam main and the return pipe carries condensation.

leading to heat emitters in the various rooms or parts of the building, and subsequently returning to the boiler. Some motive power must be provided to force the heating medium round the circuit. Early distribution systems both for water and air employed the thermosiphon or 'gravity' system for their motive power. This meant that the force available to move the water or air was limited to the difference in weight between the hot, expanded column rising from the boiler or furnace and the cooler, denser return columns. This system imposed severe limitations on planning. The boiler or furnace had to be at the bottom of the installation. The length of run was strictly limited; pipes were large and heat transfer was slow and difficult to control. Steam overcame many of these difficulties but had its own problems. Fans for air distribution systems and pumps for water have changed the situation completely. Quick, efficient distribution is possible through small diameter pipes and ducts with horizontal and downward distributions in patterns that would not previously have been possible. The main factor which governs the size both of pipes and ducts is now the maximum acceptable velocity that will not give rise to trouble from noise or corrosion. Figure 7.12 shows the principles of operation of a basic heat-distributing system.

When low pressure hot water is used the pipe circuits distributing the heat throughout the building are supplied with water from a feed and expansion cistern at high level. Figures 7.15 and 7.16 show some of the patterns of distribution which are possible with some notes on their planning implications. The diagrams show conventionalized two-dimensional arrangements. In practice more than one of these arrangements may be combined in the same installation, various plan arrangements may be

employed (e.g. the ladder system may run completely round one floor, the rise and drop being sited adjacent to each other in the same duct) or several separate units employed to serve different zones (e.g. separate ladders may be employed on different façades to make separate control possible). With very simple systems a single-pipe distribution arrangement may be employed for economy. In this arrangement one flow pipe serves radiators progressively, the water temperature in the pipe being steadily reduced. Radiators at the end will receive water at a lower temperature than those at the beginning, and will consequently have to be of a larger size for the same heat output. There may also be difficulties in the control of heat distribution. The system is, however, neat and economical in installation cost, and can give very satisfactory service in small installations. In general, only radiators can be used in connection with single-pipe systems. Convectors will not operate efficiently at the reduced flow temperatures which arise with low pressure, single-pipe hot-water systems can be overcome by using separate pipes for the flow and the return, so all heat emitters receive water at approximately the same temperature. Figure 7.16(e) shows the radiator connections for each method. The pipework of the system does cost more, and if it has to be exposed may be unsightly, but heat distribution is good, radiator sizes may be standardized or other emitters employed.

It is possible to obtain some of the benefits of each system by using a two-pipe arrangement for main distribution, but serving groups of radiators from a single pipe.

High pressure hot-water systems can follow the same basic pattern of distribution. The pressure was often maintained by steam trapped at the top of the boiler but nitrogen pressurization is now more usual; the feed cistern will be at low level in the boiler room and the water required forced into the system by means of a feed pump (provided in duplicate for maintenance and breakdown). Safety valves and air vents are required instead of vent pipes. Figure 7.17 shows a large-scale high pressure system. The nitrogen pressurization system, now often used in preference to steam, had enabled high pressure systems to be used in smaller installations, and the development of small pressure vessels requiring no attention or power supply has extended the use of the system to the smallest installation. Figure 7.18 shows the type of pressure vessel used. A metal container with a rubber diaphragm is pressurized with nitrogen and connected, on the other side to the nitrogen, to the pipework of the system. When expansion of water occurs as a result of the system warming up, the nitrogen can be

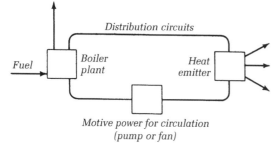

Figure 7.15 General operating principles of a heat distribution system. Control of time and temperature can be introduced at boiler, pump or emitter.

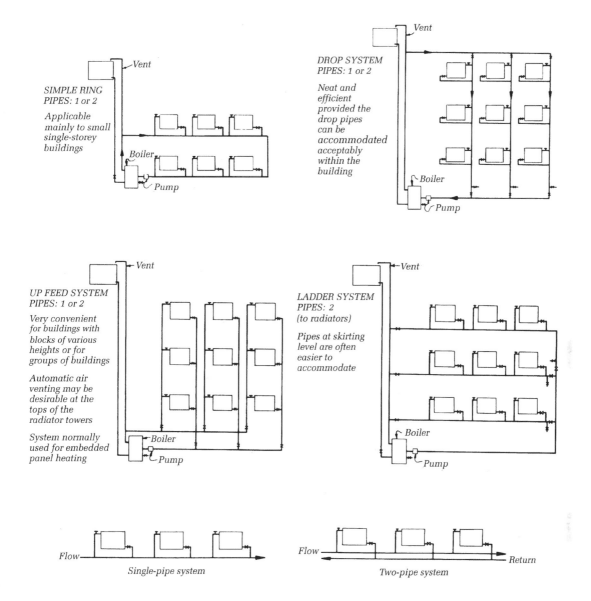

Figure 7.16 Various arrangements for heat distribution, and the difference between a single-pipe and a two-pipe system.

compressed via the flexible diaphragm. When the system cools, the nitrogen will force the water back from the pressure vessel. Figure 7.19 shows the form of the installation that results. There are many circumstances when the avoidance of feed and expansion cisterns and vent pipes, achieved by pressurization, will be of considerable architectural advantage.

With the pumped circulation which is the rule in modern installations, there is considerable freedom in placing the boiler or furnace, and it is no longer necessary to have it at the bottom of the installation. In solid-fuel installations considerations of fuel delivery and ash removal will restrict boiler positions to low levels in the building. Oil can, however, be pumped up to a small boiler feed tank, and with gas

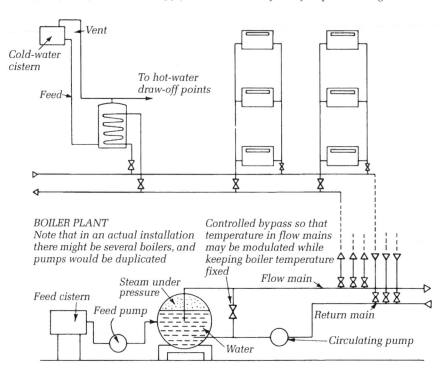

Colorifier system for hot-water supply   *Convector system for space heating*

Figure 7.17   A high pressure hot-water space heating system. Here the local hot-water supply is heated by the main space heating flow. Where the points requiring hot water are scattered this can save on capital costs compared with a separate hot-water supply circulation.

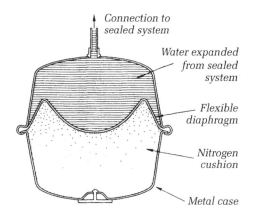

Figure 7.18   Pressure vessel for small high pressure hot-water space heating system.

there is no difficulty at all in placing the boiler at high level if desired. If the boilers can be installed and the structural loading accepted, this arrangement may sometimes have considerable advantages in freeing groundfloor or basement areas for other uses. Several factors must be taken into account when positioning boiler rooms.

Flue

Flues must normally be carried up to a level where the products of combustion will be harmlessly discharged. This often means that the flue must rise in the highest part of the building and will certainly limit its possible positions. Although it is possible, using induced-draught fans, to force flue gases through lengths of horizontal flue, where the boiler room is remote from the flue, there is clearly great advantage in having the boiler room at the foot of the flue.

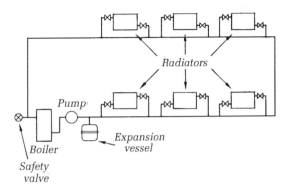

Figure 7.19 Small high pressure hot-water space heating installation.

Delivery

Solid fuel fired boiler rooms require lorry access for fuel delivery and suitable provision must be made in all cases for replacement of the plant.

Light and air

Air is necessary for combustion and to keep boiler-room temperatures to a reasonable level. Although this can be provided by ducts to subterranean boiler rooms, far better environmental conditions can be provided by natural light and ventilation. There can be no doubt that good conditions in the boiler room contribute significantly to proper maintenance of the plant.

At its simplest a boiler plant consists of a boiler and pump. Larger installations appear to be a very complex mass of pipes and plant. In fact, the basic principles involved are the same and the apparent complexity results mainly from the fact that it is usual, for maintenance and for flexible heat output while using individual boilers to their full capacity, to divide the heating load between several boilers. Pumps are provided in duplicate, for maintenance and in case of failure; and where the building is divided into separate zones for control purposes, a separate set of duplicate pumps may be required for each zone. The hot-water supply will exhibit similar duplication of calorifiers and pumps. Figure 7.20 shows the functioning of an oil-fired, low pressure hot-water space and water heating boiler plant.

The correct choice of heat emitters is crucial for economy, amenity and comfort. Some types are shown in Fig. 7.21. The main types available for use with hot-water space heating are as follows.

Pipes

In early central heating installations large pipes were run around the building. Installations of this sort can

still be seen, particularly in churches and assembly halls. The pipes were, however, awkward and obtrusive, and although pipe coils are still used in storerooms and round rooflights to avoid downdraughts, more efficient means of transferring the heat have been devised.

Radiators

Radiators present a much greater surface area for heat transfer, and the vertical arrangement of columns encourages convection currents. Most of the heat is given off as convection, but a useful if small proportion is emitted as radiation. To minimize wall staining, obstruction of useful wall space, and to provide the best thermal results, radiators are usually placed under windows. Pressed steel is now used for many radiators instead of cast iron, particularly for domestic applications. Steel radiators are light and often easy to clean. Control of individual radiators is usually by manual radiator valve, which does not permit any fine adjustment. Valves which give a flow proportional to an indicated setting are, however, now available and thermostatic radiator valves are also obtainable, although price may restrict their use. Systems employing radiators do not react swiftly to control because of the substantial thermal capacity of the radiators and the water contained in them. Improved performance can be achieved by mounting reflective panels on the wall behind the radiator.

Convectors

These consist of gilled tubes at low level in a casing which encourages the movement of air over the gills by stack effect. The amounts of metal and water are considerably less than those contained in radiators, so an installation of this sort will react more swiftly to control. Individual convectors can be controlled by the same types of valves as radiators, or in some cases by manually operated dampers closing their inlet and output grilles. It is easy to build in convectors under sills, and in built-in furniture, and arrangements can be made to combine ventilation with convectors by drawing in air from outside rather than from the room itself. The amount of radiation from convectors is negligible. It is important for efficiency of convectors that the flow temperature of the heating installation should not drop significantly.

Fan convectors (and unit heaters)

The heat transfer takes place in a battery of gilled tubes similar to that of a convector, but instead of relying on natural convection, air is blown over the heating battery by an electric fan. The output rate is usually

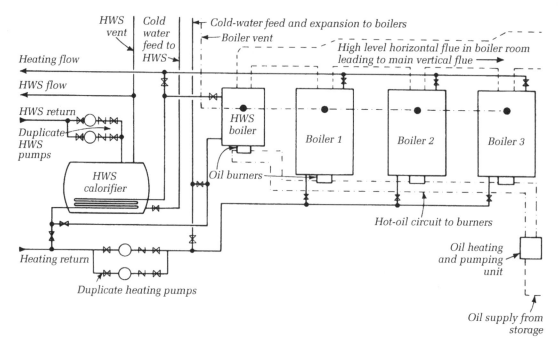

Figure 7.20 Functioning of a typical low pressure hot-water space heating and hot-water supply boiler.

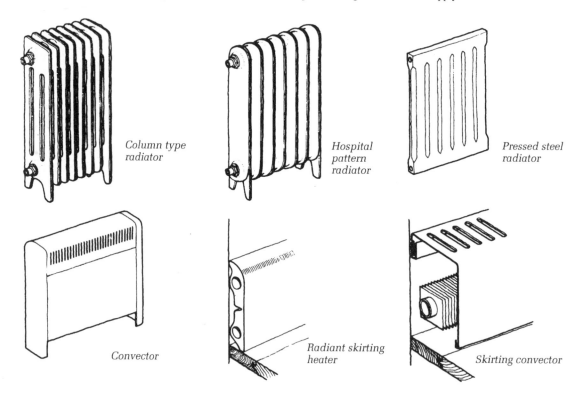

Figure 7.21 Some types of heat emitter suitable for use with low pressure hot-water space heating.

high, and this type of heater is not usually suitable for very small rooms, although manufacturers are developing smaller models. Individual control is, however, very easy since the output is governed by the fan, which can very easily be time-switched and thermostatically controlled. The delivery of air by fan assists in distributing the heat and there is considerable freedom in placing fan convectors. Advantage can often be taken of this to minimize pipe runs. Installations are often economical in first cost because of reduced pipe runs and small numbers of appliances required, and economical in running costs because of the closeness of control possible. In rooms where the heat requirement may change quickly, due perhaps to a sudden influx of people, fan convectors can give optimum comfort conditions, since some other systems which might provide better comfort in principle would not react sufficiently quickly to changing needs for heat. If fan convectors are placed on external walls it is possible by means of a duct through the wall and dampers to allow either the circulation of room air or the introduction of fresh air through the heater. This system can also be used for cooling, when it is usually given the name fan coil.

Skirting heaters

Convectors and special pipes are available shaped to take the place of a skirting-board. There may be difficulties in making connections around doors, but in many cases the specially adapted pipes or convectors can give a neat unobtrusive solution.

Panels

Continuous coils of pipe, usually 12 mm in diameter and about 60 m long, can be embedded in the construction of a building in the bottom of the structural floor (ceiling heating) or more often nowadays in a screed on top of the structural floor (floor heating). Floor heating, in particular, maintains a very even temperature gradient and gives off some 50% of the heat by radiation. Flow temperatures are limited to 50 °C, so that floor surface temperatures are kept low to avoid any discomfort. Floors are not normally warm to the touch. Boilers producing hot-water supply and serving other radiators must have a flow temperature considerably higher than this. Figures 7.22 and 7.23 show a combined floor heating radiator and hot-water supply installation in a house. Figure 7.22 shows a mixing valve used to control the temperature of the water flowing in the underfloor panels. Figure 7.24 explains the principles involved in

more detail. The construction and insulation details of the screeds for hot-water underfloor heating are similar to those for electric underfloor heating except that the thickness of the screed is standard at 50 or 62 mm since thermal capacity is not involved. Concern to save energy by close time and temperature control and installation cost have reduced the popularity of underfloor heating. One system, relatively new to Britain, provides a new approach to floor heating which offers better control and reduced installation cost. The system is based on triple thermoplastic rubber tubing. Each tube is approximately 7 mm in diameter so that the volume of water is reduced. The increased surface area compared with a single tube of comparable cross-section gives greater heat output. Flow takes place through the centre of the three tubes and return is through the outer tubes, giving a more uniform heat output than could be achieved by a single tube. The system can be embedded in a floor screed but it can also be fixed in a variety of ways in a timber floor, enabling faster response. Figure 7.25 shows a section through the tube and a typical timber floor installation detail. Embedded panels react slowly to control. Systems of pipes forming part of a suspended ceiling provide a very much quicker reaction and a number of patented systems are on the market. Care must be taken that the rooms where this type of heating is used are sufficiently high to ensure comfort for the users of the building.

There is a wide variety of designs in most of the appliances described and many other devices (e.g. radiant panels, patent suspended heating ceilings). New developments are constantly occurring. To obtain a comprehensive and up-to-date knowledge attention should be given to the contributions and advertisements in the technical journals.

7.4 Control of heating installations

The great advantages of automatic control of heating installations are leading to rapidly increasing use of more and more sophisticated control systems. Small domestic heating installations which not long ago would have been controlled only by use of the flue damper are now fitted with thermostatic control of boiler firing, time-switch control of pump operation and thermostatic control of water flow temperatures or radiator output. And with a time switch it is also possible, using motorized valves, to close off particular circuits when they are not required. In some cases control units, giving the user choices of several

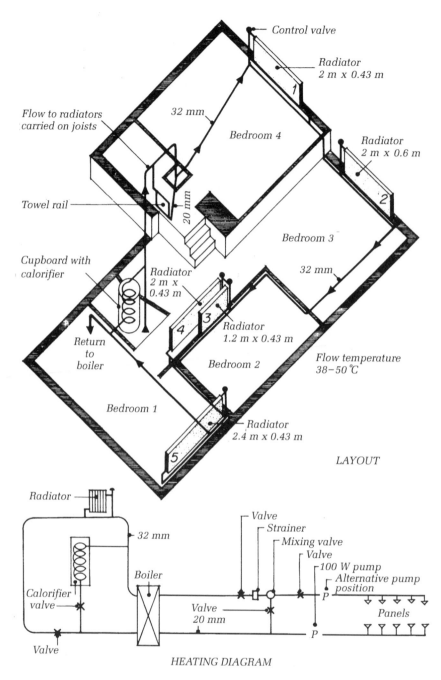

Figure 7.22   Axonometric view of pipe and radiator layout of first floor (top) and schematic of heating installation in a two-storey house having radiators on the first floor and underflow heating on the ground floor. The mixing valve enables the underflow circuit to be controlled to a lower temperature than the radiator and calorifier circuits.

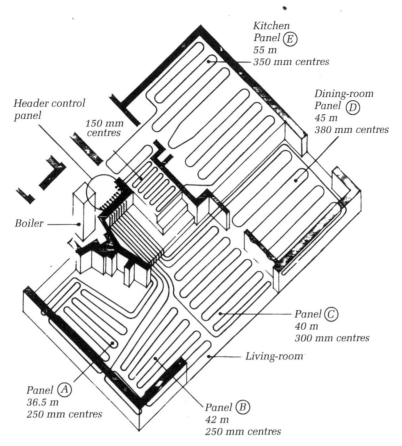

Kitchen
Panel Ⓔ
55 m
350 mm centres

Dining-room
Panel Ⓓ
45 m
380 mm centres

Header control
panel
150 mm
centres

Boiler

Panel Ⓒ
40 m
300 mm centres

Living-room

Panel Ⓐ
36.5 m
250 mm centres

Panel Ⓑ
42 m
250 mm centres

Figure 7.23   Axonometric view of pipework layout for underfloor heating (see Fig. 7.22).

possible modes of use with variable timing, are being installed in small systems.

Control is exercised in respect of two main variables time and temperature. It is now standard practice even in the smallest installation to have a time switch giving overriding control of pump operation for the heating circuits (leaving the boiler in action providing hot-water supply) or giving control both of pump and boiler firing (thereby giving overnight economy). Even very simple boilers now have thermostatic control of firing, which can be used to govern flow temperatures. This means, however, that hot-water supply and space heating will fluctuate together, which is clearly inappropriate. A better arrangement is shown in Fig. 7.24, which indicates how a thermostatically controlled three-port valve can, by allowing return water to recirculate, provide a circuit where the flow temperature can be varied independently of the boiler temperature (but not above the boiler temperature).

This is important in large installations to ensure efficient boiler firing without risk of condensation of flue gases, and in all sizes of installation it enables hot-water supply to be maintained at a fixed temperature by the direct boiler flow, while the space heating circuits can be modulated to suit prevailing weather conditions.

This method means that the whole of the space heating is controlled by the temperature prevailing at a single point in the building, and although it is an economical and effective method, some disadvantages are apparent, particularly in respect of larger buildings where very different conditions can prevail in different parts. One method of attempting to overcome this is by use of external sensing devices, usually known as external compensators, which are sensitive to air temperature, usually wind, and in some cases to other external factors. This method assumes that the heat input required to the building can be directly related to

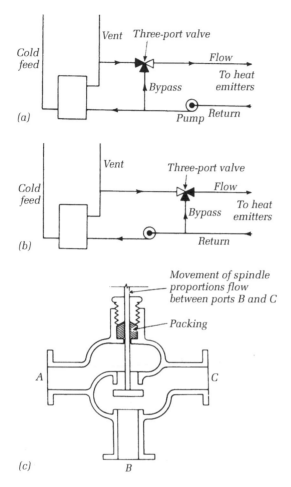

(a)

(b)

(c)

Figure 7.24 (a) This mixing valve gives a constant rate of flow round the space heating circuit with a variable flow temperature; usually the most appropriate arrangement for space heating. (b) This mixing valve gives a variable rate of flow at a fixed temperature. (c) Section through the three-port valve.

the external conditions. Clearly sunshine on one façade can upset this assumption and buildings having systems of this sort will usually have to be divided into zones having different exposures. It is also assumed, however, even when the building is zoned, that the heat input required is entirely based on external conditions. This is often not the case as significant heat gains can arise from occupants' electric lighting, machinery and other processes. In order to take these considerations into account in individual spaces, local thermostatic control of the heat emitters in the room would be required. This is

comparatively easy in the case of fan convectors where a simple thermostat can easily control the electric fan and thereby the heat output of the apparatus. Radiators and emitters where the flow of water has to be impeded present a more difficult problem but effective thermostatic radiator valves are available which can govern the water flow in the radiator in accordance with the needs of the room. Thermostatic valves cost considerably more than ordinary radiator valves. On the other hand, the closer control they give will improve fuel economy as well as comfort.

7.5 Flues: gas transport pipes

The construction, insulation and fireproofing of chimneys is dealt with in the MBS title *Structure and Fabric Part 1*, but their functioning in terms of the discharge of effluent gases forms part of the scope of this volume. It is conventional to apply the term *chimney* to the structural elements whereas *flue* refers to the gas passages and to non-structural pipes (e.g. horizontal flues in boiler rooms or vertical flue pipes supported from some other structural member).

The primary purpose of a flue is to carry the gases produced as a result of combustion together with entrained air and to discharge the necessary quantities of gas at a point where no nuisance or damage to health can result. In addition to this many solid-fuel or oil-fuel appliances require for the effective operation the 'pull' of air through the appliances which results from a proper flue draught.

The movement of gases up a flue is caused by the difference in density of hot flue contents compared with the colder outside air, which being heavier displaces the flue gases upwards. The forces generated are influenced by the height of the chimney and the temperature difference between the flue gases and the external air. Increases in either of these factors will increase the force available, known as the draught. The forces are quite small and may be increased or diminished or even overcome and put into reverse by the operation of the wind. Where the pattern of air movement (see Chapter 3) round the building gives downdraughts or produces higher pressures at the flue outlet than at the fire, or boiler, then the downdraught in the flue can result. This is normally overcome by carrying the flue up to a level which takes its outlet above the zone of varying air pressure; 1 m above the ridge level of a pitched roof or 2 m above a flat roof are figures sometimes quoted as likely to be satisfactory for small buildings, but topography, surrounding buildings and protrusions on the roofing, may affect conditions. Since the force moving the air is small,

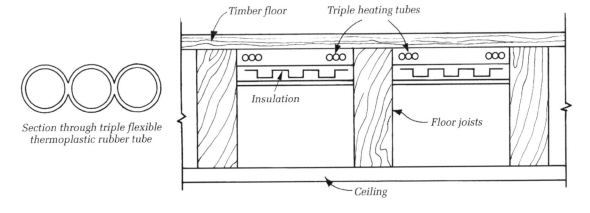

Section through triple flexible thermoplastic rubber tube

Figure 7.25 Low thermal capacity underflow heating. The tubing may be installed from the top, resting on the insulation; or it may be stapled to the underside of the floor and the insulation fixed afterwards.

resistance to flow in the flue itself such as friction due to rough internal surfaces (such as protruding cement joints between liners), sharp bends or changes of size or shape have a very significant effect in reducing draught and should be avoided. Maintaining the temperature of the flue gas is important and, where possible, flues should be internally rather than externally situated. This is particularly important in domestic situations where valuable heat savings can be made by internal flues, and in the case of oil- and gas-fired appliances which operate intermittently and consequently allow the flue to cool. Apart from the effect in reducing draught, due to the reduced temperature of the flue gases, the cooling effect of flues may cause condensation which can attack the materials of the flue and run down, creating problems at the foot. Where external flues are essential thermal insulation between the liner and the structure is very desirable. Leakage of air into the flue will cool the gases and dissipate the draught. Careful attention should be given to sound joints in the structure, and also to properly sealed pipe junctions and soot doors. Figure 7.26 shows a typical convection of a flue from a small boiler into a brick chimney.

For solid-fuel and oil-fired appliances short flues (less than about 4 m) may not in any case develop sufficient draught to operate effectively. But tall flues (over about 10 m) may develop excessive draughts, particularly when aided by the wind. This effect can be dealt with by means of draught stabilizers, which are counterbalanced flaps fitted at the base of the flue and weighted so as to open and admit air if the draught exceeds an appropriate value for the appliance. Draught stabilizers can also act as safety doors to

cater for the risk of explosion with oil-fired installations, which can suffer from sudden ignition of gas in the flue. Figure 7.27 shows a typical stabilizer.

Gas-fired appliances can normally operate without any flue draught and in many cases incorporate a hood which prevents fluctuations of flue draught from influencing the apparatus. Gas flues, therefore, act merely as a means of carrying away the products of combustion, which in this case consist mainly of air and water vapour. Conditions for the siting and the discharge of gases are not so stringent and several interesting flue variations have been developed for them. Figure 7.28 shows one such arrangement. Special unobtrusive terminals, including some in the form of ridge tiles for small installations, flues incorporated within partition blocks and, of particular significance, the balanced flue and SE duct. Balanced flue appliances have combustion chambers sealed from the interior of the building, but taking air from outside through a wall and also discharging the products of combustion at the same point. Inlet and outlet are so close that the air pressure on them is the same for any wind conditions. A typical arrangement is shown in Fig. 7.29. The SE duct enables balanced flue appliances to be used away from external walls by having a duct running through the building from bottom to top and vertical at both ends. Gas appliances take air from the duct, and return is such that any input of combustion products does not have a bad effect on the operation of appliances further up the building.

With increasing concern for heat economy and the extensive use of draught stripping, care is taken not to limit the quantity of air that can enter a room

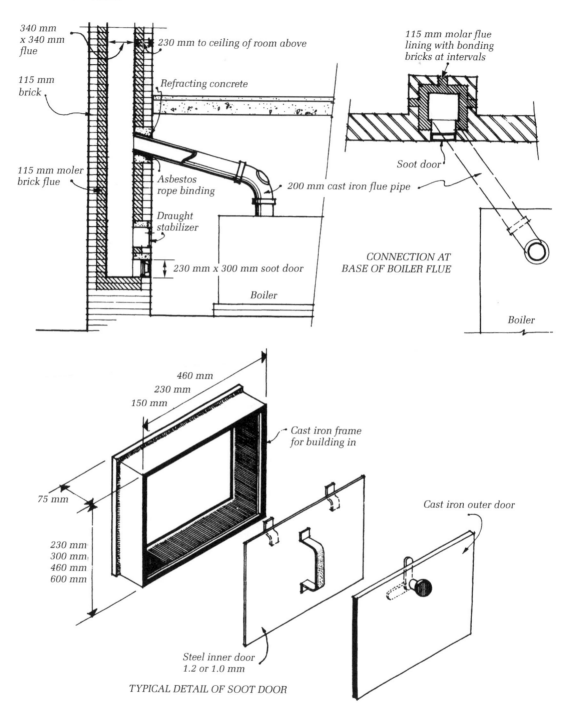

340 mm
x 340 mm
flue

230 mm to ceiling of room above

115 mm molar flue
lining with bonding
bricks at intervals

115 mm
brick

Refracting concrete

115 mm moler
brick flue

Asbestos
rope binding

Soot door

200 mm cast iron flue pipe

Draught
stabilizer

CONNECTION AT
BASE OF BOILER FLUE

230 mm x 300 mm soot door

Boiler

Boiler

460 mm
230 mm
150 mm

Cast iron frame
for building in

75 mm

Cast iron outer door

230 mm
300 mm
460 mm
600 mm

Steel inner door
1.2 or 1.0 mm

TYPICAL DETAIL OF SOOT DOOR

Figure 7.26   Connection of the flue from a small boiler into a brick chimney.

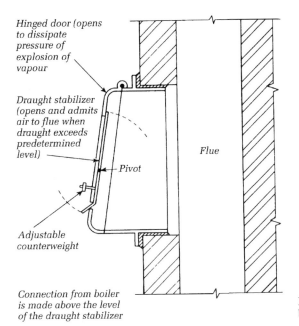

Hinged door (opens to dissipate pressure of explosion of vapour

Draught stabilizer (opens and admits air to flue when draught exceeds predetermined level)

Pivot

Flue

Adjustable counterweight

Connection from boiler is made above the level of the draught stabilizer

Figure 7.27   Draught stabilizer for oil-fired boiler flue.

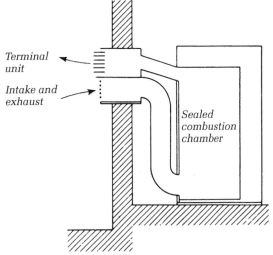

Terminal unit

Intake and exhaust

Sealed combustion chamber

Figure 7.29   Section through a gas-fired space heating boiler with a balanced flue.

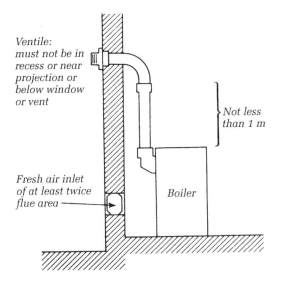

Ventile: must not be in recess or near projection or below window or vent

Not less than 1 m

Fresh air inlet of at least twice flue area

Boiler

Figure 7.28   Simple flue installation with gas-fired boiler. (Adapted from a drawing by Thomas Potterton Ltd).

containing a heating appliance. Enough air must enter to allow combustion and to supply the flue draught without causing excessive resistance. In the case of boiler rooms large quantities of air are needed and the areas required for ventilation are recommended in heating engineering handbooks. In the case of domestic heating appliances in habitable rooms, efforts have been made to duct air from outside to a point in the hearth near the appliance to provide air for combustion without drawing the cold air across the room.

In the traditional domestic open fire a large opening was provided to the flue and very large quantities of air passed from the room up the flue with consequent heat loss. Better practice was the use of a narrow throat (usually having an opening about 100 mm wide) which substantially reduced the flow. Most modern stoves overcome this problem, particularly when their doors are kept shut. For a time multistorey flats presented a special problem in terms of flues, since carrying the flues up separately for many floors was expensive. A system of branches into a main flue was designed to overcome the problems of penetration of flue gas and noise from one dwelling to another, but the development of central heating has eliminated the problem.

In large boiler rooms it is usual to find the load divided between several boilers. Traditionally the flues from these boilers were collected into one single large-diameter flue. The disadvantages of this in terms of flue efficiency are apparent and it is becoming increasingly usual to have separate flues carried up together grouped into a single structural chimney.

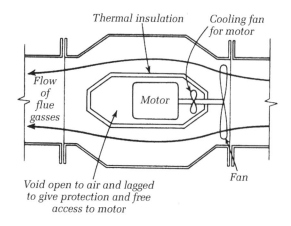

Figure 7.30 Bifurcated fan used as induced draught for a flue with inadequate natural draught.

Small flues to gas-fired appliances often have no means for cleaning but flues to solid-fuel and oil appliances should normally have access by soot floors so that inspection and cleaning is possible.

In cases where it is impossible to provide flues of adequate draught, such as inadequate height, or long horizontal runs of flue to reach chimneys remote from boiler rooms, induced-draught fans may be employed to provide adequate 'pull' by mechanical means. In small and medium-sized installations it is very usual to employ bifurcated fans. In this type of apparatus an electric motor, driving a fan fixed on the same spindle,

gives a simple and cheap arrangement. The motor is projected from the hot flue gases by dividing the duct as shown in Fig. 7.30. In domestic installations particular care must be taken to avoid annoyance due to noise from the fan.

7.6 Ventilation systems

The need for ventilation and the range of conditions where natural ventilation will not suffice have been discussed in Chapter 3.

Specific types of extractor are used in industrial premises or kitchens where the main consideration is the extraction of excess heat or fumes and close control of comfort is appropriate. Their intakes are preferably near to the sources of fumes, and fresh air is allowed to enter the space directly. This gives a cheap and effective method of preventing grossly uncomfortable or unhealthy conditions from developing rather than the provision of comfort. Forcing air in rather than extracting would enable the air to be filtered and warmed but control of the distribution would be limited and liable to short-circuit when doors were opened. Where comfort is the aim a combined system is usually employed where both input and extract are controlled. Figure 7.31 shows how it works.

When the system is required only for ventilation the extracted air will often be exhausted and the whole of the input will consist of fresh air, which in winter will have to be warmed before it is delivered to the rooms. In order to make warming-up in the morning more

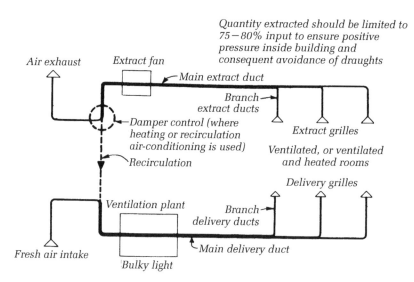

Figure 7.31 Operating principles of a typical mechanical ventilation plant.

rapid, provision for recirculating the air is very often made. Since it is necessary to warm the air there are advantages in using it as the means of heating the building and thereby eliminating any other form of heating installation. Air is not a very efficient heat transfer medium (Table 7.5) so it will be necessary, when heating is involved, to circulate more air than is required for ventilation. It is obviously undesirable to waste this air and it is also recirculated, sufficient fresh air for ventilation purposes being drawn in and mixed with the recirculated air.

The motive power for the distribution of air in ventilation systems is invariably provided by electrically driven fans. There are three main types in use.

Propeller fans

Propeller fans are suitable for situations where no great resistance to airflow has to be overcome. The free intake and discharge condition of ventilation fans situated in wall openings giving direct in/out and out/in movement are eminently suited to this type of fan, which under these conditions can move large volumes of air economically and with very low installation costs. Short duct systems can also be served, provided the resistance of the system is low.

Centrifugal fans

Centrifugal fans can develop pressures sufficient to drive air through air-treatment plant duct systems and are extensively used for this purpose. There are several types having impellers of different patterns and giving various types of performance. Output from a fan can be varied by different motors and speeds of operation, which enables a limited range of casings to serve a wide range of output. The output of fans already fixed can be easily varied on the job by altering pulley sizes. This type of fan is bulky and inevitably turns the direction of air movement through 90°. The change of direction may be convenient in many cases.

Axial flow fans

This type of fan is becoming popular. Efficiency is high, installation simple and the appearance neat, particularly in a line of ducting. Figure 7.32 shows the three types of fan.

Ventilation plant

Except in the case of simple extractor systems some treatment of the air for ventilation is usually required.

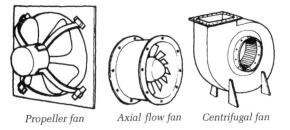

Propeller fan   *Axial flow fan*   *Centrifugal fan*

Figure 7.32   Types of fan used for ventilation systems.

This can involve merely filtering and warming the air. Figure 7.33 shows a simple ventilation plant for this purpose. In cold, dry weather the warming of the air in this type of plant will reduce the already low relative humidity still further and may cause timber shrinkage and cracking, and complaints of dry, sore throats from the occupants of the buildings. Figure 7.34 shows a more elaborate type of plant incorporating a spray chamber and two heater batteries; during the winter it is capable of controlling both temperature and relative humidity. The preheater enables incoming air to be raised to a predetermined temperature, after which it is humidified in the spray chamber, then heated to the desired final temperature and relative humidity. This process can only be effective while the outside temperature is low, permitting the final heating to be effective in reducing the relative humidity. In summer these conditions do not apply but the plant can still be used to give fully conditioned air by cooling the water in the spray chamber. This requires the provision of refrigeration plant. Heat pumps of the types described in Fig. 7.7 are normally used for this purpose.

The water contained in the spray chamber type of humidification shown in Fig. 7.34 is now believed to be a health hazard. In the very large number of existing plants careful routines of cleaning and sterilization will now be employed. In new air-handling plants steam injection is likely to be used as a hygienic form of humidification.

Several types of filter are used for ventilation plant. The simplest are porous screens of paper or fabric which intercept the dust. A layer builds up on the surface of the filter and when this is giving an excessive resistance to flow the filter is replaced.

In order to give more surface area of filter in a given space, this type is often of a concertina shape. To overcome the inefficiency associated with the increasing resistance to airflow of the type of filter described, a number of mechanisms have been developed. One example is an endless belt-oiled wire

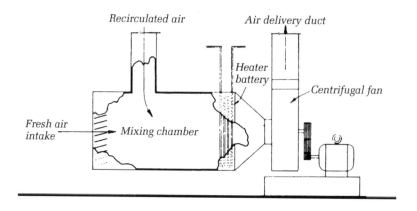

Figure 7.33 Simple mechanical ventilation plant. The heater battery can be served by hot-water and return pipes from the normal space heating system. Electric heating batteries may also be used.

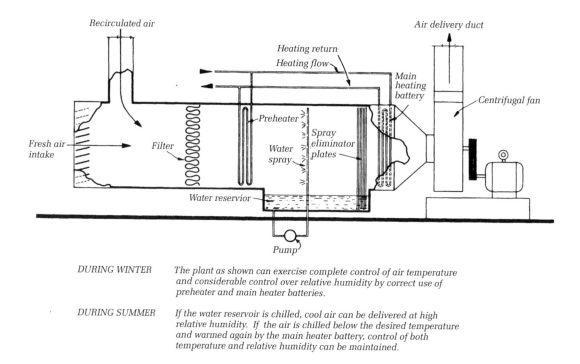

DURING WINTER   *The plant as shown can exercise complete control of air temperature and considerable control over relative humidity by correct use of preheater and main heater batteries.*

DURING SUMMER   *If the water reservoir is chilled, cool air can be delivered at high relative humidity. If the air is chilled below the desired temperature and warmed again by the main heater battery, control of both temperature and relative humidity can be maintained.*

Figure 7.34 Air-handling plant to control temperature and humidity in summer and winter.

mesh filter which is slowly and continuously rolled through an oil-bath, thereby cleaning and reoiling the fabric.

A particularly efficient type of filter having a very low resistance to flow is the electrostatic type. The air is passed through a grid of wires charged at high voltage and the particles of dust are deposited as a result of electrostatic attraction.

The plant for air treatment is large and its placing warrants special consideration in planning. It is bulky and light; it must draw in fresh air and exhaust vitiated air. The top of the building may well be very convenient. Severe structural loads are not involved, and the occupation of space is less important than it might be at a lower level. Long ducts from a central plant may represent a severe planning problem, often overcome by having a number of local plants rather than one central plant. As insurance against total failure it may be desirable to have more than one plant. Subdivision of this sort may well assist in zoning the building for control purposes. When summer air-conditioning is required and a refrigeration plant must be provided, the siting consideration for this is very different. The plant is small but heavy, and produces considerable noise and vibration. There are advantages in accommodating it when possible at the bottom of the building. When refrigeration plants are used the heat must be dissipated. This is usually done by the external air, in small installations by an evaporative cooler where a fan blows air across heat-dissipated coils as water flows over them, or in larger installations by a cooling tower. In some cases ornamental fountains and pools have been used for the purpose.

Figure 7.35 shows the relationships between the various plant areas required for an air-conditioned building.

Air trunkings

Galvanized sheet steel is the material most commonly used for air trunkings. It is light, cheap and easily fabricated. Builders' work may be used when this is convenient and when it can be made reasonably airtight, and finished smoothly internally. Table 8 of BS 5720 *Mechanical ventilation and air-conditioning* tabulates a number of materials with their application and relative efficiencies, and gives details of the construction of sheet steel trunkings.

The most efficient shape for trunkings is circular. Friction is reduced to a minimum, and the form of construction is inherently rigid. Rectangular trunkings, however, have many advantages; they are often more easily accommodated in buildings, and changes of cross-sectional area and junctions are more easily formed than in circular sections. Provided the shape of the rectangle does not become too elongated

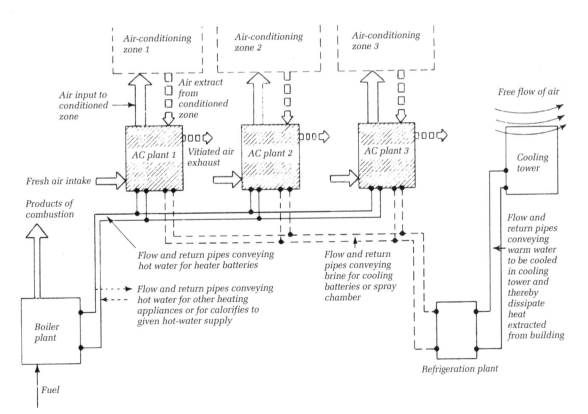

Figure 7.35   Relationships between the plant spaces required for an air-conditioning system.

(maximum ratio of sides usually taken at 1:3), they give an acceptable aerodynamic performance.

For small systems, such as lavatory extractors, the use of PVC piping is becoming popular.

Input and extract devices for air distribution systems

When air is the heat distribution medium the problem of transfer to the environment is simplified in that the air is emitted directly into the rooms concerned. Care must, however, be given to the selection of a suitable grille or nozzle to give the right distribution of air without noise or other untoward effects. The simplest form of grille is the stamped plate. It is cheap but has limited free area (about 60%), no directional control and risk of noise if the velocity of the air is high. The vane type is a substantial improvement and there are many other special types giving particular distributions. It is even possible to input and extract through the same device at the same time. Nozzles can be used to direct jets of air in special directions with the minimum of trunking. Figure 7.36 shows these grilles together with an adjustable type often used in ships and aeroplanes. The form of extractor grille is not usually critical and simple stamped or louvred types are generally used. Special patterns are made for use in floors, however, and the diagram shows a mushroom floor device suitable for use under theatre seats when a downward system of air distribution is adopted.

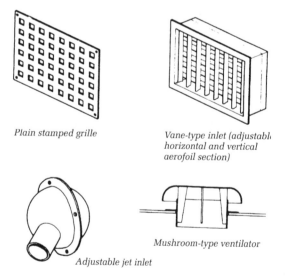

Plain stamped grille

Vane-type inlet (adjustable horizontal and vertical aerofoil section)

Adjustable jet inlet

Mushroom-type ventilator

Figure 7.36 Simple input and extract devices for mechanical ventilation.

Special types of ventilation systems

In single trunking systems all the air delivered is at a single temperature and humidity, so different conditions in various parts of the building can only be achieved by having separate plants serving the different zones. The large trunkings also present a problem, especially where the branch serving a particular floor leaves the main vertical duct. If floor heights have been reduced to a minimum, difficulties can arise with headroom. Systems which attempt to overcome these problems have been developed. They reduce trunking sizes by increasing velocity. This requires circular trunkings to ensure airtightness and freedom from drumming, and some sound attenuation at the delivery points. There are two main methods of achieving local control. In the *high velocity dual system* every space is served by a duct carrying hot air and a duct carrying cold air. The trunkings deliver to boxes where the air is mixed and where the sound level is reduced by an absorbent lining and baffles. A thermostatic device controls the balance of hot and cold delivery and a wide range of different conditions can be achieved in different rooms from the same plant and system. Figure 7.37 shows the principles of this system and Fig. 7.38 shows a typical room input unit. The *induction system* uses a single high velocity trunking delivering air through a unit rather like a convector. The water temperature in the convector will be different to the temperature of the air and local manual control of the water flow enables a range of different conditions to be achieved. The quantity of air delivered is limited to that required for ventilation. The additional air for effective heat transfer is entrained from the room air by the design of the unit. The trunking requirements for this system are more modest than for any other type and it is currently popular for office buildings. Its success depends on proper balance of the air input and water flow temperatures, and zoning for different façades will normally be desirable. Figure 7.38 shows a section through an induction unit.

Where rooms are too large to be served by induction units of the type shown in Fig. 7.38 it is possible to use the same principle and achieve economy of main air trunkings by delivering conditioned outside air from a central intake plant to small subsidiary plants which take an appropriate volume of fresh air but also extract and recirculate air from the rooms they serve. Thus the main trunkings need only be sized to carry air for ventilation whereas the relatively larger trunkings required to achieve thermal control only run between the room served and the adjacently sited local plant. Local plants can be served with both chilled and hot

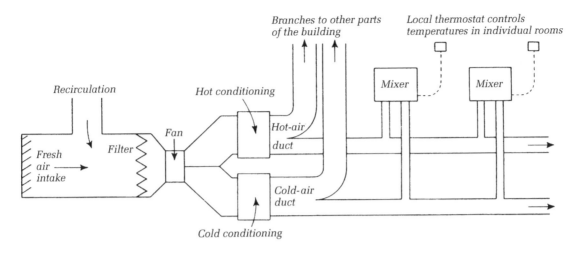

Figure 7.37 Operating principles of a high velocity dual-duct air-conditioning system.

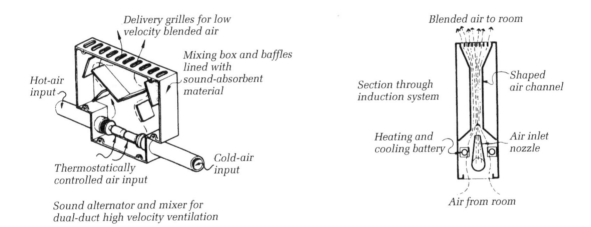

Figure 7.38 Typical room input for a high velocity dual-duct system and an induction unit.

water and thus exercise local zone control. The system illustrated in Fig. 7.42 is based upon this method.

Variable-volume systems

There is increasing interest in methods of avoiding the need for both air and water systems and returning to the simplicity of the all-air system. Apart from size of trunkings, one of the main difficulties with all-air systems is the lack of control. In most cases the system has to be divided into zones, each with a separate plant, and within any one zone no individual room control is possible. Provided that adequate air for ventilation is maintained, regulating the supply of air

to each room would enable temperatures to be controlled. In the normal air-conditioning installation reducing the supply to one room would automatically increase the air supplied to other rooms, whether they needed it or not, because the main fan could not easily be modulated. There are now at least two systems for overcoming this problem. The first depends on the use of fans with variable-pitch blades, pneumatically controlled, which enable the airflow to the system to be automatically controlled from full flow to zero by pressure sensors in the ductwork. With this method individual pneumatic thermostats can control air input to rooms by damper, larger dampers will control pressure in main branches and pressure in the whole

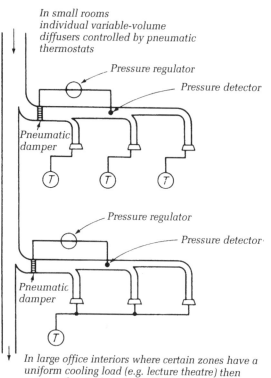

*In small rooms
individual variable-volume
diffusers controlled by pneumatic
thermostats*

Pressure regulator

Pressure detector

*Pneumatic
damper*

Pressure regulator

Pressure detector

*Pneumatic
damper*

*In large office interiors where certain zones have a
uniform cooling load (e.g. lecture theatre) then
groups of variable-volume diffusers may be
controlled by a single pneumatic thermostat*

Figure 7.39 Variable-volume air-conditioning system
using fan modulation.

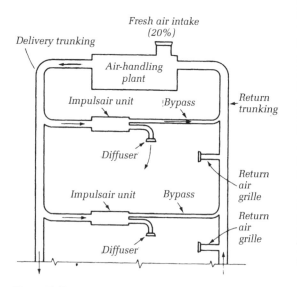

*Fresh air intake
(20%)*

Delivery trunking

*Air-handling
plant*

Impulsair unit   *Bypass*   *Return
trunking*

Diffuser

*Return
air
grille*

Impulsair unit   *Bypass*

*Return
air
grille*

Diffuser

Figure 7.40 Variable-volume air-conditioning system
using ring trunking.

system will be regulated by fan adjustment. The
second system employs a fan continuously driving air
round a ring circuit of trunking from which room
supplies can be diverted without requiring adjustment
of fan operation. Figures 7.39 and 7.40 show sections
of systems operating on those principles. In both
systems, although control can be exercised between
rooms, all the spaces served must be on the heating or
the cooling cycles. Some heating and some cooling is
not feasible.

Extract through light fittings (luminaires)

Where high levels or illumination are used
considerable heat gains result. Levels of illumination
of 500 lx present a significant problem, but at 1000 lx
the heat gain probably outweighs the losses, even in
cold weather. Since most of this heat is given off as
convection from light fittings it is sensible to extract
air in mechanically ventilated buildings through the
light fittings rather than allow the heat to make its way
into the room, when it will represent a load that must
be met by the refrigeration plant if comfort is to be
maintained. It is possible not merely to extract the heat
from the light fittings but also to recirculate it during
winter, so as to warm the incoming fresh air. In the
summer the hot air from the light fittings would be
exhausted to the outside air. Figure 7.41 illustrates the
principles involved. Figure 8.8 (page 184) shows a
light fitting specially adapted for air extract.

These ideas have been substantially developed in
recent years, particularly for large open office areas.
Large volumes have relatively small rates of heat loss
when compared with smaller volumes, and where high
levels of lighting, often 1000 lx, are employed there is
often no heating problem, even in winter, and the heat
from the lights can maintain adequate temperatures.
Cooling will be needed in the summer. Figure 7.42
shows part of a system of this sort. A local-zone air-
handling plant takes fresh ventilation air from a main
trunking supplied by a central intake plant, mixes it
with recirculated room air and delivers it to a large
open office through a system of air trunkings in the
ceiling and slot diffusers. The local zone plant draws
air from the ceiling void which acts as a plenum
chamber, equalizing pressures over the whole ceiling
and enabling equally sized apertures in all the light
fittings to extract room air evenly throughout the room.
The extracted air is then recirculated to give the
volume flow required for adequate heating or cooling.

Very much attention is now given to energy
conservation. Warm air exhausted from large air-
conditioned buildings as part of the ventilation process

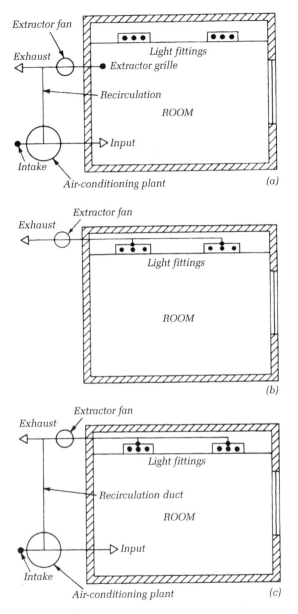

Figure 7.41 Air-conditioning configurations: (a) system not connected to lighting; (b) separate exhaust system for lighting; (c) light fittings form part of the extractor system in an air-conditioning installation – heat may be exhausted in summer or recirculated in winter.

can waste considerable quantities of energy, so several systems have been devised to save some of this energy. The simplest method is the runaround coil of Fig. 7.43 where water is circulated by pump from a heat exchanger in the arm exhaust to another heat exchanger in the cold intake. Heat is transferred from

the exhaust to the intake sufficiently well for many installations to make effective use of the principle. It is clearly possible to use a heat pump to improve the efficiency of heat recovery, as shown in Fig. 7.43(b). A most ingenious method of achieving very high levels of heat recovery from exhaust to input trunkings is the heat recovery wheel. A rotating disc, usually a honeycomb of asbestos impregnated with lithium chloride, passes through both ducts and transfers heat from the warm exhaust to the cold intake very efficiently (Fig. 7.44). At present heat recovery wheels are expensive to buy.

In many buildings it is possible for one façade to require cooling whereas the others require heating. In these circumstances it is uneconomic to input heat at the same time as extracting it from another part of the building and dissipating it via a cooling tower.

Figure 7.45 shows a heat pump system that enables the heat extracted from part of the building to be put back into other parts if they require it. If there is no need for heat in the building, the system can store it using large water cylinders. And if no further storage is needed, the excess can be dissipated via a cooling tower.

Trunking noise

Noise emanating from air systems must be kept to an acceptable minimum level. It can arise in two ways. First by vibrations from the plant being transmitted along the trunkings. This is avoided by providing in all systems a flexible canvas link joining the main trunkings to the fan. Second by noise generated in the trunkings or grilles by the passage of air. This noise can either emerge from the grille or may penetrate the wall of the duct even though there is no opening. To avoid the generation of unacceptable sound levels, air velocity must be limited. The air velocity will have to be low near openings, it can increase somewhat away from openings and increase still further if the trunking is enclosed behind sound-insulating construction. Table 7.7 indicates a range of appropriate air velocities. Drumming may occur in the case of rectangular, sheet steel trunkings unless the flat sides are stiffened at appropriate intervals. High velocity systems terminate in sound-attenuating boxes, which also reduce the velocity of the air admitted to the room.

Ventilation of internal lavatories

Ventilation systems for internal lavatories normally consist of extraction only; the fresh air is drawn via grilles or by gaps under the door from the adjoining

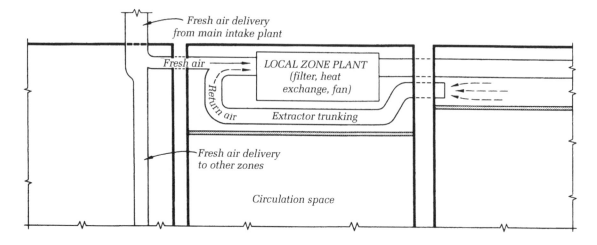

Figure 7.42   Local zone system for air-conditioning.

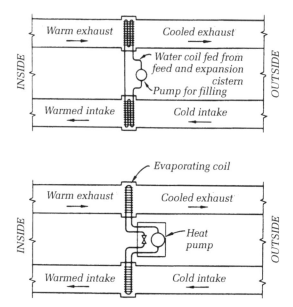

Figure 7.43   Heat exchangers transferring heat from exhaust to intake: (a) runaround coil; (b) heat pump.

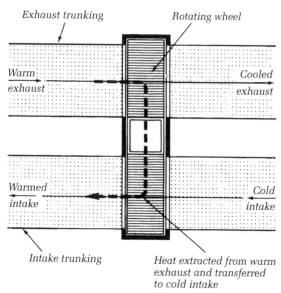

Figure 7.44   Heat recovery wheel.

corridors. They are required to be separate from the rest of the ventilation trunkings and plant, and except in cases where one extractor fan serves one domestic lavatory and bathroom, they must be provided with duplicate fans so that ventilation can still be maintained even when one of the fans fails. Several compact units incorporate two fans and some even have automatic switches.

With a system of this sort serving a number of bathrooms or lavatories, one above the other, and very often serving pairs of rooms on each floor, penetration of sound from one room to another could occur in a quite unacceptable way. This is avoided by serving the rooms using branches from the main trunkings called shunts, which should be more than 1 m long and should incorporate two bends. Figure 7.46 shows a detail of

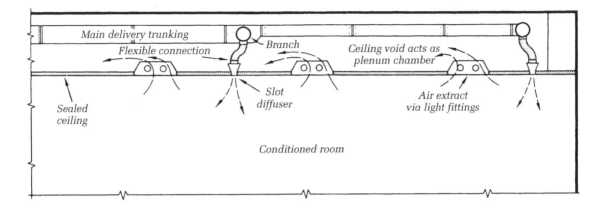

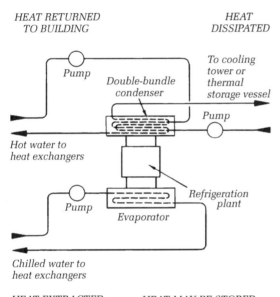

HEAT RETURNED
TO BUILDING

HEAT
DISSIPATED

HEAT EXTRACTED
FROM BUILDING

HEAT MAY BE STORED
FOR SUBSEQUENT RETURN

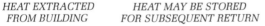

Figure 7.45   Heat pump recovery system.

this type of installation. The arrangement is also effective in preventing smoke penetration to other spaces served in the case of a fire in one room. Since the trunkings involved are usually small, plastic pipes can form a very neat and satisfactory way of forming them.

Work at the Building Research Establishment described in BRE Digest 78 (second series) *Ventilation of internal bathrooms and WCs in dwellings* has established standards for minimum ventilation rates; they are based on the occupancy rather than the

volume of the space, which is the usual basis. The BRE recommendations are given in Table 7.8.

It is recommended that the ductwork is designed to cater for 20% more ventilation to allow for balancing. A velocity of 3 m/s is usual for this type of installation.

The main trunking is often kept to a fixed size for the whole height of building served, and in all cases extract grilles will be fitted with dampers to allow the system to be balanced.

Control of fans for individual components can be by means of the light switch, preferably with a time delay switch to ensure 20 min running after use. Fans serving more than one room must be kept in permanent operation.

Concern is sometimes felt about the resultant heat losses but natural ventilation would also produce heat loss and there is no means of eliminating it. But since the ventilated rooms are often totally internal, no fabric losses can take place. There are several sources of heat gain. Occupants, lighting, hot-water supply, heating flow and any return pipe that passes by, all of them contribute heat; sometimes flues may be adjacent and drawing off bathwater raises the room temperature. It may be found that internal lavatories become uncomfortably hot. The problem should be borne in mind during design. Ventilation rates may have to be augmented to keep the temperature rise to within acceptable limits.

Approximate sizing of heating and ventilating pipes, trunkings and plant

In the early stages of building design, before a proper engineering analysis of the heating and ventilation installation is possible, it is necessary to make assumptions about the sizes of plant rooms, pipes

Table 7.7 Typical values for airspeed in trunkings

Conventional ducts	Airspeed (m/s)		
	Low background noise	Crowded or with activities	Machine or process noise
Diffusers or grilles (assume 60% free area)	1.7	2.5	3
Branches opening to diffusers	2	3	4
Main distributing ducts	5	7	10
High velocity systems[a]	Airspeed (m/s)		
	Duct in ceiling	Duct in ceiling of adjoining corridor	Duct in concrete casing	Generally
Distributing ducts	8	12	30	20

[a] High velocity systems deliver to sound-attenuating and mixing boxes; details should be obtained from manufacturers. Data on diffusers and branches to diffusers is not appropriate. High velocity ducts are normally circular (better airflow, reduced noise, less leakage). In approximate sizing some arithmetic can be saved by using charts. Extraction from the spaces served by high velocity systems is at conventional velocities.

and ducts. Figures 7.47 and 7.48 give typical sizes of boiler rooms, air-handling plant rooms, refrigeration plant and flues in relation to ranges of loadings, and Table 7.6 gives similar indicators for pipes and ducts. It is therefore possible, if the load from the building can be estimated, to reach a decision about the amount of space to allocate. In the early stages of design the estimate will be very approximate, but as design proceeds it can be made more accurate until, when the design is settled, the final engineering calculations can be made with reasonable confidence that the spaces allocated will prove to be adequate. There are several approximate methods for estimating thermal loads in early design stages:

Heating

Assume an average U value for the whole of the skin (including windows) and an average ventilation rate for the whole volume.

Estimated heat requirement =
   (U × area)
   + (volume × air change rate
      × volumetric specific heat of air)
   × temperature difference

Single-storey building U value of walls and roof 1.1 W/m² K with 50% glazing. The total requirement may vary from about 35 W/m² for 50 m × 50 m on plan to 60 W/m² for 10 m × 10 m on plan.

Multistorey building 20 m × 80 m on plan with 50% glazing and similar walls and roof. The total

requirements may vary from 30 W/m² for two floors to 27 W/m² for 12 floors.

Cooling

For perimeter zones (up to 7 m from windows):

   25% glazing 120 W/m²
   60% glazing 180 W/m²

For interior zones (more than 7 m from windows) use 75 W/m². Perimeter zones are greatly affected by orientation and fenestration and all areas by lighting levels and occupancy.

Ventilation trunking sizing

Ventilation trunkings are large and often need to be fitted into limited spaces. A slightly more detailed procedure for approximate sizing than the value given in Table 7.6 may be of use in design. The following notes describe a possible method.

Procedure for approximate sizing

It may be necessary in early design stages to make approximate estimates of sizes. The following method is suggested for this purpose. It is usually helpful to sketch a three-dimensional diagram of the proposed layout to assist in the sizing process.

Method

1 Establish the ventilation rates required in each area; see Table 7.9.
2 Calculate the volume of air to be moved *per second*

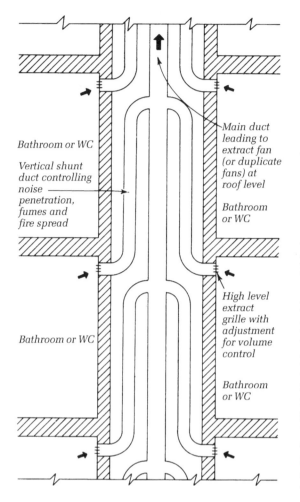

Main duct leading to extract fan (or duplicate fans) at roof level

Bathroom or WC

Vertical shunt duct controlling noise penetration, fumes and fire spread

Bathroom or WC

High level extract grille with adjustment for volume control

Bathroom or WC

Bathroom or WC

Figure 7.46 Shunt systems for the ventilation of internal bathrooms and lavatories.

Table 7.8 Minimum ventilation rates: BRE recommendations

Space	Minimum ventilation rate (m³/h)
WC compartment (3–4.5 m³)	21
Bathroom without WC (5.5–7 m³)	21
Bathroom with WC	42

in each area to achieve the desired ventilation rate. This is likely to take one of the two forms shown:

Volume (m³/s) = number of people × rate per hour per person/3600

or

Volume (m³/s) = volume of space × number of air changes per hour/3600 (see Table 7.7)

3 Size the grilles by deciding the proportion of the volume of air (step 2) which will pass through each grille or diffuser and divide by the speed. This will give the free area. (Where a manufacturer's literature is available, volume flow through grilles may often be directly established from the figures quoted.)

4 Size the branches to the diffusers and the main trunkings by adding the volumes of air passing through grilles and trunkings served by the section under consideration and dividing the total by the airspeed given in Table 7.7.

Note Sharp bends should be avoided in duct layouts, and in the case of rectangular ducts the ratio of short to long sides should not exceed 1 : 3.

Packaged plants

In recent years there has been a rapid increase in the use of completely prefabricated plants known as packaged plants. This is particularly so in the field of air handling, where packaged plants capable of being installed on flat roofs without protecting buildings are frequently employed. Access for maintenance and repair is from the roof itself, thus saving circulation space, and the plants themselves are extremely compact. For a given duty a packaged plant will be much smaller than the corresponding plant room. Several manufacturers produce ranges of sizes and their literature should be consulted. Figure 7.49 shows a typical roof-mounted packaged air-conditioning plant.

Environmental control in large buildings

The development of mechanical and electrical environmental control installations for buildings initially led to separate targets and procedures, and the environmental design for large buildings was strictly divided into its separate aspects, each of which had to perform independently. Daylighting and electric lighting were separate functions each having to achieve its own standards without any mutual compensation. Buildings were either naturally ventilated, mechanically ventilated or air-conditioned and the systems were designed to perform independently of each other. Heating could be combined with mechanical ventilation or air-conditioning, but the standards would be based on

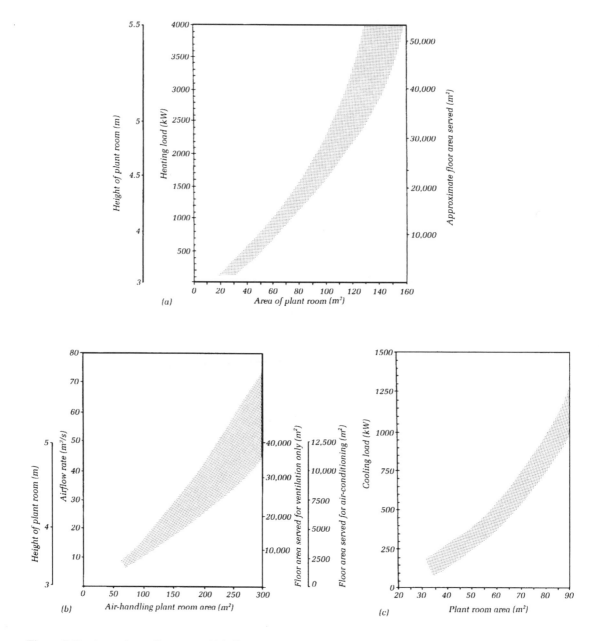

Figure 7.47   Approximate floor areas: (a) boiler plant; (b) air-handling plant; (c) refrigeration plant.

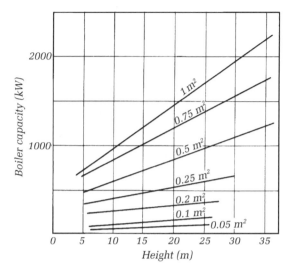

Figure 7.48 Flue sizes for solid-fuel and oil-fired boilers: approximate relationship of boiler capacity, flue height and cross-sectional area.

Table 7.9 Typical air change rates

Ventilation space	Air changes per hour[a]
Boiler rooms	15–30
Garages	
Parking	6 (minimum)
Repair	10 (minimum)
Laundries	10–15
Kitchens	20–60[b]
Sculleries and wash-ups	10–15[b]
Banking halls	6
Cinemas and theatres	6–10
Dance halls	10–12
Restaurants	10–15
Offices	4–6
Bathrooms	6[b]

[a] Extraction rate less than input rate. These rates are for ventilation rather than heat transfer, which may sometimes require higher rates.
[b] Extraction rate greater than input rate. Make extraction the standard and use negative pressure within the ventilation space, if possible.

the rate of heat loss through the external skin, with little if any consideration of the thermal capacity of the building. The design skills were similarly divided. There was little opportunity for the development of integrated design development.

Significant changes can now be observed where form, fenestration, insulation and thermal capacity are balanced with heating, ventilation and lighting, and they can interact with each other to produce an optimum environmental performance. Several UK developments attempted to combine previously separate design elements. In the 1960s an integrated approach to office lighting was proposed by BRE. It involved the use of electric lighting remote from the windows to supplement daylighting, permanent supplementary artificial lighting installations (PSALI). It offered the opportunity for more even lighting distribution and deeper offices. Its application was, however, limited. The wishes of users, the lamps and control systems and the design of fenestration current at the time did not combine to make the system popular. Other work at BRE identified the significance of thermal capacity in controlling internal temperature and led to the admittance method of design, which enabled peak summer temperatures in offices to be predicted. Work under the auspices of the Electricity Council demonstrated that the building envelope could be designed to modify the thermal effects of the outside climate and that thermal installations could redistribute and even store excess heat. This approach was given the name integrated environmental design (IED).

In the mid-1970s the IED principles were applied to the design of a new CEGB administrative building near Bristol. The form and fenestration of the building were designed to give good daylighting without excessive solar gain. Task lighting was provided rather than high overall electric lighting levels. Heat reclaim was used to redistribute waste heat to other areas. Air ducts formed in massive concrete floor slabs enabled the slab to be cooled overnight in the summer so that ventilation air could be cooled during the following day by passing it through the same ducts. Figure 7.50 shows the general form of the building. Since then considerable development has taken place.

In the United States the concept emerged of an open atrium rising through the centre of multistorey buildings. Many of the early buildings were hotels which would otherwise have taken the form of rooms served by internal corridors. And the climate in the United States meant that air-conditioning was provided. The concept of the atrium was very attractive to architects. Provided adequate control of the environment could be achieved, it offered the possibility to replace artificially lit central corridors with an attractive, well-lit interior space which could supplement daylight at the rear of offices and provide a route for through ventilation. The volume of the building would be increased but the office depth could also be increased and air-conditioning might be avoided. The method has been adopted with success

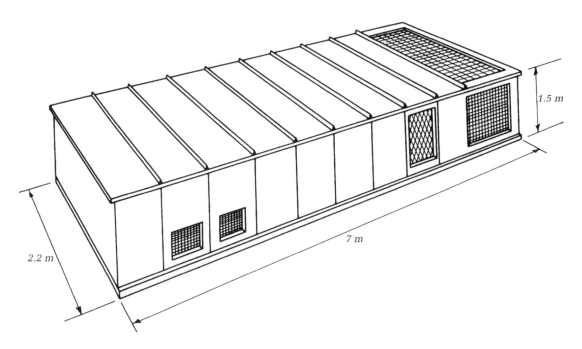

Figure 7.49   Roof-mounted packaged air-conditioning plant.

across Europe, including the United Kingdom, and the opportunity taken to use natural ventilation in conjunction with the thermal capacity of the building to improve both performance and economy. Thermal capacity and ventilation have become an integral part of the environmental design strategy.

Integration of all the elements of building into the provision of environmental comfort is currently most apparent in large commercial offices using atria. The method has also been used in academic buildings that have a similar configuration. The traditional form for offices was side-lit with a depth of about 7 m from the window on both sides of a central corridor. Provided there was adequate daylight, limited noise and limited atmospheric pollution, these offices could be lit and ventilated by natural means. Developments in commercial methods and organization began to call for larger and more flexible spaces. With traditional techniques the deeper offices that might have resulted would have required air-conditioning with its associated costs and high energy usage.

Several general principles apply to this type of design. All the elements of the building and its environmental installations must be considered as an integrated whole and properly balanced in their performance to achieve success. This includes the

form of the building, the materials of construction, the internal finishes, control of insolation, control of natural ventilation by windows and by vents at the top of the atrium, and reduction of heat gains. The method requires an integrated approach to building design and active involvement of the occupants in day-to-day running. The control strategies must be fully developed and made clear to the occupants and also the technical staff responsible for the overall operation of the building and maintenance of the installations. The logic derived from conventional buildings does not always apply to buildings of this type.

The following items are among the important detailed design considerations:

- *The materials of construction*
  The fabric of the building plays a vital part in absorbing heat during the day. It is important that the structural floors should be of high thermal capacity and exposed to form the ceilings to the offices. The floors not only absorb heat during the day but also provide a cool surface to improve radiant conditions. They must be allowed to cool at night. This is achieved by keeping high level windows open at night and encouraging the flow of cool night air. The temperatures in the floor slab

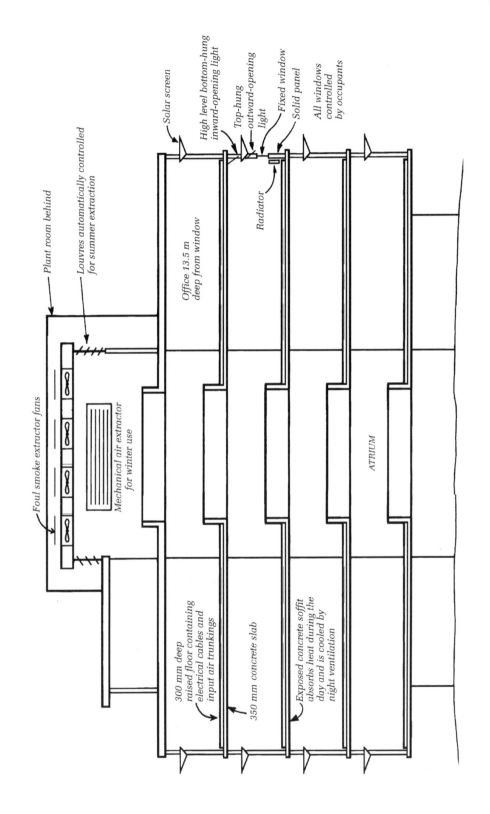

Figure 7.50 Schematic section through the Addison Wesley Longman offices at Harlow. (Architects CD Partnership, senior engineers Cundall Johnston and Partners).

will be monitored and influence the control of ventilation, particularly at night. The surface for radiant cooling must form the ceiling of the floor below.

- *Servicing zones*
  A suspended ceiling containing ventilation trunkings and light fittings has been conventional in large offices. Recently raised floors have provided access for cables to electrical equipment and computers. The use of the floor slab and ceiling for thermal control precludes the use of suspended ceilings. Fortunately though, naturally cooled offices have no need for major ventilation trunkings. The relatively small trunkings which may be provided for supplementary ventilation can be accommodated in the raised floors. The arrangement represents a significant initial cost saving.

- *Fenestration*
  Fenestration is a critical control element. Screens must be provided to control solar gain in summer. Opening lights are needed at normal level and at high level. The high level vents are particularly effective in providing airflow to the interior, without annoyance to the occupants near the windows and also for cooling the floor slab at night. The normal level windows give high rates of air change when needed. It has been demonstrated that occupants are better satisfied and will accept greater variations in conditions if they are in control of window opening. It is possible to have low level openings for daytime use controlled by occupants, and high level openings for night and interior cooling controlled automatically. However, it is vital that the occupants operating windows understand how the overall system operates. Automatic control will be required in buildings with a changing population that cannot be kept informed about operation.

- *Open plan*
  The system requires open-plan use. Some subdivision is possible for small conference and interview rooms but the number must be limited and there may be a requirement that doors should be open when the room is not in use.

- *Automatic control*
  A building management system is essential for naturally cooled buildings. It is not only the heating plant that has to be controlled. Windows and the vents at the top of the atrium also have to be controlled, taking into account not only the air temperature but the direction and strength of the wind along with the time of day. During occupancy

the aim will be to maintain comfort and air quality, at night in the summer the aim is to cool the structure. The control system must not only react to prevailing conditions but also identify time and rates of change so that coming conditions can be anticipated and an optimum balance maintained. Complex detailed strategies will be adopted to govern the extent of window opening in conjunction with time of day, solar gain and temperature in the floor slab.

- *Heating systems*
  Floor and ceiling heating are precluded by the use of the floor slab for temperature control. Most other conventional systems can be used.

- *Supplementary air input*
  Air quality problems might arise where windows are closed during winter. Air can be supplied through flexible ducts running in the raised floor space. The diffusers can be located to suit local conditions and relocated if required.

- *Fire and smoke precautions*
  Fire and smoke precautions are very critical in a large open building and should be discussed with the fire brigade at the earliest stages of design. In general the atrium provides a natural path for smoke extraction. It should extend above the ceiling of the highest office so that horizontal spread at ceiling level is prevented.

- *Insects*
  Shining into the darkness through open windows, lights may attract insects into buildings, especially during the summer. Some problems have been noted, but they are not universal and do not appear to have been serious. But the problem should be considered in buildings where summertime office activities continue until after dark.

Design procedures
There is no specific design procedure for atrium-type offices. Computer software of various types can be used to assist in design. Since much depends on air movement, computational fluid dynamics (CFD) programmes may be particularly useful. However, there are many variables in building, both internal and external, which are difficult to predict accurately at the design stage and modifications during construction can have important effects. Actual performance may differ significantly from early predictions. A very sound understanding of the nature of the phenomena involved and careful study of the performance of existing examples appear to be the prerequisites for design.

Buildings without atria

There are examples of buildings too small to enclose atria, where ventilation stacks are being used to promote airflow. This is most effective at night when the internal air is warmer than the outside and this buoyancy creates airflow. An experimental office at the Building Research Establishment uses this technique, in conjunction with ducts in concrete floors, to cool the floor overnight.

Renovation

Many offices built in the 1960s are reaching the point where major renovation is appropriate. It will not usually be possible to create atria. However, the original designs will have assumed windows which were closed at night. Replacement windows might be provided with facilities to give secure night-time opening and to achieve some of the advantages of the more modern thermal design concepts.

8 Electric lighting

Since the end of the nineteenth century virtually all buildings have been provided with electric lighting installations for use at night. During this period there has been continuous development in the efficiency and types of lamp available. It is possible to use electric lighting to supplement daylight and also to provide dramatic visual effects. The next section describes the types of lamp currently available and their application to working environments.

8.1 Types of lamp

General lighting service

Approximate performance   12 lm/W
Approximate life   1000 h

The filament lamp where an electric current heats a thin conductive filament to incandescence was the first practical source of electric light and, in the form of the tungsten filament lamp, remains in effective and widespread use. The modern lamp has a coiled filament in a glass bulb filled with inert gas. It is cheap, requires no special control equipment and gives light of an acceptable colour. It may be made with decorative or coloured glass or with a specially shaped bulb and silvering to provide a spotlight beam. It has a relatively low efficacy at converting electric power into light, so its electrical consumption and rate of heat generation are high. It is available in a wide range of sizes from 10 to 2000 W.

Tungsten halogen

Approximate performance   20 lm/W
Approximate life   2000 h

This type of lamp has a halogen filling, which combines with the tungsten evaporated from the filament and redeposits it on the filament. This increases the life of the bulb and improves the light output since the tungsten is not deposited on the inside of the glass bulb and the bulb can operate at a higher temperature. Quartz or special glass bulbs are required.

Fluorescent lamp

Approximate performance ≤ 90 lm/W but it
                           depends on phosphor
Approximate life   6000–10,000 h

In common with most of the following types, fluorescent lamps produce radiant energy from an electric discharge passing through a long tube containing gas and vapour mixture. Fluorescent lamps use mercury vapour and ultraviolet radiation is produced. This is converted to visible light by a coating of phosphors on the inside of the tube which forms the lamp. Different phosphors can produce a wide range of different light outputs. The efficacy is high in comparison with tungsten lamps and the heat output modest. Special control gear is required to start the process.

Compact fluorescent

Approximate performance   50 lm/W
Approximate life   7000 h

The length of the original fluorescent fitting limited its application and a more compact form has been devised, where the discharge path is folded by the form of the tube. In some cases the necessary control features are incorporated in the lamp itself and the compact lamp can be installed into fittings suited for GLS lamps. The efficacy is slightly reduced but the performance is still more efficient than GLS fittings.

Sodium lamps

Approximate performance   80 lm/W under high pressure

Approximate life   12,000 h

Some discharge lamps use sodium vapour. Low pressure lamps have a very high efficacy (up to 200 lm/W). But they produce yellow light, unacceptable indoors, and their use is largely confined to street and security lighting. High pressure sodium lamps have improved colour rendering, so some models can be used for display lighting or uplighting.

8.2 The purpose of lighting

Although its necessity has been traditionally unquestioned, natural lighting is associated with some serious disadvantages. The depth to which natural light can penetrate is limited, so this constrains the depth of buildings lit in this way, creating long external walls and limited use of restricted sites as well as high heat loss. Glazing itself makes a significant contribution to heat loss, presents a cold surface on the interior of the building and may be subject to condensation. In the summer the heat of the sun may give rise to overheating. At night the dark area of windows normally requires a covering. Where especially close control of lighting is needed, the variations in natural lighting will often mean that electric lamps must be used. Sky glare may result from large windows and, where the effect of direct sunlight could be disadvantageous, special control devices may be required. The north-light roofs of factories are an example of this. Patches of sunlight can affect proper vision and make accidents more likely where machinery is involved.

It might be thought that, as a result of these disadvantages, electric lighting would be used very frequently to the exclusion of daylight. There are many buildings where this is the case. Department stores are often designed in this way. Some factories are deliberately designed to exclude daylight, in many more the contribution of daylight to illumination is very limited. Many people, however, regard daylight as very important and desirable. It seems that this desire stems not only from discernible differences between artificial lighting and daylighting but also to a considerable degree from a wish to have visual contact with the exterior and to be aware of changes in the weather and other external activities. In terms of quality of illumination it must be borne in mind that, although electric lighting comes from overhead, the sideways light from windows will often give better light on vertical surfaces and a better three-dimensional quality to objects than will be obtained from the overhead lighting.

In the design of modern buildings it cannot automatically be assumed that windows are needed. A decision will be needed about the best form of illumination, in view of the considerations described, and if daylighting is selected, considerations beyond simple illuminations must be borne in mind.

In relation to electric lighting it will be necessary to decide what part the lighting will play:

- *By night* electric lighting will be the sole illumination.
- *By day* electric lighting will be the sole source of illumination (windowless building) or the main source of illumination with windows for views or to give better modelling. Small windows can make a considerable contribution to better three-dimensional appearance.
- Electric lighting can supplement daylight where rooms are too deep for adequate natural lighting.

And as well as providing overall illumination in a way similar to daylighting, electric light can be used to provide dramatic contrast effects; it is also frequently used as an emergency lighting or for security purposes at night. Rarely used spaces may be provided with minimal electric lighting sufficient for movement and safety; any work is carried out beneath portable lamps.

8.3 Lighting terminology

The terminology and quantities used for lighting are not in common use and before dealing with criteria for design and estimation some definitions are desirable. Refer to Table 8.1.

8.4 Design of lighting installations

Lighting installations may be regarded as falling into four main categories:

- overall general illumination
- task lighting with background general illumination
- lighting for dramatic effect
- specialized problems (e.g. galleries, museums, machinery)

Dramatic effects and specialized problems

Lighting for dramatic effect can take many forms. Brightly illuminated objects standing out from a dark or even totally black background form a good

Table 8.1 Lighting quantities: units and definitions

Quantity	Unit	Definition
Luminous intensity, I	candela (cd)	Intensity of small source which gives an illuminance of 1 lx at a distance of 1 m
Luminous flux	lumen (lm)	Flux that gives an illuminance of 1 lx over 1 m^2
Illuminance, E	lux (lx)	The number of lumens per square metre
Luminance, L	apostilb (asb)	Flux emitted per square metre: $L = Ev$
Reflection factor, ρ	dimensionless number	The reflected luminous flux divided by the incident luminous flux
Efficacy	lumens per watt (lm/W)	Average through life lamp efficacy

example. This particular type of design will usually require optical systems of reflectors and lenses to concentrate the light and complete control of internal finishes and decorations. The principle is sometimes employed for illumination on the working plane, particularly in restaurants. Light sources are totally recessed in the ceiling, giving light on tabletops but giving no direct illumination of walls or ceilings. In interiors with this type of dramatic lighting it is very easy for uncontrolled reflections to produce an atmosphere of gloom rather than dramatic contrast.

Another form of dramatic lighting effect, often used in showcases, is to supplement a high level of illumination by using high intensity point sources to give bright reflections from shiny surfaces.

Successful dramatic and specialized lighting depends very much upon knowledge of the effects required and the means of achieving them; they should be carried out by experienced designers.

General illumination

Traditionally the criterion for general illumination of a working environment has been the quantity and uniformity of light falling on the working plane. Although necessary for a successful working environment, this criterion is not sufficient to ensure satisfactory working conditions. Glare may give rise to

discomfort and even affect visibility. The sources and distribution of light will control the ability to discern the modelling of three-dimensional objects. The balance of brightness of the internal surfaces will control the character of the environment and should ensure that it is bright and cheerful rather than gloomy. Technical Report 15 of the Illuminating Engineering Society sets out a design method to ensure that all the criteria can be met. The following notes describe the principles involved and give a simple predictive technique which can be used for working environments.

Quantity

This is normally defined in terms of the amount of light falling on a horizontal plane at working level, known as the working plane. Quantities are given in terms of lux. BS 8206 Part 1 is an authoritative source giving illumination levels suitable for a wide range of applications. Some typical values are given in Table 8.2.

Glare

Glare is an important consideration. Excessively bright areas in the field of view can give rise first to discomfort and in acute cases to disability in seeing. Glare is a complex phenomenon dependent to some degree on the observer but also governed by the brightness, size, number, distance away and direction of the light sources. Methods have been developed to evaluate glare and it is possible to calculate a glare index for proposed installations. Suitable values for the glare index for a range of applications are given in BS 8026. In order to design lighting installations in relation to glare it is necessary to know the distribution of light output from light fittings. The British zonal method (BZ) exists for this purpose. Light fittings are divided into 10 categories based on

Table 8.2 Illumination and limiting glare for various spaces

Environment	Illumination level (lx)	Limiting glare index
Offices	500	19
Drawing-offices		
General	500	16
Boards	750	16
Auditoria and foyers	100	–
Shops	500	19
Living-rooms		
General	50	–
Reading	150	–
Sewing	300	–

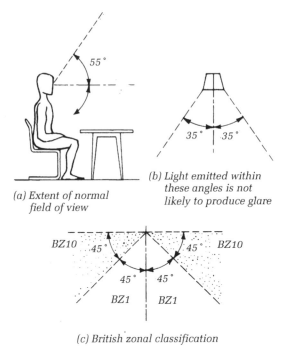

(a) Extent of normal field of view

(b) Light emitted within these angles is not likely to produce glare

(c) British zonal classification

Figure 8.1   British zonal classification and rough geometrical limits of glare. BZ1 fittings give out most of their light within 45° of the vertical. BZ10 fittings give out most of their light within 45° of the horizontal. Other numbers have intermediate performances.

the direction of light output. Figure 8.1(c) shows the principles of this classification.

An observer is not likely to be conscious of bright sources of light 55° or more above the horizontal (Fig. 8.1(a)), and some ranges of light fittings have been designed giving complete cut-off of light within this angle. Figure 8.1(b) shows the geometry of this cut-off. Glare is not likely with this type of fitting, but fittings may have to be close together to give adequate distribution of light.

Luminance ratios

It is very possible to have a room with adequate light on the working plane, low glare levels but which gives an unsatisfactory and gloomy appearance. This will usually be due to unsatisfactory luminance of the surroundings in relation to the task. Hence the term *luminance ratio* has come to represent a range of ratios of the luminances of task, immediate surroundings and background, and suitable ratios that produce interiors which appear cheerful and brightly lit have been established. The Bodmann ratio of $10:4:3$ for task,

immediate surroundings and walls and ceiling has been widely accepted as giving satisfactory conditions.

Modelling

In recent years it has been increasingly appreciated that meeting the criteria given above would not necessarily give lighting conditions which permitted good appreciation of three-dimensional form (modelling). Concepts have been developed to take this into account. Scalar illumination is the light falling on a point from all directions (not merely on to a plane). Vector illumination is used to describe the angles at which the difference of illumination on a point is at its greatest. It has magnitude and direction. The system is not used, at present, to any great extent in design and it is important to note that, in a room well lit in terms of the Bodmann ratio described above, the effect of reflected light from walls and floors will be to give less harsh modellings than would be the case with a flow of light solely from overhead.

Lamps (light bulbs)

A wide range of lamp types is produced commercially. Details of the main types are given in Fig. 8.2.

Luminaires (light fittings)

It is convenient to regard luminaires as falling into three categories: *decorative fittings* are intended to be seen rather than to give optimum lighting distribution and performance; *general utility fittings* are intended to give economical and effective illumination; and *special fittings*, usually with special optical arrangements such as reflectors or lenses, give highly directional light. It is the general utility type of luminaire which is of present concern (Figs 8.3 and 8.4).

All luminaires must be strong enough in construction to withstand handling and erection to sustain the lamps and any associated control gear (particularly the case with fluorescent lamps which require heavy ballasts and perhaps starters) and to be cleaned without damage. Access and electrical and thermal insulation is necessary for electrical cables, and earthing must be catered for. The temperatures likely to be reached in operation must be considered. Fluorescent tubes are particularly susceptible to temperature and are inefficient in operation at very high and low temperatures. Ventilation of the luminaire is therefore very common. Completely closed luminaires present another problem because thermal expansion and contraction of the air inside means they cannot be completely sealed. Dust and often flies may gain access, with consequent unsightliness and the need for frequent cleaning.

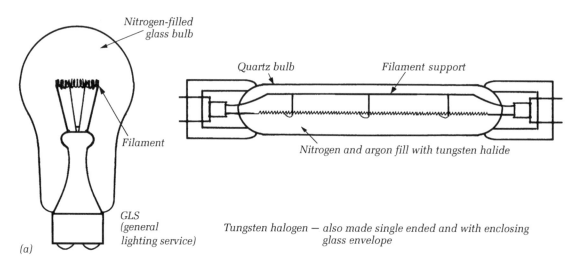

Nitrogen-filled glass bulb

Quartz bulb   *Filament support*

Filament

Nitrogen and argon fill with tungsten halide

*GLS
(general
lighting service)*

*Tungsten halogen — also made single ended and with enclosing
glass envelope*

(a)

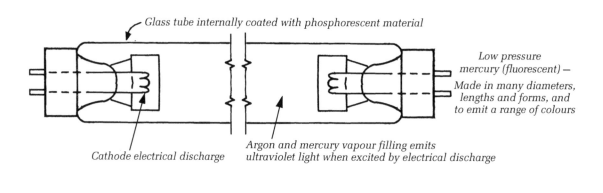

Glass tube internally coated with phosphorescent material

*Low pressure
mercury (fluorescent) —*

*Made in many diameters,
lengths and forms, and
to emit a range of colours*

Cathode electrical discharge   *Argon and mercury vapour filling emits
ultraviolet light when excited by electrical discharge*

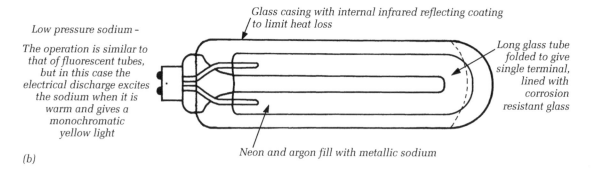

Low pressure sodium -

*The operation is similar to
that of fluorescent tubes,
but in this case the
electrical discharge excites
the sodium when it is
warm and gives a
monochromatic
yellow light*

*Glass casing with internal infrared reflecting coating
to limit heat loss*

*Long glass tube
folded to give
single terminal,
lined with
corrosion
resistant glass*

Neon and argon fill with metallic sodium

(b)

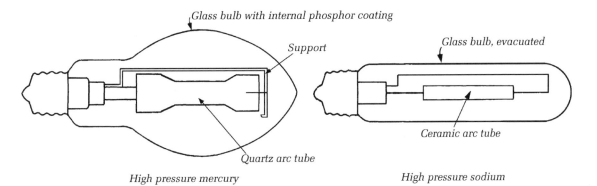

Glass bulb with internal phosphor coating

Support

Glass bulb, evacuated

Ceramic arc tube

Quartz arc tube

High pressure mercury

High pressure sodium

Figure 8.2 Types of lamp: (a) incandescent: light produced by a tungsten filament at a very high temperature; (b) low pressure discharge: light produced by the excitation of chemical by electrical discharge – special control gear is required, which may be incorporated into lamp; (c) high pressure discharge lamps: while essentially similar to low pressure lamps, the high pressures and other technical changes enable improved performance at a reduced size – Edison screw caps are normal, and in both mercury and sodium types reflector spotlight bulbs can be supplied.

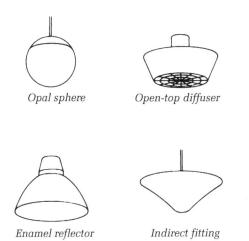

Opal sphere

Open-top diffuser

Enamel reflector

Indirect fitting

Figure 8.3 Typical luminaires for tungsten lamps.

The luminaire should obscure the lamp from direct view to reduce glare and should present, if it is to be directly viewed, a larger surface area of lower brightness. The shape and arrangement of louvres and openings, if any, will be devised to give a particular distribution of light in upwards, sideways and downwards directions. The BZ system of classification has already been described, but there is also a traditional descriptive classification of luminaires into direct (light downwards); semidirect (light mainly downwards, some up); semi-indirect and indirect which have the reverse performance to direct and semidirect. These classifications are used in the lumen method of calculation.

Figures 8.3 and 8.4 show typical luminaires for tungsten and fluorescent lamps. The enamel reflectors and trough fall into the direct category, the diffuser and louvred luminaires into the semidirect category, and the plastic diffuser and opal sphere into the general diffusing categories. It is usual now for manufacturers to quote the downward light output ratios of these luminaires. Two types are shown in Fig. 8.5.

It is not always necessary to have luminaires. Figure 8.6 shows a cornice used to support a line of fluorescent tubes to give indirect light. Some luminaires have been devised to give light while remaining completely concealed. The 'black hole' fitting shown in Fig. 8.6 is an example of this type.

8.5 Other significant factors

Heat gain

Comparatively low levels of electric lighting in buildings with good natural ventilation present little problem. But as lighting levels increase, heat input from the luminaires must be taken into account; 500 lx using fluorescent lighting is often taken as the level at which the problem becomes critical. Heat emitted from light fittings is partly in the form of radiation and partly in the form of convection.

Figure 8.7 shows typical proportions for tungsten and fluorescent luminaires. A substantial proportion of the heat is emitted by convection, and where this problem exists, it is becoming standard practice to extract air through the luminaire. This has the effect of removing a large proportion of the convected heat. The hot extracted air can then be exhausted, or as in many American installations, partially recirculated during

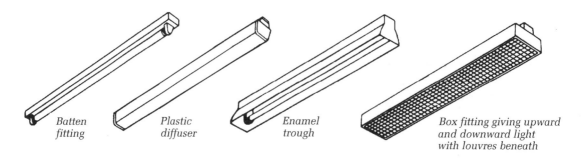

Figure 8.4   Typical luminaires for fluorescent lamps.

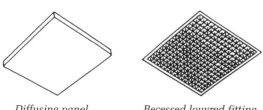

Figure 8.5   Typical panel luminaires (usually with fluorescent tubes).

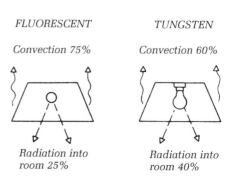

Figure 8.7   Emission of heat from luminaires.

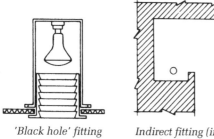

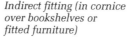

Figure 8.6   Special types of lighting arrangement.

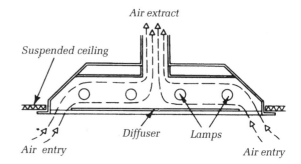

Figure 8.8   Luminaires combined with air extraction.

cold weather to warm the incoming air. Figure 8.8 shows a luminaire adapted for extract.

The radiant heat cannot be dealt with in the same way and must be balanced by an input of cold air from an air-conditioning plant. This adds considerably to the expense of the building. Note that tungsten lamps not only have a higher heat input for their light output but also emit a higher proportion as radiation. It follows that their use in the circumstances under consideration is inappropriate.

Where air is to be extracted through luminaires the design of the lighting installation must be integrated with that of the ventilator system. In view of the ducts running from luminaires, suspended ceilings are almost inevitable with this arrangement.

Cost
Tungsten lamps and luminaires are cheaper than comparable fluorescent luminaires, but in order to obtain a particular level of lighting it is possible that

more will be required. For a given level of illumination more electricity is required by tungsten installations and replacement of lamps is more frequent than for fluorescent lamps. Running costs are therefore higher for tungsten installations. However, this factor is influenced by the periods for which the installation is in use. It is apparent, therefore, that for installations which receive little use, tungsten will be more economical.

Nursery schools are an example of this. They have adequate daylighting and are closed before electric lighting is necessary. They are rarely used in the evenings because the furniture is inappropriate for adults. Lighting in cupboards, ducts and other occasionally used spaces is another.

On the other hand, where the lighting installations will be used for long periods, fluorescent installations are likely to give the most economical solution. In the case of air-conditioned buildings the cost of extracting heat from the lighting can be an important cost factor and fluorescent lighting is indicated.

In commercial buildings electric lighting forms a major element in the cost of energy, and measures to ensure economy will be important. In non-air-conditioned offices the cost of lighting can be up to 40% of the total. In air-conditioned offices the proportion will be smaller but the actual cost will be no less.

The strategy for economy of lighting energy has two main aspects. The first is careful *design* of the installation and selection of luminaires to give adequate, but not excessive, levels of lighting for the lowest electrical loading. The second is *control*. Few installations have variable intensities and the major aspect of control is switching. Occupants can generally be allowed to switch on lighting when it is needed but will rarely switch off when the use of the space is ended or daylighting levels render electric lighting unnecessary. The economy of switching on can be improved by providing local rather than large-area switching. The main approaches to these economy measures may be summarized:

- *Area control*
  - *Separate overall control* of switching for zones of buildings with different general patterns of occupancy.
  - *Local switch control* to meet the needs of changing occupancy at workstations or varying illumination requirements in different areas of rooms.
- *Switch control*
  - *Time-delay switches* for very occasionally used areas.

- *Automatic on and off switching* activated by the presence of occupants in areas likely to be unoccupied for significant periods where time-delay switches would not be appropriate.
- *Localized manual switches* rather than concentrated banks of switches.
- *Overall time switching* to predetermined patterns.
- *Switching to provide partial illumination* where appropriate for cleaners, etc.
- *Intensity control*
  - *Photoelectric control* related to daylighting levels.

Some switches combine manual and automatic operations.

8.6 Lighting calculations

Lighting installations not only have to provide light, but they must also take into account the lighting of three-dimensional objects, irregular layout of luminaires, varying colours of surroundings and give visual satisfaction without risk of glare or gloom; this makes the calculations very complex and the experience of the designer will play a critical part. However, there are simple methods which can be applied to rooms with regular layout of luminaires where the need is lighting on a working plane. This circumstance is very common in buildings.

Lumen method

The lumen method is a long-established and robust technique which is well supported by data from lamp and luminaire manufacturers. It enables the number and size of regularly spaced luminaires required to provide a required even level of lighting on the working plane. It takes into account the reflectances of room surfaces and the risk of glare from luminaires.

The method is based on a simple formula:

$$N = \frac{A \times E}{F \times UF \times MF}$$

where N is the number of luminaires
  A is the area of the working plane (m^2)
  E is the level of illumination required on the working plane (lx)
  F is the average lamp flux (lm)
  UF is the utilization factor
  MF is the maintenance factor

The formula may be solved for any one of the variables if the others are known. A check on glare can be made as part of the calculation. The number of luminaires, the area and the level of illumination required are self-explanatory. The average lamp flux is the 'average through life' value which can be obtained from the manufacturer's data. The utilization factor is experimentally determined by the manufacturer for each type of luminaire. It takes into account the pattern of light distribution from the luminaire and provides a factor based on the reflection factors of the surfaces, the spacing-to-height ratio of the luminaires and the room index.

$$\text{Room index} = \frac{\text{Length} \times \text{width}}{\text{Height of luminare above work plane} \times (\text{length} + \text{width})}$$

The maintenance factor is the average value of the lamp lumen output, the luminaire efficiency and the reflectance of the room surfaces. Accurate values of these at the time of building can be determined but they all vary with time. The average lamp lumen output will vary with the frequency with which lamps are replaced; the luminaire efficiency will depend upon the cleaning regime; and the surface reflectance will depend upon the efficiency of cleaning and the frequency of redecoration. In the absence of information on these points it is conventional to take the maintenance factor as 0.8. An increase may be made if the building is air-conditioned, a decrease if poor maintenance is anticipated.

Glare index

It is possible to have an installation giving adequate levels of lighting but causing discomfort because of glare. Glare is the visual discomfort that arises from a luminous source in the field of vision when its luminance, size and location in relation to the general background combine in a certain way. Calculations from first principles would be very complicated. However, manufacturers of luminaires publish tables which enable the risk of glare to be identified and avoided.

Additional factors

In some specialized cases additional factors must be taken into account (e.g. absorption of light by dust in the air in industrial buildings). To ensure even illumination the spacing of luminaires should not exceed $1\frac{1}{2} \times$ mounting height above the working plane (1.25 in the case of louvred or recessed luminaires).

Luminance design

Luminance design is a means of taking into account the appearance of all wall, ceiling and floor surfaces and selecting luminaires not only to give light on the working plane but also on to the walls and ceiling. For designers of buildings and interiors the technique seems particularly appropriate and a step-by-step method of carrying out this type of analysis is provided, together with a worked example, in the appendix to this chapter (pages 189–96). In order to use this method it is necessary to have the manufacturer's data on a range of possible shapes and sizes of electric luminaires.

Point-by-point calculation

Both the above methods are related to even, overall levels of lighting; where lighting distribution is to vary over a room or where it is desired to achieve areas of more intense illumination, the point-by-point method is used. The calculation is complex and requires knowledge of the light distribution from the fitting; so it is beyond the present scope. Detailed descriptions of the method are contained in most texts on electric lighting.

Combination of day and electric lighting

Estimations of the output of lighting installations, intended to be used in conjunction with daylight, are usually made by the lumen method. The level of lighting required is, however, no longer the same as that quoted in BS 8026. The lighting level provided must be related to the natural lighting conditions which prevail. The problem is discussed in BRE Digest 76 (second series) *Integrated daylight and artificial light in buildings*, which gives the formula:

Level of supplementary lighting =
10 × average daylight factor
over the area supplemented
$\times \frac{1}{3}$ [sky luminance (cd/m^2)]

as a guide to the levels of supplementary lighting required. In most working interiors a level of 500 lx gives reasonable satisfaction.

Ceiling lighting

Ceiling lighting, although popular at one time, came largely into disuse, except in ecclesiastical interiors, because of the high running cost and heat output. The development of high efficiency discharge lamps with acceptable colour characteristics make ceiling lighting a very practicable proposition. These lamps are very economical to run, but have a very high light output. It is difficult to use them in local light fittings but they are very suitable for uplighting. Uplighters can be free-standing, mounted on furniture or, if the ceiling height permits, suspended from the ceiling.

8.7 Task lighting

The general practice in electric lighting in commercial buildings for many years has been to provide overall lighting on the working plane of adequate intensity for the most exacting visual task to be undertaken. This system results in bright, well-lit interiors and enables completely flexible replanning of activities within the room. But it does have two important disadvantages.

First is the large energy requirement and consequent high cost. In brightly lit offices the energy for lighting can be almost half the total energy requirement for the building, and a significant proportion of the non-lighting requirement can be the cooling required to limit heat gains from the lighting. Although all the energy for lighting makes a general contribution to a well-lit interior, at the working-plane level much of it is being wasted in unnecessarily high lighting levels in circulation, storage and waiting areas and in lighting workspaces which are, for the time being, unoccupied.

Second is the problem which arises when, in large offices, free-standing screens or items of furniture are used to provide a measure of privacy. With general lighting a corner between such screens may be very badly lit due to the shadow of the screens.

A modest level of general illumination, supplemented by special lighting in areas where work is taking place, overcomes these problems. Light is not wasted in circulation spaces, and lights can be turned off when workspaces are unoccupied. The task lighting can be from special luminaires suspended from the ceiling or by independent floor- or desk-standing luminaires, or by luminaires incorporated into furniture. The background lighting can be provided in the normal way from a grid of ceiling luminaires, but these are also examples of high efficiency lamps being concealed in the tops of items of furniture and giving general illumination by reflected light from the ceiling. The disadvantages of the system are the difficulties of rearrangement and the need to provide a grid of cables at either ceiling or underfloor level in order to give some flexibility for replanning. The selection of suitable local light sources is also a problem. Overall more economical than general lighting, the traditional adjustable tungsten luminaire is nevertheless a relatively high consumer of energy and gives off considerable heat. The proximity to the work results in considerable variation of lighting intensity across the work area and the extreme brightness of the lamp encourages specular reflections. In recent years small fluorescent lamps and specially designed desk luminaires have, to a very considerable degree, overcome this problem.

The great majority of building users have a strong preference for general rather than task lighting but there can be little doubt that, as energy costs tend to become a more dominant element in building running costs, more use will be made of task lighting.

The standards of illumination for task lighting are the same as those for general illumination having the same purpose. The level of background illumination must be related to the level of illumination for the task, and it is usual to employ a level of one-third that of the task lighting.

The lumen method can be used for the design of background lighting whereas lamps and luminaires will be chosen for task lighting on the basis of their photometric performance to give an adequate level of lighting over the area of the task. Although the problem of general glare is much reduced, care must be taken to ensure the local luminaire screens the lamp from the view of the user and that it is positioned to illuminate the work adequately without unacceptable reflections.

VDUs

VDUs are becoming almost universal in offices, academic buildings and schools and in many other types of building. They present a special lighting problem. Their screens present a substantial area of specular (mirror-like) reflection immediately in front of the user. This presents little problem in daylight. The problem of glare if the user faces the window or reflection if he faces away from it, can be overcome by taking a position parallel to the window. In small rooms with electric lighting there is little problem since the angle of light is too high to reflect into the eyes of the user. In larger rooms, however, the problem can be acute. When designing a room for computer users, consideration should be given to downlighters or luminaires with similar directional characteristics. Uplighters do not produce very bright areas and may also assist in this application.

Emergency lighting

Emergency lighting is required in public buildings and is also being installed in many private buildings.

Intended to take over if the main lighting fails, emergency lighting is often good practice and may be a statutory requirement under health and safety legislation. The general principle is that escape routes should be determined, marked with clear signs and provided with emergency lighting; auditoria, large open areas, rooms which provide access to other areas, toilets of area over $8\,m^2$ and rooms with special risks should also be provided with emergency lighting and signs indicating the escape routes. The emergency lighting should come into operation within 5 s of failure of the main lighting system in buildings with public access and 15 s in other types.

The power for the emergency systems must be independent of the main building supply. Traditionally this has involved central battery systems with special circuits serving emergency light fittings. Luminaires are now available with their own batteries and control system. The batteries will be kept charged by the normal electrical installation and be brought into operation immediately the main system fails. The same luminaire can also provide normal lighting.

The level of lighting required for open areas where the escape route is not clearly defined is 1 lx and the same level of illumination is appropriate for stepped areas and spaces with adjustable seating. In auditoria with fixed seating the level reduces to 0.1 lx.

Escape routes themselves should be clearly marked by illuminated signs saying EXIT or EMERGENCY EXIT. The signs may be self-luminous, externally lit or internally lit. Gas discharge lighting may be used for this purpose. The size of the lettering increases with viewing distance. The illumination along the centre line of each 2 m width of escape route must be 0.1 lx.

All measurements of escape lighting are made at floor level. Care must be taken to avoid glare along escape routes and limits are imposed on the luminous intensity of objects which might interfere with the field of vision.

Calculations for emergency lighting do not include any reflected component. They can be undertaken with isocandle diagrams, which provide a plan of luminous intensity on the horizontal plane at a given distance under the luminaire. This provides a very simple procedure. Acceptable glare levels involve a calculation based on the illumination of the line of escape and the maximum luminous intensity of objects within the line of sight, which includes an angle of 20° above eye level.

Recently it has been recognized that smoke can be a major problem in escape routes and that the conventional high level siting of luminaires and signs makes them very likely to be obscured. Aids known as 'wayfinding' systems are under development and test. They use continuous bands of lighting at low level provided by electroluminescent strips or by light-emitting diodes. It appears that the continuous, low level guide may be more effective than the conventional systems.

Standby lighting

There are many activities which cannot be stopped immediately. Hospital operating theatres and the operation of complex plant and equipment are examples. In these cases the level of lighting must be fully sustained for as long as necessary to complete the activity. The period might be quite short for closing down plant, much longer for an operating theatre. Standby plant capable of maintaining the full level of lighting and power supply to equipment is required in these cases. The general level of illumination is 1 lx; this is required for large open areas with no clearly defined circulation pattern or escape routes.

Appendix: Step-by-step procedure for appearance lighting calculations*

This routine is for determining the number and type of luminaires required to provide a desired luminance pattern or balance of illumination in a space with given internal finishes. It has four main stages.

1 *Development of luminance specification*
   Specification of desired internal finishes and definition of the task to be carried out in the space. From these basic decisions a complete luminance specification is developed.
2 *Direct and indirect illumination*
   Calculation of interreflections in the space and hence establishment of the distribution of direct illumination required to achieve the desired conditions.
3 *Luminaires*
   Selection of the type(s) of luminaire that will provide a suitable light distribution.
4 *Number and layout of luminaires*
   Determination of the number of luminaires required.

Stage 1: development of luminance specification

1 Select surface finishes, colours and reflectances for the working plane and for the other room surfaces, and enter in columns 1, 2 and 3 respectively of Table A.
2 Establish the reflectance of the task (from manufacturers' literature or published data, or by comparison with shade cards of known reflectance, etc.). Enter in column 3 of Table A.
3 Establish the illumination level required on the task (from BS 8026) and enter in column 4 of Table A.
4 Calculate the task luminance (luminance = illumination × reflectance) and enter in column 5 of Table A.
5 Complete the specification by determining *either* luminance *or* illumination of the room surfaces. As a guide in working environments, use either Bodmann's luminance ratios or Jay's illumination ratios, which give the desirable levels on other room surfaces in terms of the task luminance or illumination. The recommended ratios are given below.

	Task	Immediate surround	Walls	Ceiling
Bodmann (luminance)	1	0.4	0.3	0.3
Jay (illumination)	1	1	0.5–0.8	0.35–0.8

6 Having specified reflectance and luminance, calculate illumination; *or* having specified reflectance and illumination, calculate luminance. Complete Table A for all room surfaces (luminance = reflectance × illumination).

Stage 2: direct and indirect illumination

7 Calculate the *room index*:

$$\frac{\text{width} \times \text{length}}{(\text{width} + \text{length}) \times \text{height}}$$

Height is measured from ceiling to working plane or floor, as appropriate.

8 Using the value of the room index just calculated, select the appropriate multiplying factors from Table 8.3 and enter them in Table B. Also, enter luminances from Table A in the first column of Table B. When the room index does not coincide with the tabular values use linear interpolation. To check that the multiplying factors have been entered correctly, the sum of all the factors in one column of Table B should be unity.

Table A

Surface	1 Finish	2 Colour	3 Reflection factor	4 Illumina-tion	5 Lumin-ance
Task					
Immediate surround					
Floor[a]					
Working plane					
Walls[b]					
Ceiling					

* This appendix is based on a method developed by Peter Jay and Dennis Coomber. The worked example was devised by Dennis Coomber of Peter Jay and Partners. They are reproduced courtesy of Peter Jay and Dennis Coomber.

[a] *Floor cavity reflectance* should be used instead of reflectance when working plane is above floor level. It is roughly the weighted average reflectance of floor and work surfaces, minus 0.1.
[b] Where walls are of different reflectances, use the weighted average. Reflectance of uncurtained windows may be taken as 0.1.

Table B

Light coming from	Luminance[a]	Light falling on					
		FWP[b]		Walls		Ceiling	
		MF	SL	MF	SL	MF	SL
FWP[b]							
Walls							
Ceiling							
Total indirect illumination on each surface							

[a] From column 5 of Table A.
[b] FWP = floor or working plane.

9 For each pair of surfaces, work out the indirect illumination on one surface due to light reflected from the other surface: indirect illumination = source luminance (SL) × multiplying factor (MF). Enter the results in Table B and sum the indirect illumination on each surface. Note that the values of indirect illumination in one row of the table are arrived at by multiplying in turn the factors in that row by the luminance at the beginning of the row.

10 Obtain the direct illumination required on each surface by subtracting indirect illumination (Table B) from total illumination (Table A). Complete Table C.

11 If the direct illumination required on any surface has a negative value, the luminance specification developed in steps 1 to 6 cannot be met because that surface is too brightly illuminated. In this case return to stage 1 and adjust the specification, either by altering reflectances of the room surfaces (step 1) or by choosing different luminance or illumination ratios (step 5). Changing the task, luminance or illumination alone will not help.

Table C

Surface	1 Total illumination[a]	2 Indirect illumination[b]	1–2 Direct illumination required
Floor or working plane			
Walls			
Ceiling			

[a] From Table A.
[b] From Table B.

Stage 3: luminaire specification

12 Work out the ratio R:

$$R = \frac{\text{direct illumination on walls}}{\text{direct illumination on flloor or WP}}$$

13 If luminaires are to be suspended below the ceiling, calculate a new room index for this stage of the procedure, measuring height from the place of the fittings to the floor or the working plane. Otherwise use the room index found in step 7.

14 Select the appropriate BZ classification for the luminaires from Table 8.4 using the values of R and the room index from steps 12 and 13. Linear interpolation should be used for intermediate values of room index.

15 From Table 8.5 obtain the direct ratio d for this BZ classification and room index, and note its value once again; interpolate for intermediate values of room index.

16 Calculate the ratio C:

$$C = \frac{\text{direct illumination on ceiling}}{\text{direct illumination on floor or WP}}$$

17 Calculate the required flux fraction ratio:

$$F = Cx$$

Flux fraction ratio is the ratio between the amount of light emitted from the luminaire above the horizontal to the amount below the horizontal. Some manufacturers give this value directly; in other cases the flux fraction ratio is obtained by dividing the percentage of light emitted upwards by the percentage emitted downwards.

18 Choose a luminaire of the required BZ classification and flux fraction ratio from manufacturers' literature. Note the downward light output ratio (DLOR) given by the manufacturer.

Stage 4: number and layout of luminaires

19 Determine the total lamp flux required:

$$\text{Flux} = \frac{\text{area} \times \text{direct illumination on working plane}}{\text{direct ratio} \times \text{DLOR} \times \text{maintenance factor}}$$

A good average value for the maintenance factor is 0.8; for information on the more precise determination of maintenance factors see IES Technical Report 9, *Depreciation and maintenance of interior lighting*.

20 Establish the number of lamps required by dividing the total lamp flux obtained in step 19 by the light output of the type of lamp appropriate

Table 8.3 Luminance design calculations. (Reproduced by permission of the Illuminating Engineering Society from *IES Monograph 10*)

Room index, k_r	Multiplying factors			
	On to ceiling or floor, from floor or ceiling, $_{cf}M_{cf}$	On to ceiling or floor from walls, $_{cf}M_w$	On to walls from ceiling and floor, $_wM_{cf}$	On to walls from walls, $_wM_w$
0.500	0.1998	0.8002	0.2000	0.5999
0.525	0.2129	0.7871	0.2066	0.5868
0.550	0.2257	0.7743	0.2129	0.5742
0.575	0.2384	0.7616	0.2190	0.5621
0.600	0.2508	0.7492	0.2248	0.5505
0.625	0.2630	0.7370	0.2303	0.5394
0.650	0.2749	0.7251	0.2357	0.5287
0.675	0.2866	0.7134	0.2408	0.5184
0.700	0.2980	0.7020	0.2457	0.5086
0.725	0.3092	0.6908	0.2504	0.4991
0.750	0.3201	0.6799	0.2550	0.4900
0.775	0.3307	0.6693	0.2594	0.4813
0.800	0.3411	0.6589	0.2636	0.4728
0.825	0.3512	0.6488	0.2676	0.4647
0.850	0.3610	0.6390	0.2716	0.4569
0.875	0.3707	0.6293	0.2753	0.4493
0.900	0.3800	0.6200	0.2790	0.4420
0.925	0.3892	0.6108	0.2825	0.4350
0.950	0.3981	0.6019	0.2859	0.4282
0.975	0.4068	0.5932	0.2892	0.4216
1.0	0.4153	0.5748	0.2924	0.4152
1.05	0.4316	0.5684	0.2984	0.4031
1.10	0.4471	0.5529	0.3041	0.3918
1.15	0.4618	0.5382	0.3095	0.3811
1.20	0.4758	0.5242	0.3145	0.3710
1.25	0.4892	0.5108	0.3192	0.3615
1.30	0.5020	0.4980	0.3237	0.3525
1.35	0.5141	0.4859	0.3280	0.3440
1.40	0.5257	0.4743	0.3320	0.3360
1.45	0.5368	0.4632	0.3358	0.3283
1.50	0.5474	0.4526	0.3395	0.3211
1.55	0.5575	0.4425	0.3429	0.3142
1.60	0.5672	0.4328	0.3462	0.3076
1.65	0.5765	0.4235	0.3494	0.3013
1.70	0.5854	0.4146	0.3524	0.2952
1.75	0.5940	0.4060	0.3553	0.2895
1.80	0.6022	0.3978	0.3580	0.2840
1.85	0.6101	0.3899	0.3607	0.2787
1.90	0.6177	0.3823	0.3632	0.2736
1.95	0.6250	0.3750	0.3656	0.2688
2.00	0.6320	0.3680	0.3680	0.2641
2.10	0.6454	0.3546	0.3724	0.2552
2.20	0.6577	0.3423	0.3765	0.2470
2.30	0.6693	0.3307	0.3803	0.2394
2.40	0.6801	0.3199	0.3839	0.2323
2.50	0.6902	0.3098	0.3872	0.2256
2.60	0.6998	0.3002	0.3903	0.2194
2.70	0.7087	0.2913	0.3933	0.2135
2.80	0.7171	0.2829	0.3960	0.2079
2.90	0.7251	0.2749	0.3987	0.2027
3.00	0.7326	0.2674	0.4011	0.1978
3.10	0.7397	0.2603	0.4035	0.1931
3.20	0.7464	0.2536	0.4057	0.1886
3.30	0.7528	0.2472	0.4078	0.1844
3.40	0.7589	0.2411	0.4098	0.1804
3.50	0.7647	0.2353	0.4117	0.1765
3.60	0.7702	0.2298	0.4136	0.1729
3.70	0.7755	0.2245	0.4153	0.1694
3.80	0.7805	0.2195	0.4170	0.1661
3.90	0.7854	0.2146	0.4186	0.1629
4.00	0.7900	0.2100	0.4201	0.1598
4.25	0.8006	0.1994	0.4237	0.1527
4.50	0.8103	0.1897	0.4269	0.1462
4.75	0.8190	0.1810	0.4298	0.1404
5.00	0.8270	0.1730	0.4325	0.1350
5.25	0.8343	0.1657	0.4350	0.1300
5.50	0.8410	0.1590	0.4373	0.1255
5.75	0.8472	0.1528	0.4394	0.1212
6.00	0.8529	0.1471	0.4413	0.1173

Table 8.4   Values of the ratio *R*

Classification	Associated S/H_m ratio	Ratio R for room index (k_f) of:									
		0.6	0.8	1.0	1.25	1.5	2.0	2.5	3.0	4.0	5.0
BZ1	1	0.242	0.223	0.212	0.202	0.196	0.189	0.185	0.182	0.179	0.177
BZ2	1	0.323	0.295	0.278	0.264	0.255	0.244	0.237	0.233	0.228	0.224
BZ3	1.25	a	0.362	0.344	0.329	0.319	0.307	0.299	0.294	0.288	0.284
BZ4	1.25	a	0.457	0.432	0.413	0.400	0.384	0.375	0.368	0.360	0.355
BZ5	1.5	a	0.530	0.512	0.496	0.484	0.471	0.462	0.457	0.451	0.448
BZ6	1.5	a	0.780	0.760	0.746	0.738	0.733	0.734	0.739	0.751	0.764
BZ7	1.5	a	1.006	0.977	0.958	0.947	0.942	0.943	0.951	0.967	0.987
BZ8	1.5	a	1.213	1.170	1.140	1.122	1.110	1.108	1.115	1.135	1.155
BZ9	1.5	a	1.568	1.435	1.379	1.352	1.308	1.293	1.289	1.296	1.312
BZ10	1.5	a	2.345	2.048	1.854	1.743	1.627	1.573	1.547	1.531	1.535

[a] Room index 0.6 impossible with standard luminaire spacing for classifications BZ3 to 10.

Table 8.5   Direct ratios for BZ classifications

Classification	Associated S/H_m ratio	Ratio R for room index (k_f) of:									
		0.6	0.8	1.0	1.25	1.5	2.0	2.5	3.0	4.0	5.0
BZ1	1	0.553	0.642	0.703	0.755	0.792	0.841	0.871	0.892	0.918	0.934
BZ2	1	0.481	0.576	0.643	0.703	0.746	0.804	0.840	0.866	0.898	0.918
BZ3	1.25	a	0.525	0.593	0.655	0.702	0.765	0.807	0.836	0.874	0.898
BZ4	1.25	a	0.467	0.536	0.602	0.652	0.722	0.769	0.803	0.847	0.876
BZ5	1.5	a	0.430	0.494	0.558	0.608	0.680	0.730	0.766	0.816	0.848
BZ6	1.5	a	0.339	0.397	0.456	0.504	0.577	0.630	0.670	0.727	0.766
BZ7	1.5	a	0.284	0.338	0.395	0.442	0.515	0.570	0.612	0.674	0.717
BZ8	1.5	a	0.248	0.299	0.354	0.401	0.474	0.530	0.574	0.638	0.684
BZ9	1.5	a	0.203	0.258	0.312	0.357	0.433	0.492	0.538	0.607	0.656
BZ10	1.5	a	0.146	0.196	0.252	0.301	0.381	0.443	0.492	0.566	0.620

[a] Room index 0.6 impossible with standard luminaire spacing for classifications BZ3 to 10.

Table 8.6   Lamp flux: 'average through life' lumen output (typical values at 240 V and 25 °C)

5 ft 0 in, 80 W fluorescent tubes		Incandescent lamps			
Lamp	Output (lm)	Wattage	Output (lm)	Wattage	Output (lm)
Northlight	3 250	40	330	300	4 400
Daylight	4 500	60	600	500	8 000
Natural	3 500	100	1 200	750	12 800
White	5 000	150	2 000	1000	18 000
Warm-white	5 000	200	2 750	1500	28 500

to the luminaire chosen. (Use the lighting design lumen figures given by the lamp manufacturer.) Hence obtain the number of fittings required.

21 Sketch out an acceptable layout of luminaires. The number may have to be increased to give a pattern suited to the space.

22 Check the spacing/mounting height ratios for the proposed layout and ensure they do not exceed the values given for the appropriate BZ classification in Table 8.6. If they do, the layout should be modified until satisfactory spacing/mounting height ratios are obtained. Failure to do so will cause excessive variation in the illumination on the working plane.

23 Carry out a glare check on the proposed installation, using either the Mears circular calculator or the tables in IES Technical Report 10. Ensure the glare index for the proposed

installation does not exceed the limiting glare index given in BS 8026 for the task being performed in the space.

Luminance design: worked example

Required

A general lighting scheme for a computer room. No special requirement for local concentrations of light.

Room dimensions

12.0 m × 6.0 m × 3.0 m high. The working plane is taken as 0.85 m above floor level.

Stage 1: development of luminance specification

Step 1

Preferred surface finishes, colours and reflectances are entered in a table.

Step 2

The task may be taken as reading and writing, and for task reflectance we take 0.8 (white paper).

Step 3

From BS 8026, task illumination is to be 600 lx.

Step 4

Therefore

Task luminance = 600 × 0.8 = 480 apostilb

In the example, the working plane is of more interest than the floor. Before going on to step 5, we therefore work out the floor cavity reflectance and enter it in Table A. Assuming that worktops, etc., account for one-third of the total floor area, the floor cavity reflectance will be

$$(\tfrac{1}{3} \times 0.45) + (\tfrac{2}{3} \times 0.30) - 0.10 = 0.15 + 0.20 - 0.10$$
$$= 0.25$$

Step 5

Using Bodmann's luminance ratios, the luminance of both walls and ceiling should be 0.3 times the task luminance. We therefore enter 0.3 × 480 = 144 apostilb in column 5 of Table A in each case.

Obviously the illumination on the work surfaces and on the working plane will be the same as that on the task, since we have assumed that all three are at the same height above the floor. The respective luminances are therefore:

Immediate surround   0.45 × 600 = 270 apostilb
Working plane   0.25 × 600 = 150 apostilb

Enter these figures in column 5 of Table A. In this case the luminance ratio for the immediate surround is 270/480 = 0.56; this is higher than Bodmann's recommendation, but would still be quite acceptable. (The ratio could be reduced by reducing the reflectance of the work surfaces, and if this were done the floor cavity reflectance would also be reduced.)

Step 6

Calculate the illumination on walls and ceiling, and complete Table A. For the walls:

Illumination = luminance/reflectance
             = 144/0.5 = 288 lx

And for the ceiling:

Illumination = 144/0.7 = 206 lx

The calculations in stage 2 will use only the luminances of the ceiling, walls and working plane, and we do not need to complete the line in Table A dealing with the floor. The completed Table A is shown in step 6.

Step 1

Surface	Finish	Colour	Reflection factor	Illumination	Luminance
Task					
Immediate surround	Work surfaces	Light teak	0.45		
Floor	Semtex tiles	Roman grey	0.30		
Working plane					
Walls	Plaster	Pale blue	0.50		
Ceiling	Acoustic tiles	White	0.70		

Step 5

Surface	Finish	Colour	Reflection factor	Illumination	Luminance
Task	Paper	White	0.80	600	480
Immediate surround	Work surfaces	Light teak	0.45		
Floor	Semtex tiles	Roman grey	0.30		
Working plane			0.25		
Walls	Plaster	Pale blue	0.50		144
Ceiling	Acoustic tiles	White	0.70		144

Step 6

Surface	Finish	Colour	Reflection factor	Illumination	Luminance
Task	Paper	White	0.80	600	480
Immediate surround	Work surfaces	Light teak	0.45	600	270
Floor	Semtex tiles	Roman grey	0.30		
Working plane			0.25	600	150
Walls	Plaster	Pale blue	0.50	288	144
Ceiling	Acoustic tiles	White	0.70	206	144

Stage 2: direct and indirect illumination

Step 7

Since the working plane is 0.85 m above floor level, we take the height as (3.00 − 0.85) = 2.15 m. Therefore

$$\text{Room index} = \frac{12 \times 6}{(12 + 6) \times 2.15} = 1.86$$

Step 8

Consider first the multiplying factor for indirect illumination on the ceiling due to reflection from the floor. From Table 8.3, this factor is

0.6101 for room index 1.85
0.6177 for room index 1.90

By linear interpolation, the factor for room index 1.86 is

$$0.6101 + \frac{1.86 - 1.85}{1.90 - 1.85} \times (0.6177 - 0.6101)$$

$$= 0.6101 + \frac{0.01}{0.05} \times 0.0076 = 0.6101 + 0.0015$$

$$= 0.6116$$

Using a similar interpolation process for the other factors, and taking luminances from Table A, we have the following entries in Table B. As a check on the multiplying factors, note that the column totals are 1.0000, 1.0001 and 1.0000 respectively.

Step 9

The indirect illumination on the working plane due to reflection from the walls is

Step 8

Light coming from	Luminance	Light falling on					
		FWP		Walls		Ceiling	
		MF	SL	MF	SL	MF	SL
FWP	150			0.3612		0.6116	
Walls	144	0.3884		0.2777		0.3884	
Ceiling	144	0.6116		0.3612			
Total indirect illumination on each surface							

Step 9

Light coming from	Luminance	Light falling on					
		FWP		Walls		Ceiling	
		MF	SL	MF	SL	MF	SL
FWP	150			0.3612	54	0.6116	92
Walls	144	0.3884	56	0.2777	40	0.3884	56
Ceiling	144	0.6116	88	0.3612	52		
Total indirect illumination on each surface			144		146		148

144 (wall luminance) $\times$ 0.3884 (MF) = 56 lx

and so on. Complete Table B.

Step 10
Now complete Table C.

Step 10

Surface	Total illumination	Indirect illumination	Direct illumination required
Floor or working plane	600	144	456
Walls	288	146	142
Ceiling	206	148	58

Step 11
None of the direct illumination values is negative, so the luminance specification derived in step 5 is feasible. Since some direct illumination is required on the ceiling, we shall use suspended rather than recessed light fittings.

Stage 3: luminaire specification

Step 12
Ratio $R = 142/456 = 0.311$ (using direct illumination values from Table C).

Step 13
Suppose the luminaires are suspended 0.30 in below the ceiling. Then the new room index for stage 3 will be

$$\frac{12 \times 6}{(12 + 6) \times (2.15 - 0.30)} = 2.16$$

Step 14
From Table 8.4 we have the following values of R.

Room index	2.0	2.5
BZ3	0.307	0.299
BZ4	0.384	0.375

So by linear interpolation, for room index 2.16, the R values are

$$\text{BZ3:} \quad 0.307 - \left[\frac{2.16 - 2.0}{2.5 - 2.0} \times (0.307 - 0.299)\right]$$

$$= 0.307 - \left[\frac{0.16}{0.5} \times 0.008\right] = 0.304$$

$$\text{BZ4:} \quad 0.384 - \left[\frac{2.16 - 2.0}{2.5 - 2.0} \times (0.384 - 0.375)\right]$$

$$= 0.384 - \left[\frac{0.16}{0.5} \times 0.009\right] = 0.381$$

Clearly, a BZ3 luminaire is closest to the requirement.

Step 15
By linear interpolation in Table 8.5 the direct ratio of a BZ3 fitting at room index 2.16 is

$$0.765 + \frac{2.16 - 2.0}{2.5 - 2.0} \times (0.807 - 0.765) = 0.778$$

Step 16
Using direct illumination values from Table C, we get

Ratio $C = 58/456 = 0.127$

Step 17
Flux fraction ratio F is given by

$$F = Cd$$
$$= 0.127 \times 0.778 = 0.099$$

Step 18
Select a surface-mounting BZ3 luminaire with flux fraction ratio about 0.1 from manufacturer's catalogue. The computer room will be air-conditioned and fluorescent lamps will be preferred so as to keep down the heat gain. The DLOR of a suitable luminaire is found to be about 0.40.

Stage 4: number and layout of luminaires

Step 19
The total lamp flux required is

$$\frac{(12 \times 6) \times 456}{0.778 \times 0.4 \times 0.8} = 132{,}000 \text{ lm}$$

Step 20

Lighting design lumen figures for high efficiency fluorescent tubes are

1.8 m, 85 W	5550 lm
1.5 m, 65 W	4400 lm
1.2 m, 40 W	2600 lm

Dividing these figures into the total lamp flux we find that we require 24 of the 1.8 m, 85 W tubes, or 30 of the 1.5 m, 65 W tubes, or 51 of the 1.2 m, 40 W tubes to meet the specification. Assuming that twin-tube luminaires are used, we have a choice between 12 of the 1.8 m, 15 of the 1.5 m or 26 or the 1.2 m luminaires.

Steps 21 and 22

Consider first the use of twelve 1.8 m, 85 W twin luminaires; the natural arrangement will be three rows with four luminaires in each. Mounting height is 1.85 m above the working plane (step 13); the spacing between rows will be 2.00 m. This gives a spacing/mounting height ratio of 1.08 compared with a maximum of 1.25 for a BZ3 luminaire (Table 8.4 or 8.5) and is therefore satisfactory.

Each fitting will be about 1.90 m long, and the clear space between luminaires in the same row will therefore be 1.10 m. The value of the spacing along rows is taken from the centre of one luminaire to the centre of the next, and is accordingly 3.00 m. This results in a spacing/mounting height ratio of 3.00/1.62, which is excessive, and we therefore reject the idea of using 1.8 m luminaires.

Consider instead the use of fifteen 1.5 m, 65 W luminaires, arranged in three rows of five. The spacing between rows remains the same as before and is satisfactory. Each luminaire is 1.60 m long, and there is about 0.8 m clear space between them. The spacing/mounting height ratio along the rows is therefore (1.60 + 0.80)/1.85 = 1.30. This is close to the limiting value, and by reducing the clear space between luminaires to 0.70 m we can get a satisfactory spacing/mounting height ratio of 1.24.

It would also be desirable to increase the spacing between rows slightly by moving the outer rows towards the walls. With 2.20 m between luminaires, a ratio of 1.19 would be obtained.

Step 23

Carrying out a glare check by use of the Mears circular calculator gives a final glare index of about 14. A more detailed calculation using tables in IES Technical Report 10 is as follows:

Initial glare index	19.9
Correction for downward flux	+3.2
Correction for luminous area	−6.1
Correction for mounting height	−0.8
Correction for 0.3 floor reflectance	−2.8
Final glare index	13.4

BS 8206 sets a limiting glare of 19 for computer rooms, so that the proposed installation is completely satisfactory.

UTILITY SERVICES

9 Water supply

9.1 General standards

Water supply installations are governed by bye-laws made by water authorities and statutory water companies. The government publishes model bye-laws to ensure that, wherever it is appropriate, the bye-laws of individual water undertakings are uniform. Unnecessary variations lead to confusion and expense. In 1986 a new edition of the Model Water Bye-laws was published by the Department of the Environment.

The most significant technical changes in the 1986 edition were new requirements for backsiphonage, cross-connections, cold-water storage, water economy and pipe accessibility. A very important organizational change was the recommendation that all water authorities should make and advertise their bye-laws at the same time. Independent publication at different times could result in anomalies in the pattern of water supply legislation.

It would be difficult for individual designers and installers to evaluate whether particular fittings complied with the bye-law requirements. The *Water Fittings and Materials Directory* is published by the Water Bye-laws Advisory Service and gives a list of tested and approved items. A vital guide to all aspects of the subject is British Standard 6700 *Design, installation, testing and maintenance of services supply water for domestic use within buildings and their curtilages.*

Traditionally the Model Water Bye-laws governed hot-water installations and the Building Regulations up to and including 1976 did not address heating or hot-water installations except to ensure adequate provision for air supply, flues and hearths. The introduction of pressurized hot-water storage systems in 1985 caused a problem since the enabling legislation for water supply refers only to 'waste, undue consumptions, misuse or contamination'. Safety features could not be included in the Model Water Bye-laws. The Building Regulations, however, can deal with safety and the 1986 regulations included safety requirements for unvented hot-water storage. A similar situation arises with energy conservation. As a result, the legal performance requirements for hot-water installations are now divided between the Model Water Bye-laws and the Building Regulations.

9.2 Private water supplies

Although public utility water mains can serve the majority of buildings, it is sometimes necessary for private water supply arrangements to be made. Expert advice from specialist firms can be obtained, and the local authority public health department will require to be satisfied as to the suitability of the supply.

It is important to note that, where a private source water supply has to be provided, it is usually also necessary to make private arrangements to dispose of sewage, and care must be taken to avoid pollution of the water source.

For human consumption it is best that the water comes straight from the ground, rather than from a stream or pond which is exposed to probable pollution. A dug well of big enough diameter to admit a person carrying a spade, or a borehole of small dimensions made by mechanical means, and just big enough to admit the necessary pump or suction pipe are possible methods. In either of these the upper part should be lined to exclude surface water, as this is liable to be polluted. It is best to leave the lower part unlined if it will stand safely without collapsing, but lining is usually needed through clay zones, and some form of porous or perforated liner may be needed in water-bearing strata which will not stand up without support.

Most modern wells are lined with precast concrete cylinders, and boreholes with steel tubing.

When a well is being dug the water level must be held down by pumping to enable the well-digger to work. The excavated material is shovelled into a bucket and hauled out with a rope. Great care must be taken to ensure the safety of the digger both from falling objects and from collapsing sides of the excavation. Boring operations do not involve dewatering. For most modern requirements boreholes are the most practical and convenient, but dug wells are useful in situations where the water-bearing strata do not yield water freely, and where the larger peripheral area of the well is therefore an advantage.

Dug wells for domestic or farm use are commonly 1.0–1.5 m in diameter, and boreholes 150–200 mm. For small requirements where the strata are coarse-grained gravel or soft, fissured rock, driven tube-wells are often successful. These are usually of steel tube, 30–50 mm in diameter, screwed together and having a point and perforations at the lower end.

The usual consumption of water is approximately as follows:

- 0.09–0.18 m^3 per day per person
- 0.36–0.45 m^3 per day per ordinary house
- 0.14 m^3 per day per dairy cow
- 0.05 m^3 per day per dry cow

For market-garden and pasture irrigation in season, 12 mm of 'rain' per week is generally considered sufficient provision.

Figure 9.1 shows how rain soaks into porous ground and finds its way through layers of rock or sand to feed springs, wells, boreholes and rivers. Underground 'rivers' are very unusual, and the water ordinarily lies in myriads of tiny crevices in rock layers and between particles of sand. This forms a vast underground store of water, rather like a tank filled with both water and pebbles. Usually there is a movement of water through these cavities to an outlet. A simple example of this is a gravel-capped hilltop overlying clay: springs flow at the junction with the clay, and water may be obtained from wells dug in the gravel down to the face of the clay. If the catchment area is small, and the gravel clean and readily porous, the water will run out quickly and the springs and wells will fail in prolonged dry weather. If the area is extensive and the gravel interspersed with fine-grained material, the rate of flow will be diminished and springs and wells will continue to yield small supplies even in dry times. The same things happen in the more complex conditions shown in the later diagrams. An *artesian* borehole is one which pierces a layer of clay and enters a lower porous zone

from which water rises as a *gusher* above ground level. A similar borehole where water rises part way but does not reach the surface is called *subartesian*.

9.3 Water conservation

Economy in the use of potable water has always been a key factor in the design of water supply installations. The siphon type of flushing cistern used in the United Kingdom, traditionally called a water-waste preventer, was developed and required to be used in order to limit waste. Prevention of waste is a main purpose of the UK Model Bye-laws for Water Supply. In recent years population increases and increasing standards of hygiene and equipment have led to major concerns about water economy and to studies of new ways for conservation. This is the case both in countries where water is scarce and, in those such as the UK, where rainfall is more than adequate but the provision of catchment, storage and distribution facilities present economic problems. There are several ways that offer promise of savings.

Metering

Some countries have adopted the philosophy that charges for domestic water supply should not be directly related to consumption since this might limit standards of hygiene. Where high standards of hygiene are conventional, it appears that metering of domestic supplies can reduce consumption without significant effect on health.

WCs

Supplies to WCs are the main items in water consumption. The quantities of water used at each flush vary from country to country. In the United States, flushing valves are often intended to give a 5 gallon (22 litres) flush. In the United Kingdom, 3 gallon (13 litres) flushing cisterns were widely used at one time. Nine-litre cisterns are now standard and, for increased water economy a dual-flush facility can be provided which, if no solids are present, can be operated to give a reduced 4.5 litres flush. The effectiveness of this measure depends on the users. Experience in housing areas equipped with this type of flushing cistern indicates that savings of 10% of the overall water consumption can be achieved.

In Scandinavia, in particular, WCs are in successful use with flushes of 6 litres and less, and there is every reason to believe that such designs will be used elsewhere.

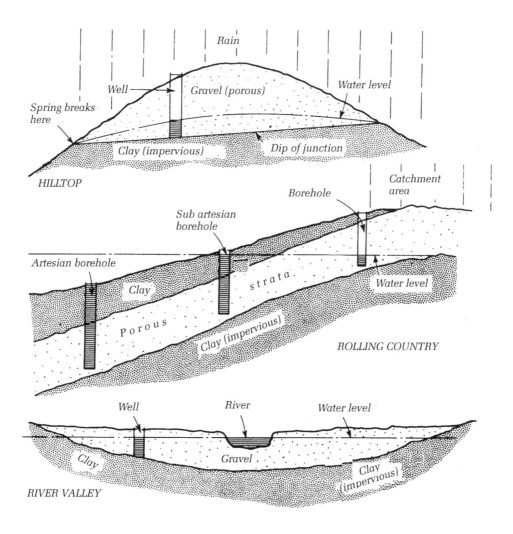

Figure 9.1 Wells and boreholes for private water supplies.

Showers

Showers, particularly those delivering a fine spray, offer great economy of both water and energy when compared to baths. They can also save space and are safer in use than baths.

Urinals

The regular automatic flushing of urinal stalls consumes substantial quantities of water. The typical automatic flushing cistern delivers 5 litres per stall once every 20 min throughout the 24 h, whether the urinals are in use or not. Many devices are now being used to limit the amount of flushing. They include electronic and pressure-operated systems for preventing flushing during periods when no use is made of the urinal.

The most dramatic development is the waterless urinal. Urinals are normally regarded as major sources of smell, requiring frequent flushing to control this. Recent experiments with bowl-type urinals have, however, demonstrated that it is possible for this type of appliance to perform satisfactorily without flushing. In experimental installations no trouble has been experienced from smell and the requirements for cleaning of traps and wastes have not exceeded those of conventional installations. Copper traps do not resist the corrosion of the undiluted flow, but plastic ones have performed satisfactorily.

Non-potable supplies

One anomaly of present water supply systems is that all the water supplied has to be drinkable but only a very small proportion is actually used for drinking or culinary purposes. Although in principle it seems uneconomic to provide pure water for all purposes, no practical and cost-effective system has yet been proposed to supersede the present methods.

9.4 Pumps: the three main types

There are very many types of pumps; broadly they fall into three main categories:

- *Displacement pumps*
  One or more plungers move backwards and forwards or up and down, each drawing water past a valve into a cylinder then forcing it out past another valve.
- *Rotary pumps*
  Meshing 'gear' wheels, or a rotor and a stator, are pressed together in such a way that water is pocketed, passed through the pump body, and forced out as the meshing elements meet.
- *Centrifugal pumps*
  A rotating impeller flings water to its periphery, where the kinetic energy is converted to pressure energy by a volute or set of diffuser vanes.

All these types are obtainable in sizes to meet almost every requirement, but each has its advantages and its limitations, and selection must take into consideration all aspects of the application – the height the water is to be elevated, the cleanliness of the water, the form of power available to work the pump, the amount of noise that can be tolerated, and the length of life required of the mechanism.

Most hand-operated pumps are of the displacement type, having a plunger which is lifted and lowered alternately, or pushed backwards and forwards in a cylinder. Another variety, called the *semi-rotary pump*, has a metal plate which fits diametrically across a round chamber and is oscillated with a backwards and forwards semi-rotary movement; the plate is fitted with valves, so is the chamber, which is divided so that the pump has the equivalent of two cylinders. This is a popular type for raising water from a well to an overhead tank.

Displacement pumps are also operated by windmills or waterwheels; gearing is provided to produce fairly slow speeds, usually between 15 and 50 reciprocating cycles per minute.

Windmills are erected on towers, usually 6–12 m high. To span times of no wind it is necessary to provide ample storage. The pump units can be small as they will be operating continuously when the wind is blowing.

The plungers in displacement pumps are fitted with *cups* made of leather or fabric, or they operate through packed glands. Usually these pumps give many years of reliable service; the cups or packing are renewed at intervals and the material costs are low.

Rotary pumps of the gear type are convenient and inexpensive, but rely on close metal-to-metal clearances, and if the water they handle is gritty they wear quickly and become noisy and inefficient. They are particularly useful for pumping oil and chemicals, where cleanliness is assured. They usually rotate at speeds of 1000 rpm or more.

Some metal-and-rubber rotaries are useful and reliable for small water-supply duties and have good self-priming characteristics. These pumps usually run at 1450 rpm – a useful electric motor speed – and are quiet in operation.

As both displacement and rotary pumps are positive in action, they give approximately constant volume at any fixed speed, irrespective of the pressure.

Centrifugal pumps usually run at 1450 or 2900 rpm and are often direct-coupled to electric motors. Ordinary centrifugal pumps have no ability to self-prime; the pipe system must be completely filled with water and all air must be released before the pump is started. But there are several devices which can be incorporated in the design to provide self-priming; most include circulation of part of the water from the delivery side, back to the suction side through a venturi tube. This draws water up the suction pipe and dispels the air in the system. Centrifugal pumps are not positive in action and the volume delivered varies widely at different pressures, even at constant speed.

Most of these centrifugal pumps can draw up water 4.5–6.0 m vertically; by special design as much as 9 m can be attained, but this is unusual (about 10 m is the theoretical maximum suction). The total head these pumps can produce depends on the peripheral speed of the impeller. Most small single-impeller pumps give total heads of 10–20 m. Larger single-stage pumps give up to 60 m. To attain greater heads several pump units are assembled in series on one spindle and are driven by the same motor.

Another type of centrifugal device is the *water-ring* pump. This has a different form of impeller which runs with close clearances between the facings of the casting sides; it pulls water in through one set of holes in the sides of the casting and forces it out through another. It is essentially a clean-water type of pump and has a relatively low power efficiency; but it is useful because it will draw water up about 8.5 m, has good self-priming characteristics, and gives total heads of 45–100 m per impeller stage when running at 2900 rpm. The volume does not vary with pressure changes as much as with a conventional centrifugal pump.

Most of these types of pumps can be made in shallow-well and deep-well forms. In every case the actual pumping unit must be either submerged or else located within the *suction limit* of the water level, for practical purposes not higher than about 4.5 m above water level. Two convenient forms of pump are now used in deep wells and boreholes:

- *The reciprocating displacement pump*
  This has its cylinder located at or below water level, suspended from ground level. Its operating rod is extended above ground and worked by a crank mechanism driven by an engine, electric motor or windmill.
- *The submersible electric pump*
  This has a special electric motor designed for working underwater. To its rotating shaft are attached one or more centrifugal-pump impellers. The water passes through each of these in series and extra pressure is added by each to build up sufficient for delivering water to the desired height. The pressure given by each impeller is a function of its peripheral velocity, hence, as most of these pumps are designed to go down relatively small boreholes, several impellers are generally needed. Pumps are available for raising over 1.3 l/s from 100 mm boreholes, and 25 l/s from 200 mm boreholes, and pressures up to 300 m are attained with standard pumps.

Three other special types of pump deserve mention; each is limited in its application.

The hydraulic ram

This device derives power from falling water. Water is taken from above a dam, piped to the ram at a lower level and discharged through a *pulse-valve* to waste below the dam. Water flowing through the ram washes the pulse-valve to a closed position, thus arresting the flow suddenly and causing a hydraulic shock; this shock jerks a small amount of water past a delivery valve into the delivery main. After the shock has passed the pulse-valve drops open and water flows again. This pulsing action is allowed to go on continuously day and night, and a hydraulic ram correctly installed can give trouble-free service for many years. Variations in design allow special types of ram to be operated by dirty water, and to pump clean water from a well or other pure source.

The jet-pump

This is a device for raising water from deep wells and boreholes. An ejector (or jet) is hung below water level and power water is delivered down a *drive-pipe* to it; this water enters the ejector near the bottom and is directed upwards through a nozzle. The water-jet from the nozzle rushes through a venturi throat and creates a vacuum, which draws in fresh water through a foot-valve below the device. This fresh water and the power water flow together up a riser-pipe to ground level. Generally, a centrifugal pump is fitted at ground level with its inlet connected to this riser-pipe: from its outlet some of the water goes to the reservoir and the rest is returned to the jet as power water. Correct proportioning of pressures, heads, volumes and pipe sizes is essential. This device is sometimes convenient, but is necessarily of low mechanical efficiency as power absorbed in pressurizing is about twice as much water as is delivered to the reservoir.

The airlift

This uses compressed air. The air is piped to a foot-piece below water level; the bubbles rise and carry water up with them. It is essential for the foot-piece to be adequately submerged, and about 50% submergence is common; that is to say, if the water level is 15 m below the discharge point, the foot-piece must be 15 m below water, making it 30 m below the discharge point. Airlifts cease to function below about 35% submergence. The higher the percentage submergence, the lower the volume of air needed to raise the air pressure. Correct proportioning of pipe sizes, depth of foot-piece, air volume and pressure is essential. Airlifts have a low power efficiency, but they are particularly useful in handling water containing sand, mud or raw sewage since it does not have to pass through any moving mechanism and wear and tear is limited to abrasion on the stationary pipe system.

9.5 Assessing water purity

It is frequently necessary to obtain information on the purity of a water supply to assess its suitability for drinking purposes, for steam-raising or other industrial processes where excessive hardness may be a disadvantage, or to forecast its action on metals such as galvanized tanks or pipes. Samples are usually collected in containers supplied by the analyst, with instructions for taking the sample.

The analyst's report will give a number of analytical figures together with an opinion on its suitability for the purpose in mind. The interpretation of the results will be influenced by the source of the water and other considerations but it will give the following figures.

Total solids

This will comprise both dissolved and suspended solids, organic and inorganic. Organic solids are mainly derived from animal and vegetable debris, inorganic solids from the earth. Water from chalk or from salt-bearing strata may contain over 1000 ppm, whereas water from upland sources may contain less than 50 ppm. (Note carefully the units employed, as it is customary nowadays to express results in parts per million, but some analysts still use parts per 100,000.)

Ammonia

More than 0.08 ppm of albuminoid ammonia in water from a shallow well, coupled with a similar amount of free and saline ammonia, usually indicates pollution by sewage, but the same amount of free and saline ammonia in water from a deep well would be due to another cause.

Oxygen absorbed

This is a measure of the amount of organic matter in a water, mainly derived from the decay of animal and vegetable matter. Generally speaking, the figure for a very pure water would be less than 0.5 ppm, up to 1.5 ppm would be reasonably pure, and anything over 2.0 ppm would be classed as impure, but these figures may be doubled if the water is surface water from an upland source.

Nitrates

More than 5 ppm of nitrates indicates that the source may be liable to pollution, although at the time of sampling it may be satisfactory.

Nitrites

The presence of nitrites normally indicates recent sewage pollution.

Chlorine

The combined chlorine is an indication of the salinity of the water. A very salty water is unpalatable. Chlorides may also be derived from urine.

Hardness

There are two kinds: hardness removed by boiling is termed *temporary*, otherwise it is called *permanent*.

Temporary hardness is deposited as fur in pipes and boilers, unless the water is softened or steps are taken to prevent its deposition. Both kinds of hardness are due mainly to salts of calcium and magnesium, and the hardness of water is usually measured in degrees Clark. Waters usually vary from 0 to 50 degrees of hardness; anything over 20 is hard, less than 10 would be regarded as soft. On this basis a soft water would not be likely to cause serious deposition of fur in 20 years.

pH

This measures the acidity or alkalinity of the water. Soft acid waters are derived from hard insoluble rocks or from peaty upland, they have a pH less than 7.0, and may corrode pipes and tanks unless passed through a cylinder packed with limestone to neutralize the acidity. Waters with a pH of more than 7.0 are alkaline and are less likely to attack metals.

Water may also contain bacteria, some derived from soil. These soil bacteria grow at ordinary temperatures and normally do not cause disease, although their presence in very large numbers is not desirable, and those which normally flourish in the intestines of animals, which grow at blood heat, may include disease-producing species. Disease-producing bacteria are called *coliform* and their presence indicates pollution of animal origin.

If the water is intended for drinking purposes, for use in a dairy or in the manufacture or preparation of foodstuffs, a bacteriological examination is necessary, in addition to the chemical analysis. Samples are taken in sterile bottles supplied by the analyst.

Treatment

The following processes are commonly used to treat water for use in buildings.

Filtration

This is normally done through sand, either by simple percolation (slow filter) or under pressure (rapid filter). It may be necessary to add a small amount of some chemical, such as alumina, to assist filtration. The filters become clogged in time, and have to be washed back with clean water. The process removes suspended matter from the water.

For water which is to be used for baths, flushing WCs, etc., no further processing will be required in many cases. Figure 9.2 shows a simple filtration and storage unit from which water can be pumped up to a storage cistern in the building itself.

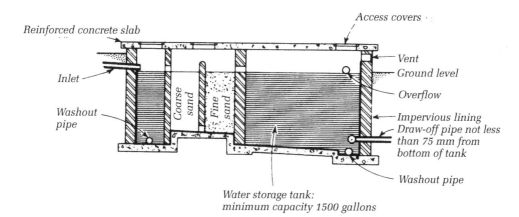

Reinforced concrete slab

Access covers

Inlet

Washout pipe

Coarse sand

Fine sand

Vent
Ground level
Overflow
Impervious lining
Draw-off pipe not less than 75 mm from bottom of tank
Washout pipe

Water storage tank: minimum capacity 1500 gallons

Figure 9.2   Simple filter and storage cistern for small private water supply installation.

Sterilization

Before water can be consumed by humans, it must be sterilized. In large installations and in public supplies chlorine is added to the water by a special plant. In small installations this is not feasible. A possible solution is the sterilizing filter, a very fine filter that can be cleaned. The filter is impregnated with silver, which has a bactericidal effect. Figure 9.3 shows such a filter. It is usually used at the kitchen sink to give additionally filtered and sterilized water at this point while the rest of the installation delivers the water from the simple filter described above.

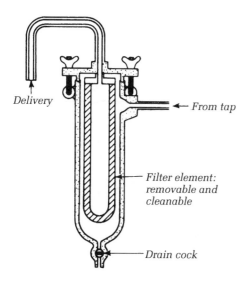

Delivery

From tap

Filter element: removable and cleanable

Drain cock

Figure 9.3   Sterilizing filter for producing potable water in a private water supply installation.

Softening

Normal hard water is not a risk to health but has a number of disadvantages. In hot-water pipes and boilers scale may be deposited and, in domestic use, scum is deposited and soap does not lather well. Some industrial processes may be affected. There are several traditional ways in which water may be softened: base exchange, lime-soda and inhibitors.

Domestic water softeners belong to the first type. The water is passed through a medium called a *zeolite*, which converts the calcium salts in the water to sodium salts. From time to time the zeolite medium requires regeneration by passing strong brine through it. A modern development of this process employs synthetic resins, etc., in place of the natural zeolites, and water of comparable purity to distilled water can be obtained for certain industrial processes where water of high purity is necessary. This is expensive and unnecessary for domestic water.

For industrial plant, water may be softened by the addition of lime to remove temporary hardness, or with lime and soda to remove both temporary and permanent. The process may be a batch process, whereby tanks of water are treated with the required amount of lime or lime-soda, the sludge allowed to settle and the water run off to use; or the addition may be made continuously by a special plant. In either case the disposal of large volumes of sludge is necessary.

To prevent the deposition of fur in boilers by hard waters, certain complex phosphates (such as Calgon) may be added, which prevent the formation of scale. These compounds are growing in number and can be regarded as inhibitors rather than true softeners, although the effect may be the same.

9.6 Mains water supply

The vast majority of buildings take their water from public water supplies. This supply will be suitable for drinking, although in some areas it may be hard and require softening for some uses, and in others the use of particular pipe materials may be prohibited because of rapid corrosion due to the nature of the water. The adequacy of a mains supply will depend on the size of the water mains, the pressure of water in them (expressed in terms of head, i.e. the height to which the water in the main would rise in a vertical pipe) and the demand on the main. The water undertaking should be consulted about a supply both to the building and for the building operation at a very early stage.

The desirable minimum size for water mains is governed by firefighting requirements. A 75 mm diameter pipe fed from both ends and a 100 mm diameter pipe fed from one end are considered satisfactory. In towns a grid of pipes served by two trunk mains enables supplies to be maintained even when individual sections have to be shut off. Figure 9.4 shows a typical arrangement.* An adequate head of water in the main is necessary to provide an adequate flow and to raise water to the top of buildings. A minimum head of 30 m is desirable for firefighting

* Supplies to isolated buildings will normally be by a single pipe.

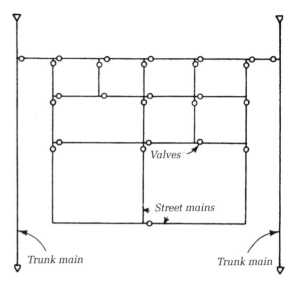

Figure 9.4   Typical urban water main grid. Two trunk mains supply the grid and any pipe can be taken out of service without cutting off the supply to the others.*

purposes, and a maximum head of 70 m is thought appropriate to limit wastage and noise in pipework. The head is provided by siting service reservoirs at suitable heights above the buildings being served. Figure 9.5 shows the principles involved. It also shows

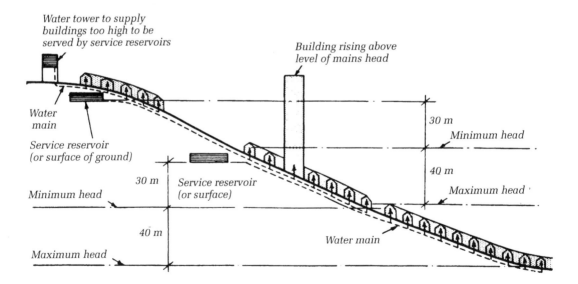

Figure 9.5   Principles for distributing water from service reservoirs. The maximum and minimum heads of water are achieved by siting the reservoir at a level higher than the building it serves. In practice the shape of the ground may make it difficult to achieve close compliance. Friction in the pipework will cause pressures to fluctuate while water is being drawn off. The water supply to the tall building will have to be pumped.

a water tower built to give a supply to buildings which cannot be served from a normal service reservoir, and a building rising above the mains head which will require its own system of pumps to raise water to the top. It is clearly not always possible to arrange service reservoirs in an ideal relationship to the buildings they supply, and although it will be rare to find heads of over 70 m, many areas will have less than 30 m. The full head of water from the reservoir is only available when no flow is taking place in the mains. The actual head at a given point will normally be reduced by the friction losses due to flow. This will vary with demand from day to day and from time to time. Water undertakings carry out a continuous programme of pressure measurement and will usually be able to give information about the pressure variations to be anticipated at any point.

Although a position beneath the footpath has often been advocated, water mains, in common with other main services, are usually to be found under the carriageway. Connection of the main to the site is almost invariably made by the water supply undertaking which has the right to dig up the road for this purpose. An ingenious box-like mechanism bolted to the main enables the connection to be made without interrupting flow in the main itself. Figure 9.6 shows the general arrangement for a typical domestic connection, where the communication pipe and the service pipe would be 12 mm in diameter. Note the

minimum cover of 750 mm below ground for the service pipe to avoid risk of freezing. The normal maximum depth is 1350 mm. This part of the work is normally carried out at the commencement of the building contract so that a supply of water for buildings can be obtained by the contractor. A larger diameter may be required in large buildings or where sanitary appliances are supplied directly from the service pipe. In some cases, where street hydrants are not close enough to the building, fire brigade hydrants will be required on the building site, necessitating a 100 mm diameter service pipe.

Traditionally in the United Kingdom the charges for water supplied to dwellings were based on a proportion of the rateable value. Charging on a basis of usage was resisted as a tax upon hygiene. With rapidly increasing use and limited supplies this approach is changing and universal metering seems to be an inevitable development. For dwellings the water meter is a small unit in the rising main after it emerges into the dwelling (Fig. 9.7).

In industrial buildings metering has been conventional. The meter is usually in a pit near the boundary. On large sites where the service pipe also serves fire hydrants it may well have a diameter of 100 mm. In this case the meter can be installed in a bypass of size appropriate to normal usage. The fire brigade will open the valve to bring the large-diameter pipe into use when required.

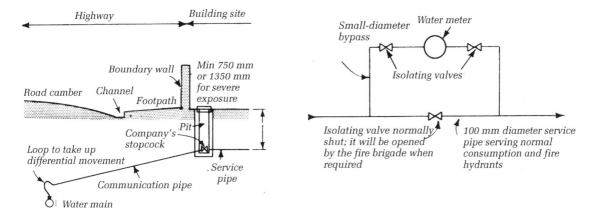

Work up to and including the stop cock is normally carried out by the water supplier

Figure 9.6 Typical details of the communication pipe connection from the town water main and the service pipe on the site. This part of the installation is often carried out in the early stages of building to provide a supply for construction.

Figure 9.7 How to install a water meter having an appropriate size for normal consumption and where the service pipe has to have a diameter of 100 mm for firefighting.

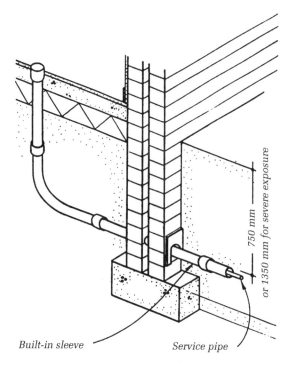

Built-in sleeve   *Service pipe*

750 mm
or 1350 mm for severe exposure

Figure 9.8   Entry for water service pipe.

9.7 Cold water: storage and distribution

The pipework from the company's stopcock onwards will usually be installed after the carcase of the building is complete. Figure 9.8 shows a pipe sleeve built into the foundation and groundfloor slab to allow the flexible service pipe to be installed when the fabric is complete. Provision for the service pipe to enter 750 mm below ground and to rise inside away from the external wall is usually made by building in drainpipes at the time of laying the foundations and groundfloor slab. Some features of pipe distribution within the building are common to all installations since they deal with avoidance of damage from frost. Water supply pipes may not be built into external walls and should not pass through or near the eaves where cold draughts are difficult to avoid and if they are not within the heated volume of the building pipes should be lagged.

A stopcock and a drain cock to enable the installation to be shut off and emptied must be provided immediately the service pipe enters the building, and drain cocks to empty any section of pipe not drained by the main drain cock or by draw-off points. The Water Supply Bye-laws now require that

no pipe should be embedded or built in such a way that it cannot readily be accessed.

Apart from these considerations the size and pattern of distribution of pipework will depend very much on the provision for water storage. Not all countries require water storage and it is possible in some parts of the world to have all parts of the water supply installation directly under mains pressure. In Britain, until very recently, hot-water supply installations (except instantaneous heaters) could not be subject to mains pressure. In the south and east, where water resources are relatively limited, it was usual to supply all appliances from storage except those supplying drinking water. The New Model Water Bye-laws permit both hot- and cold-water systems to be subject to mains pressure, provided proper precautions are taken to prevent contamination or the risk of explosion.

The main advantages of water storage are as follows:

- It provides a reserve against failure of the mains supply.
- Sudden demands are met from the storage cistern which then fills slowly, thereby making the demand on the main more even. This gives
  - economy of water mains and in the size of service pipe
  - reduced possibility of mains pressure dropping to nothing, which could lead to backsiphonage of water from sanitary appliances into the main
- Reduced pressure on the installation which minimizes noise and wastage and enables appliances (e.g. hot-water cylinders of reasonable gauge) to be used.
- Heating and hot-water supply apparatus can be vented to the storage cistern, thereby minimizing safety-valve requirements.

And the disadvantages:

- Space and support must be provided for the storage cistern. In high buildings a proportion of storage at ground level is becoming usual to avoid loads on the building and to save space at high level.
- Storage cisterns may become dirty, particularly if not provided with a cover. Drinking water is, however, supplied direct from the main.
- The reduced pressure means that distributing pipes have to be larger.

Economy

If the storage cistern can be accommodated without expense (say in the roof space) and the service pipe from the main is long in relation to the distributing pipes, a system employing a small-diameter service pipe and storage may be the more economical.

Noise

The reduced pressure from storage cisterns as compared with mains pressure will usually give quieter operation, reduce water hammer and particularly ball-valve noise. (Substantial water storage is not essential to achieve these objects which could be accomplished by a small 'break tank', but the two are normally associated.)

Reserve for mains failure

Some areas are subject to gross pressure fluctuations and even the cessation of flow at times during normal operation. Most places enjoy very consistent supply. The consequences of mains failure must, however, be taken into account. In some industrial situations provision of water storage could form a very worthwhile insurance against disruption of manufacturing processes.

Tables 11.1 and 11.2 give standards for water storage and sizes of typical cisterns. For domestic premises $0.23\,m^3$ actual capacity of storage is required for hot and cold and $0.12\,m^3$ actual capacity for hot-water supply only.

Domestic water supply installations

Figures 9.9, 9.10 and 9.11 show a typical domestic water supply installation. Figure 9.9 shows the service pipe (subject to mains pressure) serving the storage cistern and kitchen sink with stopcock and drain cocks. Figure 9.10 shows the cold-water distributing pipes (pipes subject to pressure from cistern only) and Fig. 9.11 the hot-water distributing pipes. It also shows the flow and return pipes to the hot-water cylinder that would be required to achieve hot-water supply from a domestic boiler. Where water storage is employed only for hot water, it is possible to supply all the cold taps from the service pipe. This may save on pipework, particularly if 12 mm pipe is used throughout, which is very usual. If a 12 mm pipe is used from the communication pipe, some restriction of flow may be experienced when several fittings are used simultaneously. Noisy operation of ball valves and taps is very likely with this arrangement. Figure 9.12 shows a typical installation. The diameter of the service pipe at entry, 25 mm, should be noted. This is to allow for simultaneous use of all taps without restriction of flow.

The pipe sizes shown in these diagrams, although typical of domestic installations, relate to the building

shown and are not necessarily applicable in other cases. The size depends on the flow required, the length of pipe and the head from cistern to draw-off (see pages 249–57 for pipe-sizing methods). The general principles demonstrated in these diagrams apply to larger and more complex installations which, with appropriate sizing, will be similar in their nature.

The connection of bidets presents a special problem of hygiene. If they are fitted with taps which discharge sufficiently far above the rim to form a type A air gap (see page 000), they can be supplied from a cistern and cylinder in the normal way. However, if they employ an ascending spray, the typical arrangement, special precautions must be taken to avoid backsiphonage. The cold-water pipe that serves the bidet must not supply any other appliances except WCs and urinals. It is less easy to isolate the hot-water system. An acceptable method is to provide a separate, vented hot-water supply for the bidet. The hot-water distributing pipe should leave the vent at least 300 mm above the level of the bidet rim and be provided with a check valve at the junction. The arrangement is shown in Fig. 9.13. It is also possible to provide isolation by a separate break cistern.

Figure 9.14 shows the details of installation and insulation of a typical cold-water storage cistern. The cistern should be included within the insulated perimeter of the building. If it is installed above an insulated ceiling, the ceiling insulation should be omitted within the area of the cistern insulation so that heat from the building may penetrate easily. There is some risk of condensation forming on the sides and bottoms of cold-water cisterns. This problem is particularly likely to arise in a linen cupboard if hot-water cylinder and cold-water cistern are both installed there. Trouble can be avoided if an asbestos cement sheet is fixed below the cistern. The drips will be intercepted and will then evaporate.

The traditional material for cold-water cisterns has been galvanized sheet steel. There is, however, rapidly increasing use of other materials, particularly plastics, to make domestic cisterns.

Very large storage cisterns are often built up from panels of iron or steel about 1 m square. In some cases the cisterns are formed of concrete lined with bitumen.

Cold-water storage cisterns often present difficulties of placing. They may well be too large to go upstairs and through doors or hatches. The problem is particularly acute when new cisterns have to be fitted in place of old ones which were placed before construction was complete. In these cases several

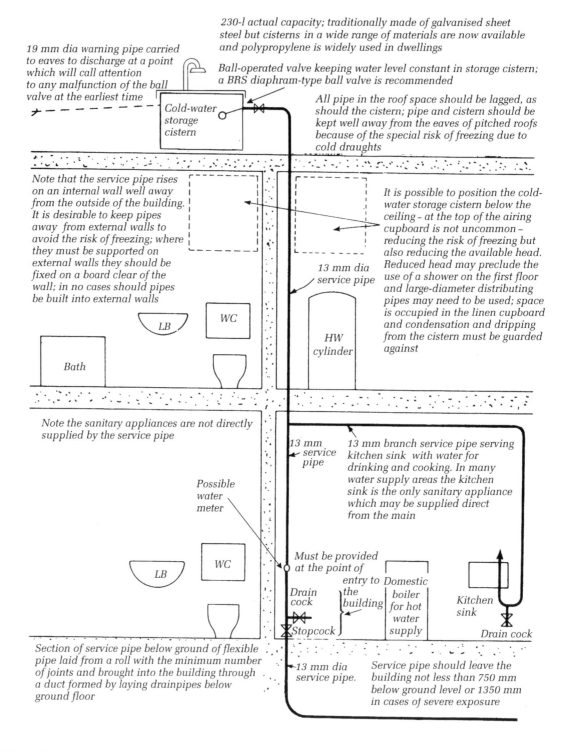

230-l actual capacity; traditionally made of galvanised sheet steel but cisterns in a wide range of materials are now available and polypropylene is widely used in dwellings

19 mm dia warning pipe carried to eaves to discharge at a point which will call attention to any malfunction of the ball valve at the earliest time

Ball-operated valve keeping water level constant in storage cistern; a BRS diaphram-type ball valve is recommended

Cold-water storage cistern

All pipe in the roof space should be lagged, as should the cistern; pipe and cistern should be kept well away from the eaves of pitched roofs because of the special risk of freezing due to cold draughts

Note that the service pipe rises on an internal wall well away from the outside of the building. It is desirable to keep pipes away from external walls to avoid the risk of freezing; where they must be supported on external walls they should be fixed on a board clear of the wall; in no cases should pipes be built into external walls

It is possible to position the cold-water storage cistern below the ceiling - at the top of the airing cupboard is not uncommon - reducing the risk of freezing but also reducing the available head. Reduced head may preclude the use of a shower on the first floor and large-diameter distributing pipes may need to be used; space is occupied in the linen cupboard and condensation and dripping from the cistern must be guarded against

13 mm dia service pipe

LB

WC

HW cylinder

Bath

Note the sanitary appliances are not directly supplied by the service pipe

13 mm service pipe

13 mm branch service pipe serving kitchen sink with water for drinking and cooking. In many water supply areas the kitchen sink is the only sanitary appliance which may be supplied direct from the main

Possible water meter

LB

WC

Must be provided at the point of entry to the building

Domestic boiler for hot water supply

Kitchen sink

Drain cock

Stopcock

Drain cock

Section of service pipe below ground of flexible pipe laid from a roll with the minimum number of joints and brought into the building through a duct formed by laying drainpipes below ground floor

13 mm dia service pipe.

Service pipe should leave the building not less than 750 mm below ground level or 1350 mm in cases of severe exposure

Figure 9.9   Typical service pipe installation to a dwelling where storage for hot and cold water is provided. Notice how the pipe rises on an internal wall to avoid frost. Also note the stopcock emtpying plug and the kitchen sink served direct from the main (all other taps are served from storage).

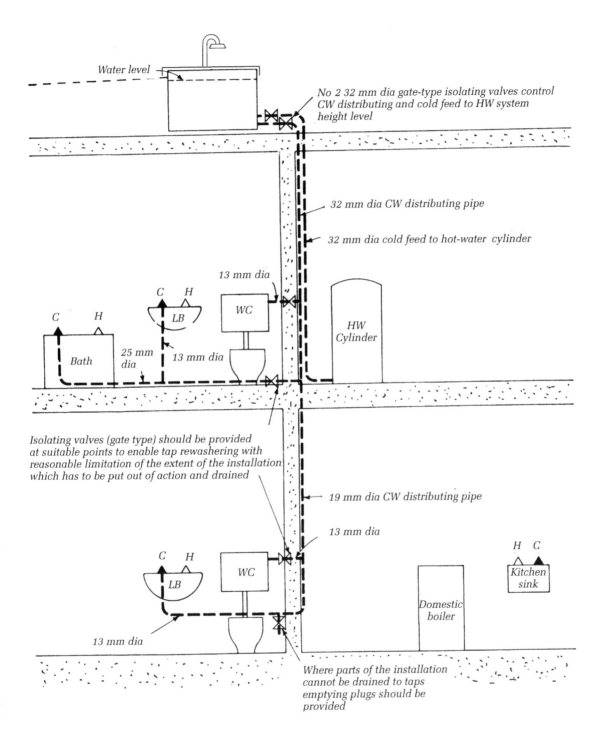

Figure 9.10 Typical distribution system for cold-water supply. Read this diagram in conjunction with Fig. 9.9.

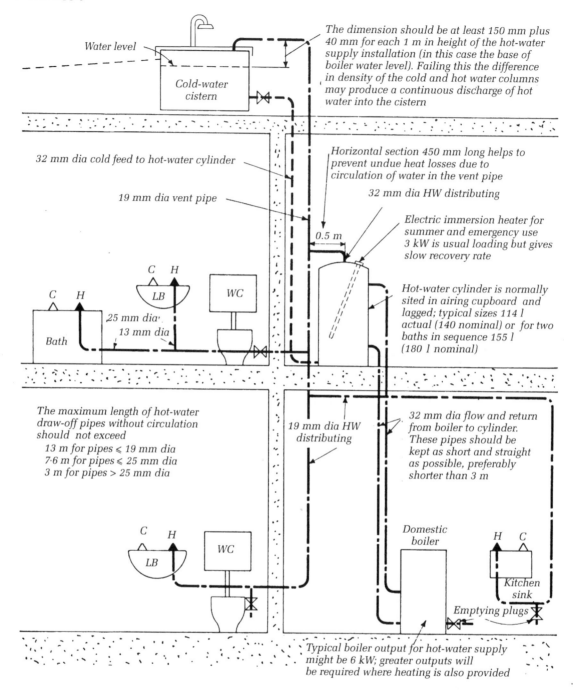

The dimension should be at least 150 mm plus 40 mm for each 1 m in height of the hot-water supply installation (in this case the base of boiler water level). Failing this the difference in density of the cold and hot water columns may produce a continuous discharge of hot water into the cistern

Water level

Cold-water cistern

32 mm dia cold feed to hot-water cylinder

Horizontal section 450 mm long helps to prevent undue heat losses due to circulation of water in the vent pipe

19 mm dia vent pipe

32 mm dia HW distributing

0.5 m

Electric immersion heater for summer and emergency use 3 kW is usual loading but gives slow recovery rate

C   H

LB

WC

C   H

Bath

25 mm dia

13 mm dia

Hot-water cylinder is normally sited in airing cupboard and lagged; typical sizes 114 l actual (140 nominal) or for two baths in sequence 155 l (180 l nominal)

The maximum length of hot-water draw-off pipes without circulation should not exceed

13 m for pipes ⩽ 19 mm dia
7·6 m for pipes ⩽ 25 mm dia
3 m for pipes > 25 mm dia

19 mm dia HW distributing

32 mm dia flow and return from boiler to cylinder. These pipes should be kept as short and straight as possible, preferably shorter than 3 m

C   H

LB

WC

Domestic boiler

H   C

Kitchen sink

Emptying plugs

Typical boiler output for hot-water supply might be 6 kW; greater outputs will be required where heating is also provided

Figure 9.11   Hot-water supply distribution pipes. Water heating is provided by a domestic boiler with flow and return pipes leading to a patent indirect system.

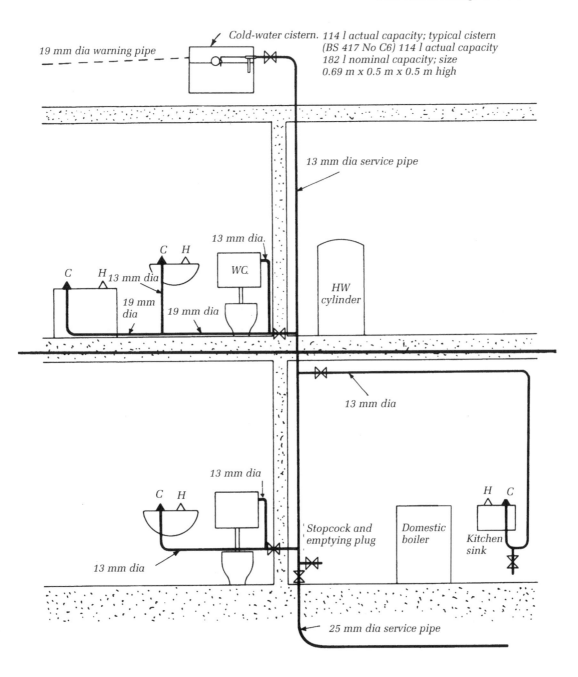

19 mm dia warning pipe

Cold-water cistern. 114 l actual capacity; typical cistern
(BS 417 No C6) 114 l actual capacity
182 l nominal capacity; size
0.69 m x 0.5 m x 0.5 m high

13 mm dia service pipe

13 mm dia.

C   H

WC.

HW
cylinder

C   H 13 mm dia

19 mm
dia   19 mm dia

13 mm dia

13 mm dia

C   H

H   C

Stopcock and
emptying plug

Domestic
boiler

Kitchen
sink

13 mm dia

25 mm dia service pipe

Figure 9.12   Typical service pipe installation to a dwelling where storage is required for hot-water supply only.

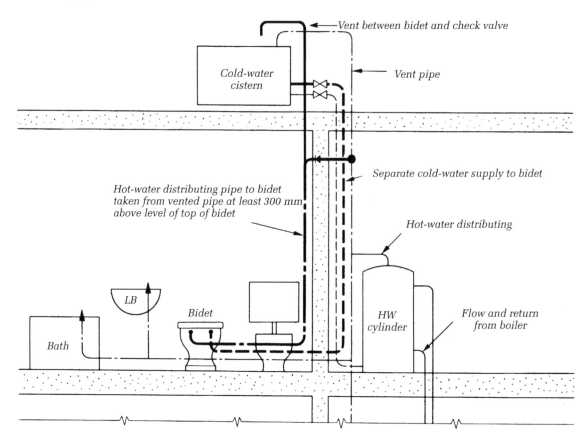

Figure 9.13   Connection of bidet to hot and cold water supply (ascending spray type with sanitary appliances at lower level).

cisterns may be linked together to provide the storage required. The supply and the draw-off should be at opposite ends of the group to ensure a flow of water through all the cisterns.

It is usual in large installations to provide at least two cisterns with separately valved ball valves and outgoes, so that one may be put out of action for cleaning or repair without affecting the supply to the building. Figure 9.15 shows a typical arrangement. Although, with proper maintenance, the arrangement shown in Fig. 9.15 should be satisfactory in relation to legionnaires' disease, it is thought that the risk of infection is reduced by having the water flow through the two cisterns in sequence rather than in parallel. In the case shown the cisterns should have pipe links so that flow can proceed from one to the other, and isolating valves to control the input and output and enable sections of pipe not in use to be isolated and drained. The cistern sizes should be minimized to encourage active water flow.

Supplies to buildings rising above the level of the mains head

New problems arise in high buildings. It will probably be necessary to pump water up to the top of the building since the head in the main is not likely to be adequate, and if the building is very tall it may be necessary to divide the distribution into zones to keep the water pressure within reasonable limits. It is not satisfactory to have pumps in continuous operation, nor is it practicable to have them switch on and off for every draw-off of water. Figure 9.16 shows one way to overcome this problem by using a pneumatic vessel. A cushion of air under pressure is maintained in the top of a pressure cylinder; when a tap is opened the air is able to expand by forcing water out of the cylinder and through the pipework. This process can continue until the water level drops to a predetermined point, when the pumps will be switched on to raise the level again. Drinking water is drawn off from the pressure vessel, although within the reach of the mains head the

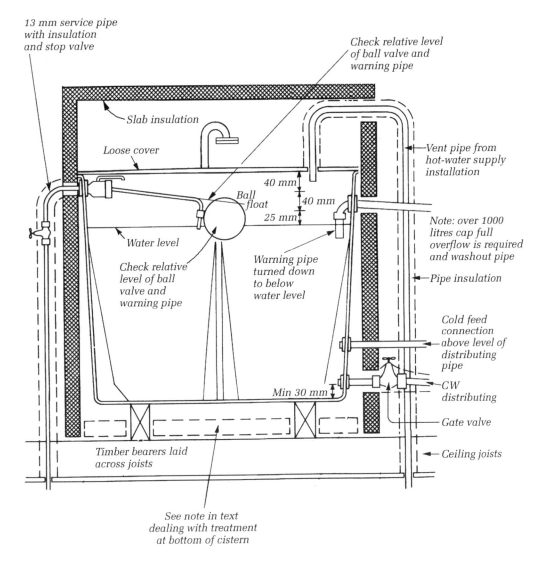

Figure 9.14   Typical cold-water supply cistern.

drinking water is supplied direct from the main. Precautions are taken to ensure purity. The air pumped in is filtered to prevent dust and insects gaining access and the capacity of the vessel is kept reasonably small to prevent stagnation. For dwellings the volume of water between high and low water levels would be no more than 4.5 litres per dwelling. For flats of up to about 15 floors a simplified system is possible. Figure 9.17 shows the arrangement. It differs from the previous system principally in supplying minor drinking-water draw-offs by means of an enlarged section of service pipe above the level of the highest

flat. This enlarged section, coupled with the special air vent above, enables water to flow to drinking-water taps without the pump being operated except when the whole enlarged section becomes empty. As soon as this occurs a float switch brings the pumps into operation until the pipe is full. The ball valve prevents overflowing into the storage cistern unless the level there drops when the pumps will also be brought into operation by a float switch. In this case, too, the drinking-water taps are supplied direct from the main wherever possible. In buildings which rise above 15 floors, zoning is necessary to reduce the pressure in the

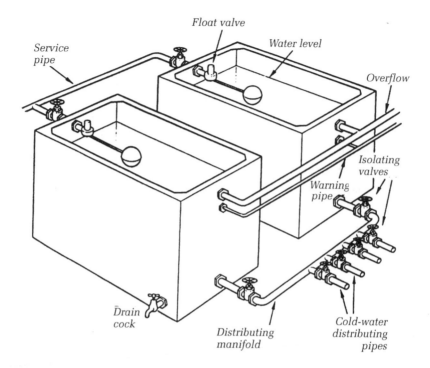

Figure 9.15 Typical water storage installation with two linked cisterns.

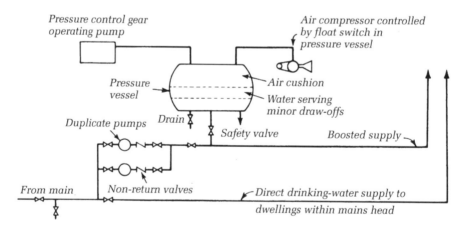

Figure 9.16 Pneumatic booster for raising water in buildings extending above the level of the mains head.

distributing pipes at the bottom of the building. The usual zone height is about 30 m. Figure 9.18 shows one method of dealing with the problem. The main water storage is now at ground level. This avoids a load of perhaps hundreds of thousands of kilograms at the very top of a high building. The systems for drinking-water supply and cold-water distribution are now completely separate. Pumps raise the water in stages of 60 m in both cases, but with cold-water cisterns distributing to stages of 30 m. Drinking water is distributed on the same system using storage cisterns but in this case the cisterns are sealed against insects and dust, the

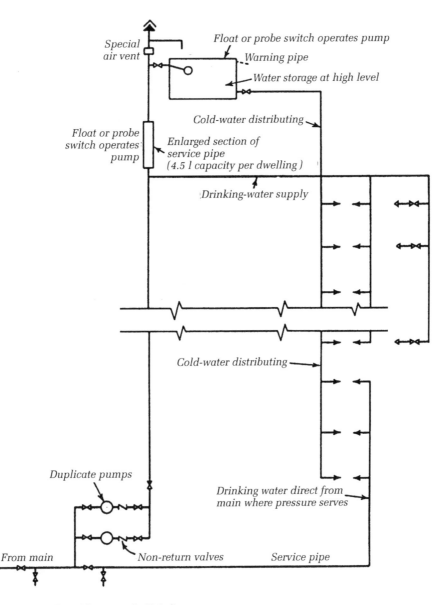

Special air vent

Float or probe switch operates pump

Warning pipe

Water storage at high level

Cold-water distributing

Float or probe switch operates pump

Enlarged section of service pipe (4.5 l capacity per dwelling)

Drinking-water supply

Cold-water distributing

Duplicate pumps

Drinking water direct from main where pressure serves

From main

Non-return valves

Service pipe

Figure 9.17 Simple system for raising water in high flats.

warning pipe and air vent are fitted with filters for the same purpose and stagnation is avoided by limiting the cistern capacity to 9 litres per dwelling. Figure 9.19 shows typical cistern arrangements. Pressure-reducing valves can have the same effect in reducing pressure as intermediate cisterns and are increasingly used. Part 6 of Ministry of Housing and Local Government Design Bulletin 3 *Service cores in high flats* covers cold-water services and gives a detailed account of the problem along with several possible solutions.

The temperature in internal pipe ducts is likely to be at least several degrees above ambient temperature because of heating and hot-water supply pipes, flues, etc. Cold water passing through long lengths of pipe may be noticeably warmed. This can become unacceptable for drinking water, particularly in office buildings where long lengths of pipe in warm ducts may be serving drinking fountains that allow only a small flow. In such cases it is desirable to have an insulated flow and return circuit serving the drinking-

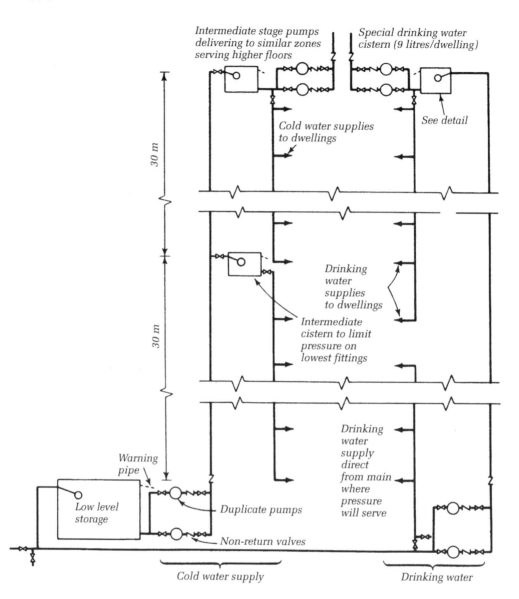

Intermediate stage pumps delivering to similar zones serving higher floors

Special drinking water cistern (9 litres/dwelling)

Cold water supplies to dwellings

See detail

30 m

Drinking water supplies to dwellings

Intermediate cistern to limit pressure on lowest fittings

30 m

Drinking water supply direct from main where pressure will serve

Warning pipe

Low level storage

Duplicate pumps

Non-return valves

Cold water supply

Drinking water

Figure 9.18 System for raising and distributing water in high buildings, showing the ground level main storage, separate drinking-water circuit and the possibility of raising the water up another stage of the building.

water points, with water pumped through a drinking-water cooler.

9.8 Hot-water supply

Hot water can be produced by a wide variety of appliances involving the whole range of available fuels. The methods can be classified as central and local systems and the appliances as instantaneous or storage. Central systems are usually of the storage type where water in a storage vessel is heated from the space heating boilers. Such types of systems are found in all types of buildings from dwellings to large office or commercial buildings. In local systems the water heating appliances are sited near groups of sanitary appliances. Electricity is widely used because of the ease of supply and the freedom from flues, although gas is often employed. Selection will be based on the

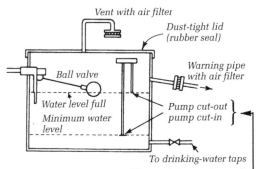

Vent with air filter

*Dust-tight lid
(rubber seal)*

Ball valve

*Warning pipe
with air filter*

Water level full

*Minimum water
level*

*Pump cut-out
pump cut-in*

To drinking-water taps

Probe switches to operate pump (float switch may be used)

Figure 9.19 Storage cistern for drinking-water supply showing filtration of entering air, sealing of the top and a limited volume to ensure fresh water (4.5 litre per dwelling).

combination of performance, capital costs and running costs. Figures 9.27 (page 224) and 7.20 (page 152) show typical small and large central water heating systems.

The characteristic of the central system is the pipework linking the central supply with the draw-off point. In its simplest form a single pipe (dead leg) serves the outlets. Between draw-offs the water in the pipe will cool and the next draw-off will involve running the cold water to waste before it reaches the tap. This creates user dissatisfaction, waste of water and waste of heat. To overcome wastage the model bye-laws impose maximum lengths for dead legs.

In order to keep water consumption within reasonable limits dead legs of pipework serving hot-water taps are governed to maximum lengths:

- 12 m for pipes not exceeding 20 mm diameter
- 7.6 m for pipes not exceeding 25 mm diameter
- 3 m for pipes exceeding 25 m diameter
- 1 m for pipes serving spray taps

This limits the amount of cold water which has to be run to waste before hot water is delivered. The reason for the length limit is water saving, but it also serves to keep within bounds the time delay and consequent inconvenience which would result from the use of very long pipes.

Draw-off points too distant for a dead leg to be used can be served by a secondary circulation where a loop of pipework carries a continuous flow of hot water. Figure 9.20 shows the principles involved. The secondary circulation avoids waste of water but loses heat continuously. The Building Regulations do not appear to require insulation of secondary circulations unless they run outside the heated volume of the

building. During the winter the heat loss from the pipework will contribute to space heating; in the summer the heat is not only wasted but may contribute to overheating. Good practice calls for insulation of secondary circulations.

A recent innovation which can provide hot water at taps without secondary circulation is the use of electric heating tape. If this is attached to the pipework of a dead leg and insulation is provided, the water can be maintained at the required temperature during the periods of demand. The savings in pipework can be set against the cost of the electrical work. The electric heating tape can also be used to raise the temperature of the water in the pipe to 65 °C for short periods to combat legionnaires' disease.

A central system which has to have a secondary circulation will require a substantial pipe installation and the heat losses from the pipework, even where reduced by lagging, will be continuous, including the summer months. Although local systems save pipework, usually being served by the cold-water pipe system, and eliminate heat losses from distributing pipes, they do require several heaters, which often use a more expensive fuel than a central system. Large and continuous hot-water demands, particularly hot if they are close to the central plant, are therefore best dealt with by central installation, whereas scattered hot-water points, particularly if their use is very intermittent, can often be more economically served by local water heaters. In many cases it will be difficult to decide which method is more economical.

Instantaneous water heaters are normally restricted to use as local heaters because of the limited flow of hot water which can be produced. The thermal capacity of water is high and to raise 8 litres of water per minute by 55 °C (the flow for a lavatory basin tap and the usual temperature increase required) requires a heat input of 17 kW. This is a rate equivalent to the full space heating requirement of a sizeable house. The output which can be achieved from an appliance of reasonable size is therefore very limited. A large gas instantaneous heater may deliver a flow of 10–20 l/min, whereas the flow from a 30 kW electric spray tap may be only 1 or 2 l/min. Figure 9.21 shows an instantaneous gas water heater. Gas instantaneous heaters are used for the whole hot-water supply of small dwellings, but apart from this, instantaneous water heaters are used locally for the specific appliances they serve. Instantaneous gas water heaters can be equipped with balanced flues.

Thermal storage heaters use a vessel to accumulate hot water. It is possible with this form of heater to use a modest heat input, and sudden heavy draw-off can be

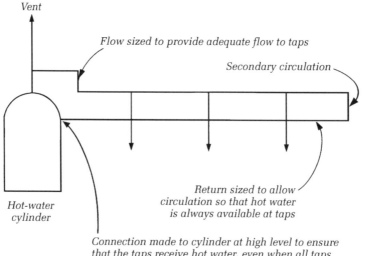

Figure 9.20   Hot-water secondary circulation.

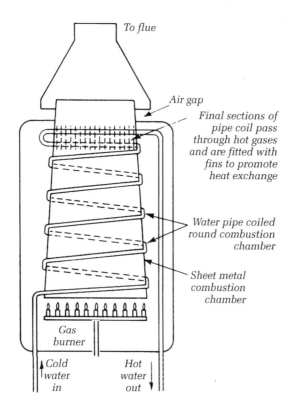

Figure 9.21   Instantaneous gas water heater.

catered for by the stored hot water. It is necessary to balance the capacity of the heater and the heat input (recovery rate) so that the likely demands can be met. Figure 9.11 shows a typical central thermal storage system for a house. The water is heated in a boiler on the ground floor and the hot water accumulates by thermosiphon action in the cylinder on the first floor. Fresh cold water is introduced into the cylinder from a cistern at high level and makes up any water drawn off from the hot taps. At one time a decision had to be taken whether the system should be *direct* (water passing through the boiler can also be drawn off from the taps) or *indirect* (water circulating through the boiler and heating the cylinder as a separate system from the water drawn off at taps). The direct system is subject to furring or corrosion of pipes, whereas the indirect system, although universal in large installations, is expensive for small houses. The problem has been overcome by the development of patent indirect cylinders which effectively separate the boiler flow (primary circulation) from the stored hot water while allowing the boiler to be filled and vented through the cylinder without the need for an additional cistern and pipework. Figure 9.22 shows a section through a patent indirect cylinder of this type, the Primatic, demonstrating how the primary flow is kept from contact with the stored water by means of entrapped air pockets.

The system shown is provided with an immersion heater for use when the boiler is out of action or during

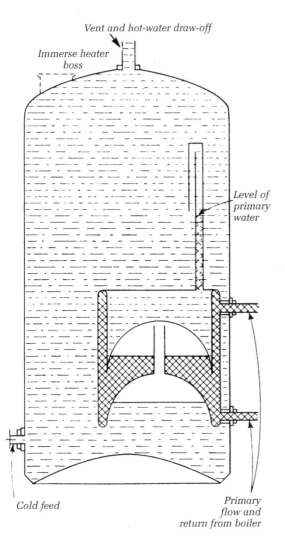

Vent and hot-water draw-off

Immerse heater boss

Level of primary water

Cold feed

Primary flow and return from boiler

Figure 9.22 The Prismatic patent indirect hot-water storage cylinder. The primary flow is shown cross-hatched. It is possible to observe the air bubbles that separate it from the water in the cylinder. It is also possible to accommodate expansion or contraction of the water, even the effects of pump operation, without allowing the primary flow to mix. Filling and venting, however, take place through the cold feed and vent.

and often contrasts unfavourably with the very much quicker heating achieved by the boiler, which will invariably have a higher rate of heat output. This is particularly the case when heating is also provided by the boiler. Except in the coldest weather the boiler will have unused capacity which can contribute to water heating and will thus speed the recovery rate of the cylinder.

In large buildings, particularly blocks of flats, several indirect hot-water cylinders or pressure-type water heaters may be required. The separate cold-water supply to an individual hot-water cylinder coupled with hot-water draw-off from the top of the cylinder means that, even in the case of failure of the water supply, the cylinder cannot be emptied. A similar performance is required from installations with more than one cylinder. Figure 9.23 shows how this may be achieved by taking the connections to the cylinder from the distributing pipe at a level higher than that of the cylinder served. Figures 9.24 to 9.26 show various types of electric water heaters.

Figure 7.20 shows the arrangements in the boiler room for water heating and water storage in large installations with pumped secondary circulation. The term *calorifier* is usually used in this case to describe the hot-water storage vessels.

Figure 7.17 shows how hot-water supply may be achieved using the heat from the boiler room, but without any separate hot-water secondary circulation. An indirect cylinder is heated by water from the space heating pipework. The cold-water storage and distribution arrangements are similar to the domestic ones in Fig. 9.11. In the summer, when the heating is not operating, an immersion heater in the cylinder is used. This is a particularly economical arrangement when small groups of sanitary appliances are widely scattered in a building. It is quite feasible to have several hot-water supply systems of this sort in different parts of the building served from one boiler room.

There are several different ways of providing hot-water supply from a domestic boiler. The general layout of appliances and pipework is shown in Fig. 9.16. Water is heated by a boiler on the ground floor. Gravity circulation transfers heat to the cylinder on the first floor via flow and return pipes. Hot water accumulates in the cylinder and is drawn off for use at hot taps. The differences between systems arise from the degree of separation between water circulating in the boiler and in the cylinder and from the way in which fresh water is introduced into the cylinder. The methods may be divided into four types.

the summer, when the boiler, which often also supplies space heating, may be turned off. The usual loading for immersion heaters used in this way is 3 kW. This gives a very slow recovery rate. The 3 kW heater will take 3 h to raise 150 l of water by 55 °C. This slow rate of heating a cylinder may cause considerable annoyance

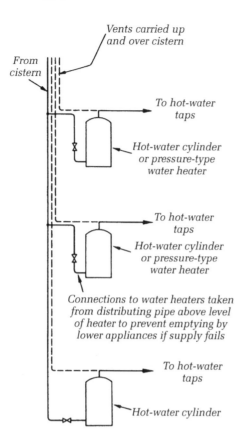

Figure 9.23 Cold-water supply to several hot-water cylinders ensuring that the cylinders cannot be emptied if the water supply fails.

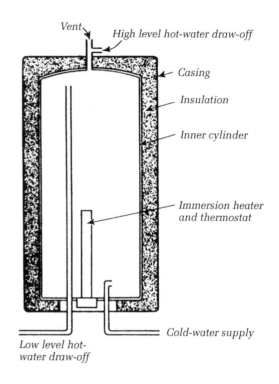

Figure 9.24 This pressure-type electric water heater requires its vent pipe to be carried back to the cold-water cistern (as Fig. 9.11). It cannot supply many taps at levels below and above the heater. The delivery of hot water is controlled by taps on the distribution pipes served by draw-off points. Several taps may be served.

Direct system

The direct system (Fig. 9.27(a)) is rarely used in new work, but very many systems are still in operation. Fresh cold water is introduced into the cylinder from the cold-water cistern. This water also circulates through the flow and return pipes to the boiler in order to be heated. Installations of this type were economical of pipework but were subject to furring or corrosion of boiler and pipes, which substantially reduced the efficiency and the life of the installation. In some circumstances the hot water drawn off could be discoloured or rusty.

Indirect system

The disadvantages of the direct system can be substantially overcome by separating the primary water circulating in the boiler and flow and return pipes from the water to be drawn off at taps. This is achieved by providing a separate cold-water cistern to supply the boiler and containing the primary flow in the cylinder within a coil of pipework or an annular cylinder to allow heat to be transferred from the primary flow to the water in the cylinder (Fig. 9.27(b)). Since the water in the boiler and primary circuit is not changed, corrosion is much reduced. The progressive deposition of scale is reduced and only takes place in the hot-water cylinder, where its effect upon performance is much less than in primary pipework and the boiler. This method is universally employed in larger installations but is expensive for domestic use.

Patent indirect cylinder system

For domestic-scale installations special cylinders (Fig. 9.27(c)) have been developed which allow the economy of installation of the direct system while maintaining effective separation between the primary

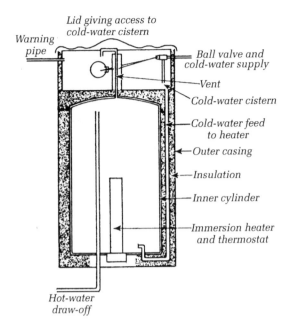

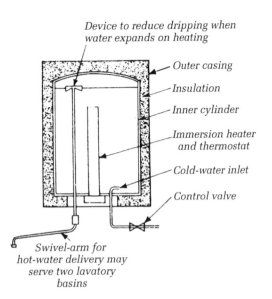

Figure 9.25 This cistern type of electric thermal storage heater is convenient for use in flats since it may be served from the main and requires only a warning pipe, not a vent. It may be used to supply many taps. The pressure at the taps will usually be low and all pipework must be kept below the level of the cistern. The vent pipe to the heater is now internal and in many areas the appliance can be connected to the service pipe. But a warning pipe must be provided and the head available from the cistern is very limited.

Figure 9.26 This free-outlet type of electric water heater requires neither a vent pipe nor a warning pipe. Venting is through the open delivery arm. Admission of cold water causes hot water to overflow down the delivery tube. It is limited to one point, two if a swivel delivery arm is used.

flow and the hot water for draw-off. Figure 9.22 shows a section through a Primatic cylinder. This allows the boiler and primary pipework to be filled and vented, and also allows expansion of the primary water while keeping primary and draw-off water separated.

Unvented storage hot-water supply systems

It has been common practice in many countries to have hot-water supply systems under mains pressure. There are several advantages. High level cisterns are not required. Hot and cold water are delivered at the same pressure, simplifying the use of showers and mixers. Smaller pipes can be used. The disadvantages are that storage cylinders have to be strong enough to resist the mains pressure and overheating must be prevented in order to avoid explosion. It is concern about the risks of explosion which has prohibited their use in the United Kingdom. The new Model Water Bye-laws 1986 now permit the use of pressurized hot-water systems subject to several precautions:

- The system must be a proprietary one and approved by an authorized body.
- The system must be installed by a competent person.
- The system must include
  - temperature control and non-resettable thermal cut-outs
  - temperature relief valves
  - provision for expansion

Figure 9.27(d) shows the features of a typical pressurized hot-water supply system.

Solar water heating

Solar space heating systems have had limited success in the United Kingdom. Water heating, however, can take advantage of sunshine at its most effective, outside the heating system. Part L of the Building Regulations, dealing with thermal performance, does recognize and provide means to assess the contribution to energy saving made by solar heating systems.

A basic system can be very simple. A small feed and expansion cistern, a hot-water cylinder and a solar collector together with flow and return pipework are all that is required. Most installations also have a small pump to improve circulation. Except for the collector,

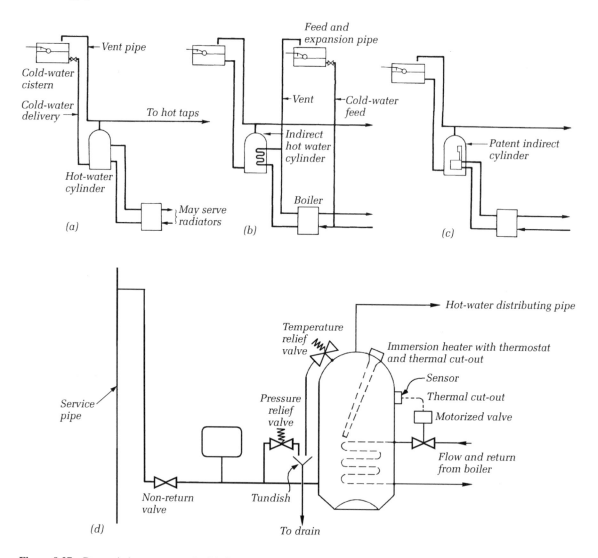

Figure 9.27   Domestic hot-water supply: (a) direct storage system (no longer installed but many are still in use); (b) indirect storage system; (c) storage cistern using patent indirect cylinder; (d) unvented system showing safety precautions.

the components are standard plumbing equipment. The collector itself usually takes the form of a copper sheet with waterways attached, backed by insulation and protected externally by one or two layers of glass, which allow solar radiation to penetrate freely but are almost opaque to low temperature radiation from the plate itself. Figure 9.28 shows a simple installation and Fig. 7.5 shows a collector unit. Precautions must be taken against freezing, usually by adding antifreeze to the circulating water. Also to be avoided is reverse circulation, when the collector loses heat to the cold night sky.

Solar heat alone cannot be relied on to provide continuous supplies of hot water. An immersion heater can be used for additional heating. Where boilers are used for space heating it is possible to provide hot-water supply combining both sources. Figure 9.28 shows a system of this type. The dual-coil water cylinder enables the solar circuit to preheat the water and the boiler-fed coil brings it to a final temperature. It is also possible to use two separate cylinders; then the cylinder with the solar coil supplies preheated water to the boiler-heated cylinder.

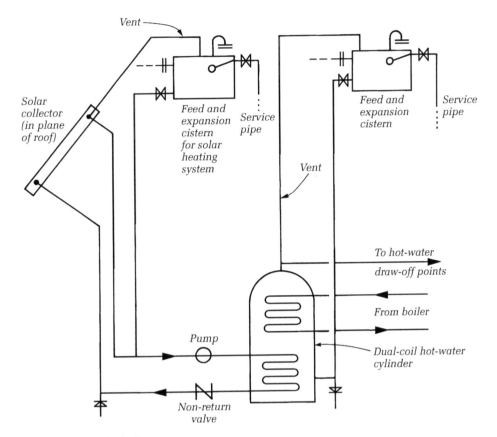

Figure 9.28   Schematic of solar hot-water system.

Backsiphonage

Backsiphonage can take place when the pressure in supply pipework drops to below atmospheric pressure. This can take place because of excessive demand or failure of supply. When it does happen it would be possible, in the absence of suitable precautions, for the contents of sanitary appliances or other apparatus connected to the pipework either directly or by means of a hose, to be drawn back into the supply pipework. The circumstances may arise in the water mains or entirely within a building, where appliances at higher levels may give rise to backsiphonage due to draw-off at lower levels.

Backsiphonage risks have been classified in three different categories of risk. *Class 1* covers a continuous risk of serious contamination (e.g. bidets, WC pans, laboratories, hospital equipment, industrial processes). *Class 2* covers occasional risk to health (e.g. most domestic draw-off points other than WCs, vending machines, car washes, humidifiers). *Class 3*

covers no significant risk to health, but water quality could be affected.

The main precaution against backsiphonage is the air gap. This separates the supply pipework from the pipework of the installation. The type A air gap has unrestricted overflow and a sufficient gap between the delivery pipe and the water surface to prevent backsiphonage under any pressure conditions. Figure 9.29(a) shows a type A air gap. Sinks and lavatory basins are normally provided with this form of air gap since the tap outlets are above the flood level of the appliance.

Figure 9.29(b) shows a type B air gap. This is similar to a cistern with a ball valve and warning pipe; it can be achieved with a BS 1212 ball valve and a warning pipe positioned to give an adequate air gap (20 mm in the case of 12 mm supply pipe). Provided the cistern feeds pipework only by gravity and its base is 15 mm at least above any fitting supplied, this type of installation can be used for some class 1 risks such as WCs.

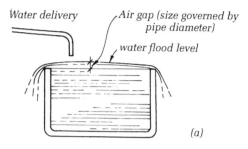

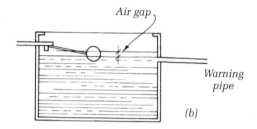

Figure 9.29 Air gaps: (a) type A has an unobstructed overflow; (b) type B has a restricted overflow.

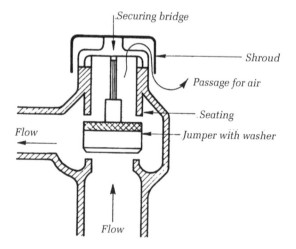

Figure 9.30 Automatic air vent: when the system is under pressure the jumper is raised and seals the air gap; when the pressure falls the jumper falls, allowing air to enter and preventing siphonage.

In some cases a type B air gap may have to be supplemented by an automatic air vent that will admit air to the pipework if the pressure drops below atmospheric. Figure 9.30 shows an automatic air vent.

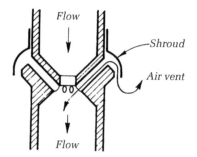

Figure 9.31   Pipe interrupter.

It would normally be located at the highest point of an installation 300 mm above the level of the ball valve. Typical applications would be where common systems serve more than one dwelling or where the supply is subject to frequent interruption. In these cases a check valve at the point of entry of the service pipe may also be required.

In some circumstances pipe interrupters may be required. They are normally used downstream of the last valve. A typical example is a WC flushing valve which delivers to below the flood level of the WC. A pipe interrupter in the flush pipe will allow free flow from the valve to the WC but would prevent any backsiphonage from taking place. Figure 9.31 shows a pipe interrupter.

Where no central boiler plant exists or where the sanitary appliances are not conveniently sited, gas or electric water heaters may be used. Figures 9.24 to 9.26 show various types of electric water heaters. The pressure-type heater (Fig. 9.26) is essentially similar to the hot-water cylinder and requires the same

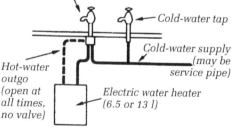

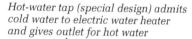

Figure 9.32   This undersink water heater with free outlet can be connected to a service pipe. No vent pipe or warning pipe is required.

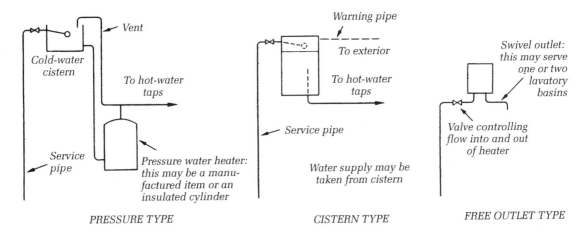

Figure 9.33   Water supply arrangements for different types of water heater.

connections, including a vent pipe which must be carried back to the cold-water cistern. In blocks of flats this can give rise to expensive pipework, so the cistern-type water heater (Fig. 9.25) has been developed to obviate the need for the vent. This type of heater can normally be supplied by the service pipe rather than a special supply from the cistern. No vent is required but a warning pipe must be provided. In operation this type of heater is similar to the pressure type except that the head available from the integral cistern is limited. It will usually be difficult to use showers, and all draw-off points and pipework must be kept at a level lower than the cistern.

The free outlet type of heater (Fig. 9.26) can be supplied from the service pipe provided its capacity is less than 13.5 litres (larger capacities may be permitted but only if effective antisiphon devices are fitted). This type of heater is particularly convenient since no vent or warning pipe is required but only one outlet can be served. This outlet is often made to swivel so that two lavatory basins may be served. By the use of special tap fittings this type of heater can be adapted to be fitted below the level of the appliance served. Figure 9.32 shows the arrangement. Figure 9.33 shows schematically the piping requirements for pressure, cistern and free-outlet water heaters. An instantaneous heater is shown in Fig. 9.34. The electrical loading for this type of heater is usually 3 kW and the output of hot water only 1 or 2 l/min. The outlet normally delivers a

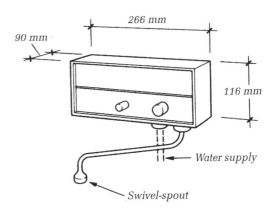

Figure 9.34   The Heatrae Carousel instantaneous electric water heater.

spray and is used only for washing hands or for simple showers.

There are gas appliances to match each of the electrical heaters. Some of the smaller ones may be permitted to discharge their products of combustion into rooms which, if acceptable, saves the expense of a flue (see Chapters 7 and 16 for gas flue considerations). Table 9.1 summarizes the main types of electric and gas water heaters together with notes on their piping and flue requirements.

Table 9.1   Types of gas and electrical water heaters

Heater type	Gas		Electricity	
	Type of appliance	Special requirements	Type of appliance	Special requirements
Instantaneous	Geyser Single point (may be able to provide boiling water) Multipoint	Flue not normally required if heater is in a room Flue needed 75–125 mm	Electric spray taps (3 kW) Limited flow	
Thermal storage	Storage heaters[a] Small 1–5 gal (serve single basin and sink) Multipoint 12–100 gal	Vent pipe and flue required	Free outlet (1 or 2 sanitary appliances) sometimes from main Pressure[a] (may be built up by use of immersion heater and copper cylinder)	Sometimes connected directly to service pipe Vent pipe required
	Cistern multipoint	Flue and warning pipe required	Cistern (must lie above highest draw-off point)	Warning pipe required
Supplementary	Circulator (fixed) adjacent to and provides heat for storage cylinder) May be used to supplement boiler	Usually installed in cupboard where flue is required	Immersion heater (provides heat for storage cylinder to supplement boiler as for summer use)	

[a] Must be normally supplied from cistern.

10 Sanitary appliances

The use of water in buildings for sanitary and other purposes is made possible and convenient by the provision of sanitary appliances of appropriate form and having water supply and outgo arrangements sized and placed to assist the function which the appliance serves. All sanitary appliances should have impermeable surfaces and preferably also the body of the material should be impervious; the shape should be appropriate to its use and free from crevices or features which would make cleaning difficult. It is important that water supplies should run no risk of contamination from the proximity to the foul water in the appliance, particularly where direct supply from the mains is involved. Appliances are usually designed that taps discharge above the highest flood level of the water, so they cannot become submerged and make backsiphonage into the supply piping possible.

10.1 Materials for sanitary appliances

Ceramics

Ceramics give very durable glazed surfaces in a wide range of colours. There are three main classes:

- Earthenware is relatively cheap.
- Fireclay is strong and resistant to knocks; it is widely specified for public buildings and institutions.
- Vitreous china is completely impermeable, unlike the other materials, which have a glaze on a permeable body.

Most designs of sanitary appliances are made in one material only and, if a particular design is desired, the material of manufacture is inevitably established.

Cast iron

Cast iron may have a white or coloured vitreous finish fired on. It is mainly used for very large appliances which would be impracticable to manufacture and too heavy to move when made in ceramics.

Pressed steel

In recent years pressed steel has been substituted for cast iron. Its finish can be vitreous enamel. Noisy when used for sinks, so sound-deadening materials are used on the underside. One-piece drainers and sinks have great hygienic advantages when compared with separate sinks and drainers.

Stainless steel

The natural surface of stainless steel, although perhaps less satisfactory visually than vitreous enamels or ceramics in good condition, is able to withstand impact and abrasion. One-piece sinks and drainers have similar hygienic advantages to pressed steel.

Plastics

White and coloured Perspex is increasingly being made into sanitary appliances. Its surface is liable to become scratched. Many other plastics are used in WC seats, cisterns, etc.

Terrazzo

Special designs, not involving large numbers of units in the materials described above, incur considerable expense because of the nature of the manufacturing process and the patterns and moulds required. Special appliances can, however, be made reasonably simply from terrazzo and very large items such as plunge baths can be formed in situ in this way.

10.2 Types of appliance

WCs

A water closet consists of a pan containing water and receiving excrement and a device for providing a flush of water. Normally the pan and flushing device are separate but in some modern fittings they are coupled together. The pans are almost invariably of ceramic material.

There is a very long history to this type of appliance. However, the first device having features clearly similar to present practice was developed by Sir John Harrington in the reign of Elizabeth I. Flushing water closets did not, however, come into general use until piped water supplies became available. Many developments in design took place in the nineteenth century. One culminated in the valve closet, which was highly effective and quiet in operation but has been completely superseded because of the problems of water supply maintaining the mechanism and because the required wooden casing was difficult to keep clean. Some closets of this type are still in operation. The non-mechanical developments led to the type of pan most widely used at present: the *wash-down*. Figure 10.1(a) shows a section through a WC pan of this type

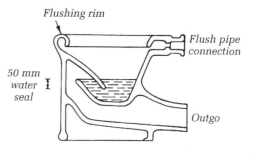

Figure 10.2   Section through siphonic WC pan.

with a P-trap. Figure 10.1(b) shows other outgo configurations which may aid planning and neatness of pipe installation. Clearing the contents of this type of pan depends on a powerful flush of water. A method that ensures clearing of the pan and does not require such a fierce flush is the siphonic closet (Fig. 10.2). The operation of the flush causes a flow in the outgo, whose shape induces a siphonic action that removes the water and other contents from the pan. The pan is refilled and the seal re-established by the last of the flush.

Although universally accepted, the form and particularly the height of WC pans does not accord with medical opinion, which regards as low a level as possible as physiologically desirable. A compromise between established practice and the ideal was the health closet, which sloped downwards from front to back, theoretically giving a low sitting position and support to the thighs. The user, however, tended to slip backwards, thereby fouling the back of the pan. Seats with stops to prevent sliding were developed but this type of closet is rarely employed. Recently detailed research into the physiological and hygienic problems of WCs has taken place in America. The outcome was a new pattern of WC, which may be produced commercially.

The floor space around and behind conventional WC pans is difficult to clean. To make this easier, particularly in institutions and hospitals where hygiene is important, *corbelled wash-down closets* are available. Two types are made: one requires a heavy wall into which the corbel extension of the pan is built; the other may be used with light partitions consisting of a metal frame fixed in the floor screed. The partitions are the same thickness as the wall, from where they project to support the WC pan.

The *squatting* or *Eastern* closet (Fig. 10.3) is rarely, if ever, installed in this country, although its use in Europe is widespread. It is flushed in the same way as a conventional WC. It is not suitable for old

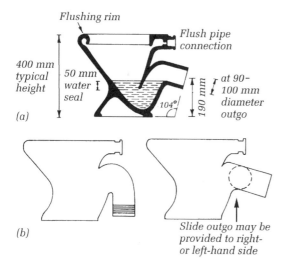

Figure 10.1   Wash-down WC pan: (a) section showing P-trap; (b) S-trap and P-trap pans may have left-hand or right-hand outgoes.

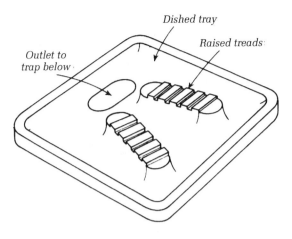

Outlet to
trap below

Dished tray

Raised treads

Figure 10.3 Squatting or eastern closet.

or infirm people but it gives a good physiological posture and when installed in a tiled enclosure enables very easy cleaning by means of a hose. It is very surprising that some of these closets are not provided in public lavatories, where they could give very much more hygienic conditions than are usual at present.

Several extra features should be borne in mind when selecting WC pans.

Flush   The flush should scour the whole inner surface of the pan and should be sufficiently powerful to discharge the contents of the pan. In most cases the water is distributed round the pan by a flushing run but in some designs the run is dispensed with. The performance in this respect of both wash-down and siphonic pans should be observed before selection.

Water area   The surface area of the water in the pan should be as large as possible so that all excrement falls directly into the water rather than on to the dry sides of the WC, where it would tend to stick even after flushing.

Back   The back of the inner surface of the pan should be vertical or nearly so to avoid fouling.

WC seats

The traditional wooden seats have been replaced by plastic ring seats which are very much more hygienic. Seats with flat undersides are desirable and open-fronted rather than complete ring seats are best for all applications. Cover-flaps, if desired, can usually be provided from the same hinges as the seat.

WC flushing devices

In order to clear the pan effectively, even siphonic WCs require a rapid flush. Two methods are conventionally used. One is to store water in a small cistern and discharge this rapidly into the pan through a relatively large-diameter flush pipe. This enables the pan to be isolated from the water supply installation, the flush to be automatically controlled and small-diameter pipework to be used to replenish the cistern. The other is to use water directly from the water supply installation and to control rate and duration of flow by automatic valves.

Flushing cisterns   The water level in a flushing cistern is controlled by a 12 mm ball valve with a 10 mm warning pipe. A very simple plug control of the outgo can be achieved and is used in many countries. But in the United Kingdom a simple siphonic device is invariably used to deliver the flush. The object is to ensure that water cannot leak continuously and unnoticed into the pan, thereby giving rise to major wastage. The object is achieved since the siphonic action requires that the delivery tube to the flush pipe rises above water level. Thus under normal conditions no leakage can take place. If the ball valve fails to stop the flow of water when the correct level is reached, the warning pipe will call attention to the problem before any leakage into the pan takes place. Originally these cisterns were called water-waste preventers (WWPs). It is now becoming conventional to refer to them as flushing cisterns.

The earliest type of WWP or flushing cistern was made of cast iron with a round sump at the centre of its base. The flush pipe, usually of 32 mm diameter, extended upwards to above water level through the centre of the sump. Its diameter was increased progressively near the top to give the entry to the flush pipe the form of a narrow funnel. A cast iron cone covering the flush pipe was housed in the sump. In operation the cone was raised and then dropped. The form of the cone forced water to rise as the cone fell. Water then flowed into the funnel leading to the flush pipe; this set up siphonic action and discharged the contents of the cistern rapidly. The cistern was normally located with its base some 20 m above floor level to assist in the force of the flush. This arrangement is known as *high level*.

A neater and quieter form uses a piston mechanism to activate the siphonic action. Figure 10.4 shows this form of flushing cistern, which is often located so that its top is only 1 m above floor level; it is known as *low level*.

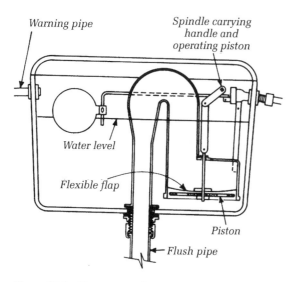

Figure 10.4   Piston-operated water-waste preventer.

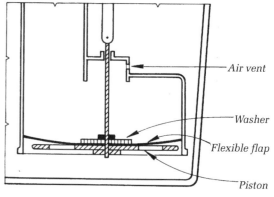

Figure 10.5   Typical flushing mechanism for dual-flush. To discharge the full contents of the cistern the operating lever is held until the flush is complete. This holds the washer on top of the cylinder, thus sealing the air vent. When the piston is allowed to fall, immediately after operation, the siphonic action will be broken as soon as the water level drops to the air vent.

High level cisterns are connected to the pan by galvanized *telescopic flush pipes* which have two sections sleeved together, allowing adjustment for differing heights. Low level cisterns have L-shaped pipes, usually enamelled, which are cut to suit the actual dimensions. Some manufacturers provide flushing cisterns which are fixed directly to the pan; they are known as *close-coupled cisterns*.

A variety of materials may be used for flushing cisterns. Cast iron is now rare. Plastics are widely used. More expensive cisterns can be obtained in ceramic materials whereas vitreous-enamelled steel is used when economy is at a premium.

The capacity of flushing cisterns in the United Kingdom is normally 9 l. Most water authorities insist that this is not exceeded. WCs consume a very large proportion of the total water supply and must be a major target in any attempt at water conservation. Many Scandinavian cisterns have a 6 l capacity and, with an appropriate pan, successful experiments have been made with cisterns down to 4 l capacity. Another cistern available in the United Kingdom is the *dual-flush cistern*. It enables the full 9 l flush to be used when solid matter is present. When this is not the case a smaller, 4.5 l flush may be used. Figure 10.5 shows the flushing mechanism for this type of cistern.

Normally the supply and warning pipes leave the cistern horizontally on opposite sides. Special cistern patterns with bottom entry are available where particular neatness of pipework is required.

In multistorey buildings warning pipes present a special problem. Normal discharge to the outside

would not be noticed. Common warning pipes are unpopular because of the difficulty of tracing the source of any flow. To overcome the problem warning pipes are now often led to baths and special fittings have been marketed to assist in this. One solution is to use a high level cistern with warning spouts delivering into its own pan.

The small-diameter delivery means that there is inevitably an appreciable delay before the cistern can be flushed a second time. In some cases, particularly in women's lavatories in industrial buildings, this gives rise to problems. Consecutive flushes can be achieved by using flushing cisterns. These consist of galvanized steel troughs running right across a row of WC compartments. There is one ball valve and one warning pipe. The delivery pipe diameter can be calculated in the normal way, taking full account of diversity. Each WC has a flush pipe and a siphonic flushing device which delivers a 9 l flush. Figure 10.6 shows the operation of the device. The quantity of water in the trough is very much greater than in a single WC cistern and advantage is also taken of the diversity of use between WCs.

Flushing valves   In many countries flushing valves deliver a fixed quantity of water directly from the pipework to the WC pan. These valves have several disadvantages. The pipework must be large enough to deliver water at the full rate required, the valve may go out of adjustment and deliver excessive quantities of water, unnoticed leakage may take place into the pan

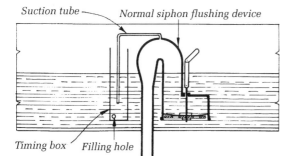

Figure 10.6 Siphonic flushing device for flushing trough. Water is drawn into the siphon through the suction tube, reducing the level in the timing box. The siphon is broken when the air enters the section tube. The duration of the siphonic action and therefore the amount of water delivered is governed by the diameters of the timing box and the filling hole and by the level of the suction tube.

and the direct connection of pipework into the pan itself may give a risk of backsiphonage. Figure 10.7(a) and (b) shows the form and operation of a flushing valve.

Slop hoppers and housemaids' closets

Slop hoppers are appliances similar to high level WCs where the pan is adapted to support a bucket or pan on a metal grid. Bib taps can be provided for filling or flushing the bucket and buckets can be emptied into the hopper and the contents flushed away. The 100 mm diameter outgo avoids problems of blockage. A sink may be associated with the slop hopper and sinks are sometimes used for disposal of excrement in hospitals where bedpan washers are not available.

The water supply for the flushing cistern is 12 mm diameter, for the taps 18 mm or 25 mm diameter hot and cold, and the outgo is 100 mm diameter.

Urinals

Urinals present the most critical nuisance problem found in sanitary appliances. They are installed in male communal and public lavatories. Their cost compares unfavourably with a high level WC suite, but cleaning is very much simplified, some space can be saved and more users can be dealt with than would be the case with WCs. Traditionally a urinal consists of a slab with a channel at its foot or a series of 'stalls' which include projecting dividing pieces. Ceramic materials are normally used. Flushing is provided at the rate of 4.5 l per stall at about 20 min intervals from an automatic flushing system and distributed through a system of sparge pipes or spreaders. Figure 10.8 shows slab and stall urinals. Since frequent use may be anticipated in situations where urinals are installed and flushing is at intervals, the flow from urinals is very concentrated relative to other sanitary appliances. In addition the faces of slab and stall urinals are wetted by urine that is not immediately washed away. These circumstances may lead to deposits in the outgo and to smells in the room containing the urinals. In addition the joints may leak and the channel lower than floor level may conflict with structural requirements. The whole urinal may have to be raised on a step to

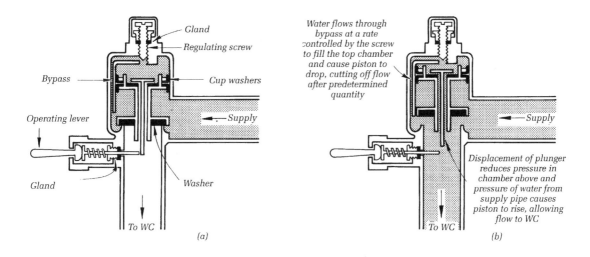

Figure 10.7 Flushing valve: (a) at rest; (b) in operation.

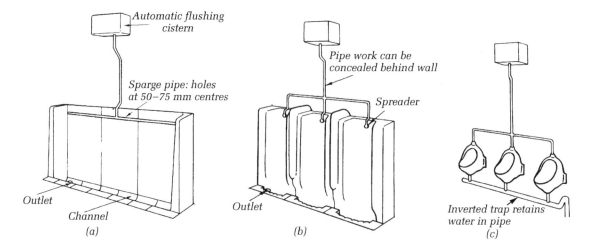

Figure 10.8   Urinals: (a) slab type with atuomatic flushing cistern, sparge pipe and channel; (b) stall type; (c) bowl type with inverted trap discharge.

accommodate the channel. An outlet with a grating to catch cigarette ends, etc., opening to a 62 mm or 75 mm diameter trap is normally provided to every six or seven stalls. Because of the concentrated flow and the nature of the cleaning fluids often employed, special care must be taken in selecting trap materials; urinal traps can be obtained in fireclay and in vitreous-enamelled metal. In expensive installations devices are often provided to minimize splashing of feet and legs. They take the form of raised ceramic screens next to the channel or glass panels at low level. The usual width of urinal stalls is 0.61 m, but it is questionable whether full use can ever be made of an installation based on this spacing.

One-piece urinals of modern materials, such as stainless steel, obviate joints in units of up to three or four stalls but are not widely used. But bowl urinals tend to be widely used in places where reasonably responsible use may be anticipated. Figure 10.8(c) shows a range of urinals of this type. They are substantially cheaper than stalls, have a smaller surface area for wetting and drying and have no cement joints. The inverted trap arrangement shown in the diagram retains some of the flushing water and thereby dilutes the first part of the fresh flows. The level of fixing requires careful thought; small boys may be inconvenienced by this type of appliance.

Bidets

Bidets (Fig. 10.9) are rare in this country, although their use is thought to be increasing. They are of ceramic materials and are provided in association with WCs for perineal washing. 12 mm diameter hot and cold valves control the temperature and rate of water supply which can in most cases be directed either round the flushing rim (thereby warming it if desired) or to an ascending spray. Outgo is via a 32 mm diameter waste and trap. It is desirable that a pop-up waste should be used rather than a plug and chain. The flushing rim and spray could become submerged if the waste becomes blocked and, in view of the nature of their use, bidets represent a particular risk of contamination for water supply. Special water supply provisions are usually called for to avoid any possibility of contamination of water. Normally the problem can be satisfied by taking special connecting pipes for both hot and cold from the distributing pipes at least 2 m above the bidet, although separate distributing pipes or even a local cold-water cistern

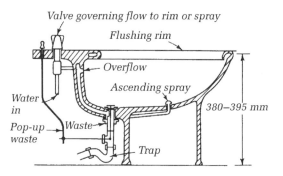

Figure 10.9   Section through bidet showing flushing rim, ascending spray and pop-up waste.

may sometimes be called for. Bidets are treated as waste appliances for drainage purposes. Figure 10.9 shows the connection of a bidet into a domestic water supply installation.

Baths and showers

Cast iron finished with vitreous enamel is the most usual material for baths. Glazed fireclay baths are available and have some advantages but they are expensive, heavy and cold to the touch unless they have been warmed by hot water. Their use is therefore confined to institutions where use is fairly continuous. In recent years enamelled pressed steel, fibreglass reinforced plastics and Perspex have been employed and the use of Perspex is becoming well established. For special designs baths were sometimes cut from marble. Terrazzo is a more economical material likely to be used at present for in situ special baths. Both marble and terrazzo suffer from the same limitations as fireclay.

The traditional roll-edge, free-standing bath still used in institutions, is shown at the top of Fig. 10.10. The Magna type shown at the centre is almost universally used for present-day domestic installations. Its flat top with straight raised edges can be built against walls, although watertight joints are difficult to achieve, particularly when the bath stands on a wooden floor, and a cover panel can be simply installed to conceal the underside and feet. The Sitz bath (lower part of Fig. 10.10) is economical of space and water and has advantages for old people since the user can maintain a normal sitting position. All these baths can be obtained in a variety of colours and with a variety of special details, such as hand grips, soap recesses and decorative features. There is a clear tendency to use smaller baths for economy of space and hot water.

Water supply to baths is usually by means of hot and cold pillar taps at the end of the bath or on the long side near the wall. In domestic circumstances 18 mm diameter taps are used but in institutions 25 mm diameter taps are normal to increase the speed of filling. Valves are sometimes used instead of taps with the water delivered to the bath by a spout. Provided both hot and cold water are supplied from a storage system and backsiphonage cannot occur, it is possible to have an inlet spout near the bottom of the bath. This is described as a steamless inlet since hot water introduced at low level will give off less steam than would be the case if the water fell from a tap. Special inlet fittings can be obtained combining hot and cold valves, spout, diverter valves and hand-shower fittings.

Wastes are 38 mm diameter in domestic situations

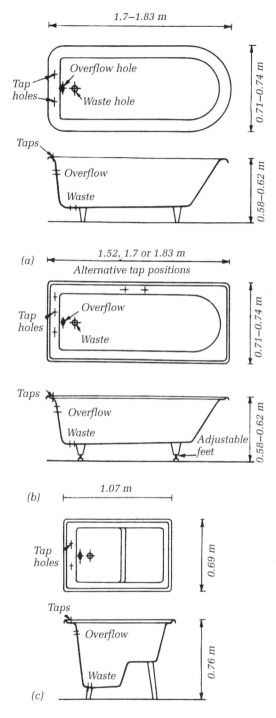

Figure 10.10   Typical baths: (a) traditional roll-top free-standing bath; (b) Magna type; (c) Sitz bath.

and 50 mm in institutions. Deep seal traps will sometimes require pockets to be left or cut in the floor because of the limited clearance under the bath. An overflow is normally provided. These overflows used to be taken through outer walls to discharge into the open. But the consequent cold draughts, entry of insects and inconvenience if water flowed out of the pipe are now avoided by connecting the overflow into the trap (Fig. 10.11). Plugs and chains are normally used to control outflow. Pop-up wastes may be used but they increase the length of the waste, making the accommodation of the trap more difficult. The standing waste is a tube rising from the waste outgo to overflow level down which water will flow. By raising or turning the tube the contents of the bath can be discharged into the waste. This device is more easily cleaned than the concealed overflow but is not so neat in appearance and does not appear to be widely used.

Many baths have hand or fixed shower fittings. Operation of the shower may cause some splashing, often controlled by plastic curtains falling into the bath. Where showers are used independently of baths, they are usually formed by a shallow ceramic tray 1 m square in a tiled compartment. A 35 mm diameter waste discharges water from the tray.

Two types of delivery heads are used for showers, the traditional rose type delivers water through a disc

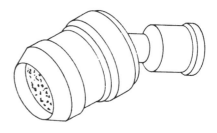

Figure 10.12   Adjustable 'umbrel a' type shower spray.

pierced with holes and is frequently fixed overhead. The adjustable umbrella type (Fig. 10.12) has come into use relatively recently. This type of spray consumes very much less water than the rose type and is conventionally installed at chest level. These points avoid the previous disadvantages of showers: high rates of hot water consumption and inevitable discharge of water overhead, very unpopular with women. The hot and cold water supplies to shower sprays must be similar in pressure and taken from storage cisterns. The simplest form of control is by manual adjustment of valves on hot and cold supplies. Fluctuation of pressure in one of the supplies, due to the opening of other taps, can alter the temperature of the shower and may even cause scalding. Mixing valves provide a superior method of control. Non-thermostatic valves normally have antiscalding provisions, which shut off the hot water if the cold water fails. Thermostatic valves automatically adjust themselves to give a fixed temperature of flow. Figure 10.13 shows a thermostatic mixing valve with associated isolating valves. In addition to their use in individual shower compartments, shower sprays can be used for walk-through installations. Single mixing valves can deliver water to a range of shower sprays. A minimum head of 1.5 m is necessary to operate a shower spray and this will be increased if a mixing valve is employed. This requirement can influence the placing of the cold-water cistern in domestic buildings.

Washbasins

A wide variety of sizes and designs are available. Figure 10.14 shows the features of a typical basin, which may range in size from barely large enough to wash the hands to 500 mm × 640 mm or more. Basins are normally made of ceramic materials, although others such as plastics are available and used where weight is important. Water supply is controlled by 12 mm diameter pillar taps for hot and cold. The waste

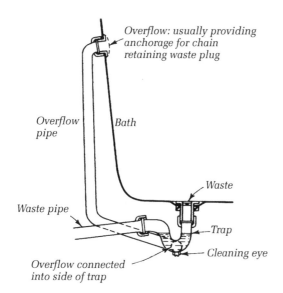

Overflow: usually providing anchorage for chain retaining waste plug

Overflow pipe

Bath

Waste

Waste pipe

Trap

Cleaning eye

Overflow connected into side of trap

Figure 10.11   This bath overflow delivers water into a trap below the water level, preventing cold draughts and ingress of insects.

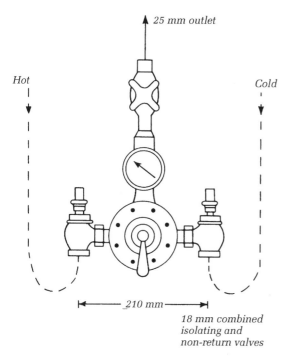

Figure 10.13 Thermostatic mixing valve serving a range of showers.

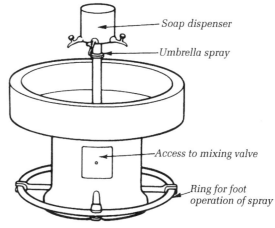

Figure 10.15 Wash fountain for industrial hand washing. Current water supply bye-laws require a separate spray and tap for each place at the fountain.

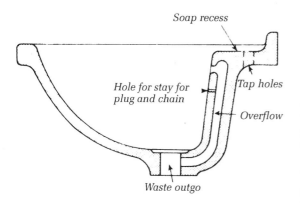

Figure 10.14 Section through typical lavatory basin.

outgo has a 32 mm diameter and incorporates an integral overflow as shown in the diagram. Control of outgo is usually by plug and chain but pop-up wastes are available. Self-draining recesses to hold soap are formed in the shelf. This type of basin has

disadvantages. The overflow and plug and chain waste are not easy to keep clean, the position of the taps makes it difficult to wash the hands in running water and the flows may be too cold or too hot. To overcome these problems a type of basin has been introduced that has only one tap, which delivers a spray of water at the right temperature for washing, and has no overflow or plug and chain. This basin is very well suited to hand washing but cannot be filled with water for the other uses which form an important part of domestic requirements. Its use is therefore usually confined to communal and public lavatories. Basins can be supported in several ways. Some basins may be cantilevered from the wall using lugs. Brackets built in or plugged to the wall give the same type of support. A ceramic pedestal support is often used; it conceals the pipework but makes access more difficult. A metal frame fixed to the wall and having legs and rails for towels may also be used.

Rectangular washbasins can be joined into ranges by cover strips. A space between each basin is neater and easier to clean and allows more comfortable use of the basins. In industrial situations economical and easily maintained hand-washing arrangements can be made using long troughs with spray taps or circular wash fountains that employ an umbrella spray. Figure 10.15 shows a ceramic wash fountain. Stainless steel models are available. In common with most other sanitary appliances the taps serving lavatory basins must discharge above the flood level of the appliance to obviate any risk of backsiphonage. Figure 10.16 shows the nature of the clearance required.

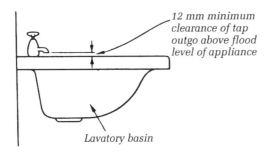

12 mm minimum clearance of tap outgo above flood level of appliance

Lavatory basin

Figure 10.16   Water supply to obviate the risk of backsiphonage.

Sinks

The Belfast sink emerged as the most successful of a variety of ceramic sinks. A range of sizes can be obtained, 610 mm × 460 mm × 205 mm is popular for domestic use. Figure 10.17 shows a Belfast sink. This type of sink is usually equipped with a draining board most often made of wood. This is not easy to keep clean, particularly at its point of junction with the sink and pressed steel sinks with integral drainers which give no lodgement for dirt have largely superseded the Belfast sink for domestic and similar use. Stainless steel and vitreous-enamelled steel are popular and vitreous-enamelled cast iron is available. Patterns with single and double drainers and single and double sinks can be obtained. One disadvantage of the sheet steel sinks is the noise which crockery makes. They are supplied with sound-deadening material applied to the underside. Teak sinks reduce crockery breakage and their use has been popular in large kitchens but the hygienic advantages of stainless steel are leading to its widespread use in these circumstances too.

The water supply to Belfast sinks is by bib taps usually 12 mm diameter for domestic sinks and 18 mm

or 25 mm for large sinks in institutions. Stainless steel and vitreous-enamelled sinks employ either separate 12 mm diameter pillar taps for hot and cold or mixer taps with swivel-arms. Kitchen sinks are supplied with cold water direct from the main and where mixer taps are used they must be of the biflow pattern, which does not allow the hot and cold water to mingle in the tap. Some special fittings to assist in washing-up such as flexible hand-sprays are available in addition to conventional taps. Wastes are 32 mm diameter for domestic sinks and 38 mm diameter for large sinks. Control is by plug and chain.

Scales of provision of sanitary appliances

BS 6465 : Part 1 *Sanitary installations* gives tables recommending scales of provision of sanitary appliances.

Taps and valves

The terms *tap*, *cock* and *valve* are loosely employed in common building usage. They are distinct in their functions, however, and should be distinguished.

Cocks consist of a body holding a bored plug, the hole through which can be aligned with the pipe, thereby allowing water to pass. A quarter-turn can fully open or fully close the cock. They are not generally permitted in water supply installations since the sharp turning off can give rise to water hammer.

Valves control flow in a pipeline. The two types most frequently used in building pipe installations are

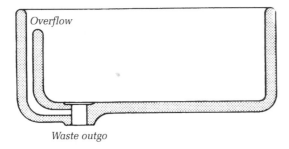

Overflow

Waste outgo

Figure 10.17   Section through typical Belfast sink.

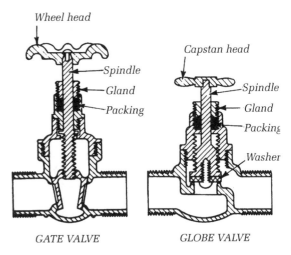

Wheel head

Spindle
Gland
Packing

GATE VALVE

Capstan head

Spindle
Gland
Packing
Washer

GLOBE VALVE

Figure 10.18   Two types of valve.

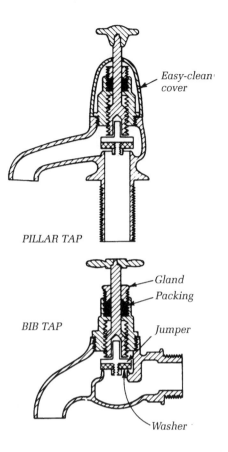

PILLAR TAP

Easy-clean
cover

BIB TAP

Gland
Packing
Jumper

Washer

Figure 10.19 Two types of tap.

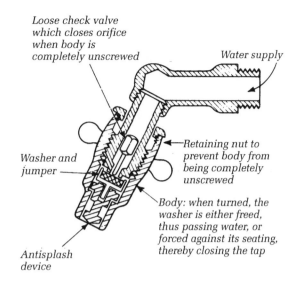

Loose check valve
which closes orifice
when body is
completely unscrewed

Water supply

Washer and
jumper

Retaining nut to
prevent body from
being completely
unscrewed

Body: when turned, the
washer is either freed,
thus passing water, or
forced against its seating,
thereby closing the tap

Antisplash
device

Figure 10.20 The Supatap is a bib tap.

globe valves and gate valves. Figure 10.18 shows both types. Globe valves are used on water supply pipework where adequate head is available and complete cut-off is essential. Gate valves offer less resistance to flow when open. Full-way types are available. They are employed when head is limited and particularly on heating pipework. Mains stopcocks are normally globe valves with washers held by loose jumpers. It is assumed that the loose jumper will act as a non-return valve if conditions for backflow ever arise.

Taps are fittings permitting draw-off of water. Figure 10.19 shows a bib tap appropriate to wall fixing over a sanitary appliance and a pillar tap which is fixed in the sanitary appliance itself.

There are many kinds of special taps for different purposes. Spring-loaded taps operated by a push button are much used in factory lavatories in an endeavour to save water. They should be of the non-concussive type, which prevents 'hammer' noise. Many water authorities do not like self-closing taps. There are many kinds of non-splash taps and some

designed so the washer can be replaced without having to turn off the water supply. Many 'combination sets' are marketed with special lavatory basins and baths, wherein hot and cold supplies are valved and the outlet is combined. With such fittings the cold cannot be connected to the mains, owing to the possibility of contamination of the main supply if it is connected directly to the hot supply. Mixing valves serve the same end but the better types are now controlled by one handle, which turns from the off position through cold up to hot. This prevents the possibility of scalding and is almost essential for showers. Thermostatic mixing valves are now obtainable and are very useful in effecting economy in hot water in big institutions as they can be set at any required temperature. Most of these valves require the hot and cold water supply pressures to be balanced.

A recent development which should become a normal accepted fitting in offices, works and other places where there is hand washing is a spray tap scientifically designed to give a cone of fine spray, adequate for washing but using comparatively little water. One such is marketed under the name of Unatap. Under independent tests by the Building Research Establishment, great economy of water use has been shown. When combined with a mixing valve, or on a hot supply of set temperature, the saving leads to a saving in fuel.

Another type of tap development, the Supatap (Fig. 10.20), has a check valve that cuts off the flow when

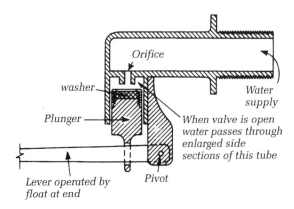

Figure 10.21   The Croydon ball valve.

Figure 10.23   The BRE ball valve.

the washer is removed. Washers can therefore be removed without turning off the water supply.

Ball valves which supply water and maintain the level in water storage cisterns and flushing cisterns should, in terms of the definitions above, be called float-operated taps. The term *ball valve* is, however, universally applied. There are several types. Figure 10.21 shows the Croydon type of ball valve which was widely used and is still sometimes installed. It cannot

be fitted with a silencing tube, which is a serious disadvantage. Figure 10.22 shows the Portsmouth type, governed in materials and manufacture by BS 1212. Notice the silencing tube. The Portsmouth valve is available in a range of sizes and in each case different orifice sizes allow for low, medium or high pressure supplies. Trouble with slow filling of flushing cisterns is sometimes due to the inadvertent fitting of high pressure valves in situations where there is a small head. Both the above types of ball valve are subject to trouble due to deposits of salts from the water impeding the operation of moving parts and wear and noise from the nature of the water flow through the valve. At the Building Research Establishment, research into these problems led to the development of the diaphragm valve (Fig. 10.23). This valve gives substantially quieter operation and better reliability in operation. The design of the nylon orifice reduces noise and wear; the plunger and pivot, which close the orifice, are kept dry so that no deposits occur to impede operation. Although the BRE valve represented a significant advance, it did not meet the increasing concern felt for prevention of backsiphonage. Although the silencing tubes of ball valves have small holes to prevent backsiphonage, this does not provide a high enough standard to meet the requirements of the Model Water Bye-laws. A further development of the BRE valve which satisfies the backsiphonage requirements is shown in Fig. 10.24. Instead of a silencing tube descending into the cistern, an outlet tube at high level delivers a number of small streams of water against the side of the cistern, thus achieving complete freedom from risk of backsiphonage together with quiet operation. The orifice is flooded during operation and this is said to reduce the noise level even further.

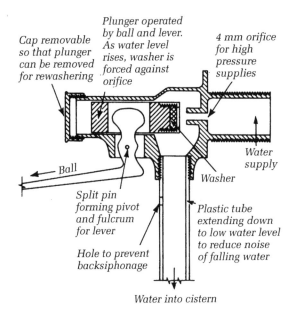

Figure 10.22   The Portsmouth ball valve (BS 1212).

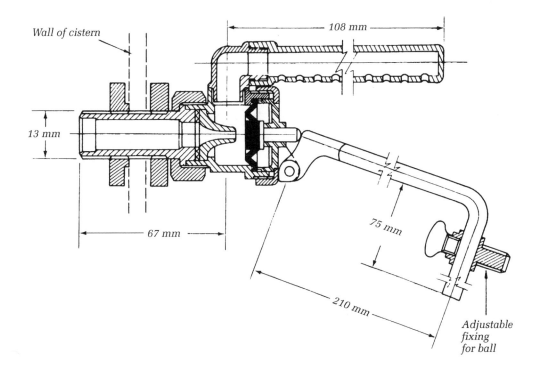

Figure 10.24   The Peglers diaphragm float valve is made to BS 858.

BS 858 covers this type of valve. In large buildings with large-diameter service pipes delivering to the storage cisterns the buoyancy of a float is not adequate to balance the pressure of the main operating against a substantial area of water. The equilibrium ball valve shown in Fig. 10.25 overcomes this problem and the float is only called upon to overcome friction.

The company Ideal-Standard has now introduced the Torbeck, an all-plastic equilibrium-type float valve.

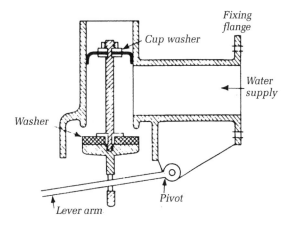

Figure 10.25   In this equilibrium-type ball valve the water pressure on the cup washer balances the pressure on the washer.

11 Pipes

Historically employed for pipes in buildings, lead owed its universal appeal to its availability, durability and ease of working. Industrial developments have produced a range of new materials which have almost entirely superseded lead for pipes. Some of the advantages which new materials may enjoy are improved economy, better appearance, lighter weight, ease of installation, reduced toxic risk and the ability to sustain higher water pressures. Cast iron, copper, wrought iron and steel have been in use for many years. More recently asbestos cement, plastic, thin-walled copper tube, stainless steel and coated tubes have come into widespread use.

11.1 Materials: properties and applications

Lead and lead alloy

Lead is obtained from the mineral galena, a compound of lead and sulphur (about 86% lead). After smelting the lead contains small quantities of antimony, tin, copper, gold and silver. These are removed by refining so that good commercial lead possesses a high degree of purity, over 99.99%. Lead is very soft. It is ductile and malleable, and does not appreciably work harden, so does not need annealing. It is highly resistant to corrosion by the atmosphere; a protective film is formed on the surface by the action of oxygen, carbon dioxide and water vapour.

Lead is very dense, $11,325 \, kg/m^3$, and has a low melting point, $327 \, °C$. The quality of lead is improved for many purposes by alloying with very small quantities of other metals; silver–copper–lead alloy has increased tensile strength and much improved creep resistance.

Uses

Lead is not now permitted for new water supply pipework or for repairs to existing systems and is rarely used for other building pipework applications. It is, however, present in many existing buildings in drainage pipework and renovations may be required.

Merits

Lead is easily worked, cut, bent cold, soldered, hence its popularity for short or complicated connections. Its slight flexibility makes it useful for connections where there may be slight movement, e.g. between rigid soil pipes (cast iron) and brittle sanitary fittings. It has a tendency to creep, i.e. it expands on heating, but tends to deform on cooling rather than contract to its original form, so it needs support when used horizontally or for hot liquids. It is now expensive; alternatives are lighter and easier to handle.

Jointing

Jointing is by soft soldering (Fig. 11.1) or by lead burning – gas welding in which the lead of the joint is first fused then strengthened by the addition of molten metal from a lead filler strip. For jointing to sanitary fittings and drainware, lead pipe has to be strengthened by use of brass sleeves or brassy sockets, to which it is soldered. For connection to metal fittings, lead pipe is usually soldered to brass tailpiece and connected to the fittings by coupling or union nuts, thus facilitating disassembly.

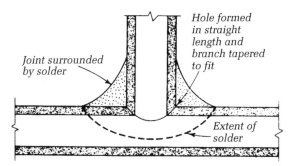

Joint surrounded by solder

Hole formed in straight length and branch tapered to fit

Extent of solder

Figure 11.1   Wiped lead joint.

Fixing

Fixing is by lead tacks soldered to the pipe. Tacks at 0.76 m centres are recommended for horizontal pipes; about 1.2 m centres are recommended for vertical pipes.

Asbestos cement

Asbestos cement is a manufactured material with three main applications: pressure pipes for water mains, soil and ventilating pipes and rainwater pipes and gutters.

Merits

Asbestos cement is economical and resistant to decay. It is used below ground for water mains as it is resistant to many soil conditions and to the bacteria which attack cast iron. In some cases water or soil conditions may require protective measures. Flexible joints are available for underground pipes. Above ground painting is not necessary except for decorative purposes.

Standards

Piping for water mains is governed by BS 486 *Asbestos cement pressure pipes*. There are three grades capable of withstanding different pressures and diameters from 50 mm to 600 mm. Soil and vent pipes are covered by BS 582; 50, 62, 75, 87, 100 and 125 mm diameters are made.

Jointing

Pressure pipes can be fitted with flexible joints. Soil and vent pipes have socket and spigot joints which should be sealed with caulked lead wool or asbestos jointing compound.

Fixing

When fixed above ground ring pipe-clips are used below each socket.

Copper

Uses

Cold-water service pipes below ground, all cold- and hot-water supply pipework, heating system pipework, particularly in small buildings, and waste pipes.

Merits

Extremely durable, strong, resistant to mechanical damage, easily worked, good flow characteristics.

Standards and specifications

BS 2871 specifies four grades:

- **Table W**: small-diameter tubes for microbore heating, supplied in coils.
- **Table X**: normal above ground uses, supplied in straight lengths of 6 m; outside diameters 15–159 mm (previously BS 659 *Light gauge*).
- **Table Y**: soft temper, easily bent, for underground use, supplied in 20 m coils.
- **Table Z**: thin-walled, hard drawn pipes (previously BS 3931); cannot be bent or used below ground; supplied in 6 m lengths.

BS 864 : Part 2 specifies suitable capillary and compression fittings for jointing.

Jointing

Jointing is by welding or by capillary or compression joints. Compression joints can be manipulative or non-manipulative. Manipulative joints are used below ground since the deformation of the pipe ends renders slipping or pulling out of the joint unlikely. Figure 11.2(a) shows a typical manipulative joint. Non-manipulative joints (Fig. 11.2(b)) are easier to form and are used above ground with BS 659 tube.

Capillary solder joints (Fig. 11.3) have a very neat appearance. They are formed by introducing clean, square, fluxed pipe ends into closely fitting joints. Solder is either applied or is contained in the fitting and when heated by blowlamp spreads throughout the joint by capillary attraction. Endfeed joints, to which solder must be applied, are also available.

Welding is much used for larger pipes and repeat work, such as a unit serving combined soil and waste to a bathroom or kitchen on one floor of a block of flats. Bronze welds are strong and easily made; autogenous welds can be buffed up so they hardly show. Connections to other materials are usually made with special fittings, available in a wide range according to the type of proprietary joint chosen, or by welding.

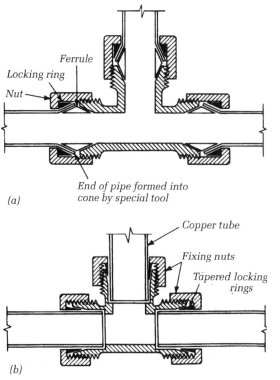

(a)

Ferrule

Locking ring

Nut

End of pipe formed into cone by special tool

Copper tube

Fixing nuts

Tapered locking rings

(b)

Figure 11.2 Compression joints for copper tubing: (a) manipulative; (b) non-manipulative.

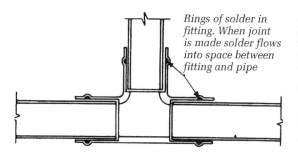

Rings of solder in fitting. When joint is made solder flows into space between fitting and pipe

Figure 11.3 Capillary solder joint for copper tubing.

Fixing
Fixing is usually by pipe-clips or holderbats. As copper pipe is so rigid, fixings need not be as frequent as with lead. BS 6700 *Design, installation, testing and maintenance of water supply* contains a table of recommended spacings.

Steel pipe

Uses
Hot water and heating installations are the principal uses, but steel pipe is also used in cold-water distribution and waste and ventilation pipework with special fittings.

Merits
Steel is cheap, very strong, both for water pressure and in resisting damage. It is susceptible to corrosion so should always be galvanized, except for closed circuits such as heating installations. Corrosion may be excessive where water is very soft. Steel is easily welded. Smaller diameters are cold bent, but much work is forged.

Standards and specifications
Steel pipe is made to BS 1387. It is available in sizes, nominal bore up to 150 mm, and in three classes: A has thin walls and is banded brown; B has medium walls and is banded yellow; and C has thick walls and is banded green. C quality is usually required for rising mains and work underground; B is used for distribution and hot-water pipes where head is not too great; A is for waste, ventilating and overflow pipes. Some authorities require C quality for all work.

No standard dimensions can be given for steel pipework for wastes, stacks and branches as they are usually multibranch assemblies designed to suit the job, with branches or bosses welded in to take screwed joints in the exact positions to suit fittings. This pipe can be supplied with spigot and socket joints, which are preferably caulked with a cold caulking compound. The multibranch units and pipes should be galvanized after fabrication.

Fittings in wrought iron, used for steamwork, and gas and water over 75 mm diameter, are covered by BS 143 and in malleable iron, for domestic work, by BS 2156. BS 2156 pipes should be galvanized.

Jointing
Jointing is by screwed and socketed joints (Fig. 11.4), or by welding. Pipes, particularly the larger diameters, can also be obtained with flanges, which facilitate bolting together, to valves and to flanged fittings. Asbestos fibre washers are used in the joint. Bends should be forged, where possible, or cold bent to allow easy flow. Frictional loss and loss by elbow and other sharp fittings can restrict the flow considerably.

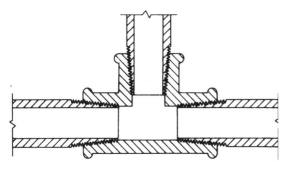

Figure 11.4 Screwed joint for steel tubing.

Fixing

Fixing is by pipe-clips or holderbats. The pipe is so rigid that fixings can be at wide intervals.

Enamelled iron pipes – mostly rainwater goods and special fittings for laboratories – are made of heavy gauge plate, pressed and welded and vitreous enamelled. These have great durability and are maintenance free. Jointing is by socket and spigot with caulking if required.

Cast iron

For nearly two centuries the strength and durability of cast iron pipes made this material dominant in waste and soil installations and in underground drains subject to possible movement. Sockets for jointing were cast as part of the pipe. Branches required a special branch unit with integral sockets. Joints were made by inserting a plain pipe end into the socket and sealing the gap with molten lead. This was a highly skilled operation. Many different units were required to provide ranges of branch configurations and junctions with sanitary appliances and other types of pipe. Although capable of lasting the lifetime of the building, the result was cumbersome and unsightly as well as very expensive. When plastic pipes capable of providing soil and waste rainwater installations became available they were very widely used, although they did not provide the same strength, durability and fire resistance as the cast iron pipes.

Recent technical developments in cast iron pipe production and design, pioneered in Europe, have changed this situation. Improved casting techniques achieve a more consistently dimensioned pipe with a thinner wall and neat couplings rather than the spigot and socket joints previously required. Wastage is reduced since cut lengths of pipe can be used. In addition to the couplings, special connectors are provided to make neat joins between branches having different diameters. The pipes have protective coatings both internally and externally. In the past there were separate specifications for above-ground and below-ground cast iron pipes. The new pipes are suited to both purposes with considerable savings in manufacture, distribution and detailing.

There is no British Standard for this type of pipe since a European Standard is being prepared. Manufacturers of this type of pipe should, however, have British Board of Agrément Certificates. The use of the traditional pipes for new work has almost ceased. However, existing installations will last for many years and require renovation and adaptation. The existing diagrams in this book have been retained to show traditional installations and Fig. 11.5 shows the new system. It will still be possible to obtain special orders of old patterns for renovation work.

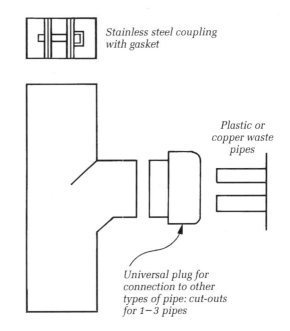

Stainless steel coupling with gasket

Plastic or copper waste pipes

Universal plug for connection to other types of pipe: cut-outs for 1–3 pipes

Figure 11.5 This thin-walled cast iron pipe has no sockets and its pipe ends are plain.

Plastic pipes

The application of plastics to piping is rapidly developing and British Standards have changed to meet the needs of new developments and metrication.

Plastic pipes are generally light, easily worked, frost resistant, economical and, depending on the material, may be available in a range of colours. Some types degrade in sunlight and several types are not suitable for flow temperatures over 20 °C. Acceptable pressures should be checked against pipe diameters and head of water. Proper spacing of supports is also important above ground. Many suppliers give proprietary names to their products. The principal materials are outlined below.

Polyethylene

One of the earliest plastic materials to be introduced for pipe installations. It is flexible, tough and very light. It can be bent cold but will spring back unless warmed to make the bend permanent. The material softens when heated and cannot be used for applications with flows at temperatures higher than 20 °C. Bore sizes range from 15 to 400 mm. There are two main types defined by British Standards:

- **BS 6572**: this pipe is blue in colour and is degraded by sunlight. Its main application is for underground supplies. In appropriate circumstances small-diameter pipes can be laid by mole plough. If the ground is impregnated by gas from leaking mains, water standing in the pipe may become tainted. Granular bedding should be used to avoid damage to the pipe when backfilling trenches.
- **BS 6730**: this pipe is black and is intended for use above ground. It requires support.

Jointing is normally with compression fittings, which should be corrosion resistant if used below ground. Figure 11.6 shows an ingenious proprietary push-fit joint. Thermal fusion can also be used with proprietary equipment.

High density polyethylene (cross-linked)

Additional chemicals added at the time of manufacture, together with high temperature and pressure produce a flexible pipe which has a wide temperature tolerance (+110 to −140 °C).

uPVC (unplasticized polyvinyl chloride)

[BS 3505] Rigid pipes suitable for temperatures up to 20 °C. Widely used for cold-water supply and particularly for water mains. Pipe sizes range from 10 to 315 mm. Compression, solvent and push-fit O-ring seal joints are used (Figs 11.7 and 11.8).

cPVC (chlorinated polyvinyl chloride)

cPVC is a similar material to uPVC. However, it is more resistant to softening and can be used with hot

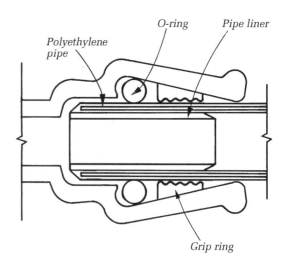

Figure 11.6   The Talbot push-fit coupling for polyethylene pipe.

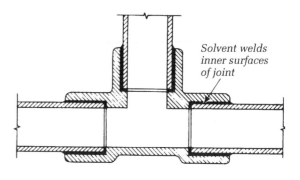

Figure 11.7   Solvent weld joint for PVC pipe.

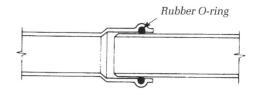

Figure 11.8   Rubber ring joint for PVC or copper pipe.

water. The range of pipe diameters is very much more limited.

Polypropylene

Polypropylene has an excellent resistance to chemicals and ultraviolet radiation; it can withstand temperatures up to 100 °C. Pipe diameters range from 20 to 110 mm. Polypropylene is mainly used in the process industries.

Polybutene
[BS 7291 (insulation); BS 5955: Part 8 (installation)]
Recently marketed, polybutene provides good performance across a range of properties, including good heat, stress and impact resistance as well as low creep and high flexibility. It is grey in colour and does not require protection from ultraviolet radiation if installed internally. It is suitable for both cold- and hot-water supplies, and for domestic central heating. Several manufacturers offer proprietary systems based on the material. The range of diameters and fittings depends on the system. Jointing is by compression fittings or by fusion. At least one manufacturer offers proprietary fittings with electric heating elements built in so that thermal fusion can be achieved by the short application of an electric current.

ABS (acrylonitrile–butadiene–styrene)
Strong, highly resistant to corrosion and with operating temperatures from −40 to 70 °C, ABS pipes are available with outside diameters from 16 to 225 mm. Its main applications are in the process industries and for cooling pipework in buildings.

Stainless steel
[BS 4127] Stainless steel is very similar to copper. The pipe diameters and dimensions conform with BS 2871 and the joints used are usually those made for copper tubes (BS 864 : Part 2). But the pipes can be brazed and fittings for adhesive jointing are available. Copper pipe bending machines are used for bending stainless steel, but the stainless steel is stiffer than copper, so bending requires more care.

11.2 Pipe fixings

Pipes fixed on walls should be secured clear of the surface by a suitable clip. Figure 11.9 shows several types ranging from the economical saddleband to the more expensive and elegant pressed and cast bronze clips. Fixing should be firm to avoid noise but not so rigid that thermal movement is prevented.

Where pipes run through walls or floors a sleeve of large-diameter pipe should be provided and the entry or exit points covered by pipe-clips as shown in Fig. 11.10.

The recent requirement that no pipe should be embedded or built in so that it cannot readily be accessed could lead to difficulty and expense in the provision of ducts and covers. It appears that pipes can

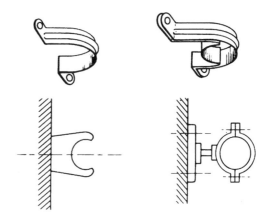

Figure 11.9 Pipe clips.

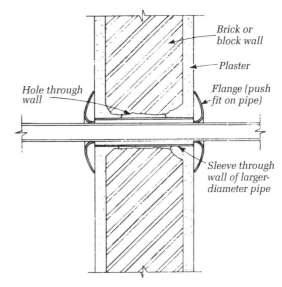

Figure 11.10 Detail showing sleeve and flanges for pipe passing through wall or floor.

be laid in chases covered by normal screed or plaster finish, but only where leakage would become immediately apparent and joints kept to a minimum.

Several manufacturers of flexible pipe have developed a system to overcome the problem, described as 'pipe in pipe'. In this system a flexible conduit pipe is built in during the course of construction; water pipe is threaded through when the plumbing is installed. The system is supported by a range of fittings to provide support at bends and to provide neat terminals at the point where the water pipe emerges.

11.3 Estimating pipe sizes for water supplies

For both cold- and hot-water supply installations it is necessary to be able to estimate the volume of water to be stored and the appropriate sizes of pipe to deliver the right flow of water to taps.

Storage for cold-water supply

For dwellings the Model Bye-laws for the Prevention of Waste, Misuse and Contamination of Water Supplies require 230 l actual capacity of storage where supply for both cold and hot is by cistern, and 114 l actual capacity for hot-water supply only. Cistern sizes are usually quoted in relation to the full capacity, making no allowance for space occupied by ball valves and warning pipes. Consequently, galvanized steel cisterns of 318 l and 182 l nominal capacity would be required. Table 11.1 gives the range of sizes for galvanized sheet steel cisterns according to BS 417. It is becoming increasingly usual, in domestic application, to use plastics or asbestos cement cisterns, which have a longer life than galvanized steel in most circumstances. Except for domestic cases, storage is based on the anticipated use by the population of the building. Table 11.2, from BS 6700 *Design, installation, testing and maintenance of water supply*, gives recommended storage capacities per head of population to give 24 hours' reserve in a variety of building types.

Storage for hot-water supply

Adequate storage must be provided to meet peak demands for hot water, but the heat input itself is spread over a longer period (known as the recovery period). For domestic installations it is usual to use a cylinder of 114 l actual capacity (140 l nominal) for dwellings where only one bath will be taken at one time and 155 l actual capacity (180 l nominal) where two baths will be taken in quick succession. In larger buildings suitable volumes of storage in relation to population are also established. Table 11.3 gives a schedule from the *IHVE Guide to Current Practice*. CIBSE now recommends that hot-water storage sizes should be calculated. The method used, although more complex, is essentially similar to the graphical method shown in Fig. 11.11. However, the original table may still be useful in giving an immediate indication of approximate quantity. Table 11.4 gives sizes of hot-water cylinders in accordance with BS 1565 and BS 1566. Cases may be encountered when these figures do not apply and some buildings may impose

Table 11.1 Cold-water cisterns: sizes from BS 417, Table 1

BSS no.[a]	Capacity (l)		Dimensions (m)		
	Nominal	Actual to water line	Length	Width	Height
C 1	45.5	18	0.45	0.31	0.3
C 2	68	36	0.61	0.31	0.38
C 3	91	55	0.61	0.41	0.38
C 4	114	68	0.61	0.43	0.43
C 5	136	86	0.61	0.46	0.48
C 6	182	114	0.69	0.51	0.51
C 7	182	114	0.61	0.61	0.48
C 8	227	159	0.74	0.56	0.56
C 9	272	191	0.76	0.58	0.61
C 10	318	232	0.81	0.66	0.61
C 11	318	227	0.91	0.61	0.53
C 12	364	264	0.91	0.66	0.61
C 13	455	336	0.97	0.69	0.69
C 14	455	327	1.22	0.61	0.61
C 15	569	423	0.97	0.76	0.79
C 16	910	710	1.17	0.89	0.89
C 17	1 137	840	1.5	0.81	0.91
C 18	1 592	1 228	1.5	1.14	1.14
C 19	2 274	1 728	1.83	1.22	1.22
C 20	2 730	2 138	1.83	1.22	1.02
C 21	4 548	3 367	2.44	1.5	1.5

[a] Note that grade A and grade B are described in this Standard. Dimensions are identical, but grade A is made from heavier gauge steel and should generally be specified.

Table 11.2 Cold-water storage: volume to cover 24 h interruption of supply, as recommended by BS 6700

Building type		Volume (l)
Dwellings	90	per resident
Hostels	90	per resident
Hotels	135	per resident
Offices	37	per head
Offices with canteens	45	per head
Restaurants	7	per head per meal
Day schools	27	per head
Boarding schools	90	per resident
Nurses' homes and medical quarters	115	per resident

space restrictions that make the full volume of storage difficult to achieve. In these cases it is possible to use a graphical method to estimate the right balance of the volume stored and the recovery rate (Fig. 11.11). In this method the estimated hot-water consumption is entered on the top part of the chart on the basis of the figures given and totalled for each hour. Commencing after the slack period during the night, when the water will have become fully heated, the amount of water consumed each hour will be deducted from the amount of cold water that can be heated (recovery rate); if a

Table 11.3 Hot-water storage capacity and boiler power as recommended by the Institution of Heating and Ventilating Engineers

Type of use	Storage capacity per person (l)	Boiler power per person (W)
Dwellings		
(medium rental)	32	750
Boarding schools	23	750
Day schools	4.6	90
Offices	4.6	120
Factories	4.6	120
Hotels, first class	46	1 200
Hostels	32	750
Hospitals, general	27	1 500
Nurses' homes	46	900

Table 11.4 Copper indirect hot-water cylinders: examples from BS 1566

Type	Capacity (l)	Height (mm)	Diameter (mm)
3	114	1 050	400
7	117	900	450
8	140	1 050	450
9E[a]	206	1 500	450

[a] Preferred for off-peak electric heating.

Table 11.5 Sanitary appliances: rates of flow recommended by BS 6700

Sanitary appliance	Rate of flow (l/s)[a]
WC (flushing cistern)	0.11
Lavatory basin	0.15
Lavatory basin, spray tap	0.03
Bath tap	
19 mm	0.30
25 mm	0.60
Shower (umbrella spray)	0.11
Sink taps	
13 mm	0.19
19 mm	0.30
25 mm	0.40

[a] The same value applies to hot and cold water.

negative answer is obtained this is deducted from the volume of hot water available and the result plotted on the chart at the foot of Fig. 11.11. Positive answers will be added to the volume available up to the full capacity of the hot-water cylinder or calorifier. If the curve of hot water available formed in this way is completed for the whole day any likelihood of the reserve being exhausted and hot taps running cold will be demonstrated. Since the hot water remains at the top of the cylinder and the cold water at the bottom – there is comparatively little mixing, particularly if the cold-water inlets are properly designed – it may be assumed that the calculated amount is an accurate estimate of the actual amount, even though part of the cylinder is filled with cold water. If the reserve of hot water is reduced at any time to below 20–25% of the total volume, the system should be regarded as inadequate. Where a system is demonstrated to be inadequate by the method described, two means of correction are available: the volume of storage can be increased or the heat input (recovery rate) increased and the new values checked by repeating the method.

Pipe sizing: cold- and hot-water supply

Adequate rates of flow are required at taps. Table 11.5 shows the rates recommended in BS 6700 *Design, installation, testing and maintenance of water supply*. Very small installations will be designed to allow all the taps to operate simultaneously and the figure in Table 11.5 can be used to give the flows required in the pipes. In large installations it would be unrealistic and uneconomic to design in this way since the probability of simultaneous use of all the taps will be very remote. The problem of arriving at a realistic figure for design flow is discussed in an article by Burberry and Griffiths (Service engineering: hydraulics, *Architects' Journal*,

21 November 1962, 1185–91); it is also considered in AJ Information Sheets 1163 to 1174. But here suffice it to say that simple numerical values may be allocated to sanitary appliances, values which represent the contribution to loading of the system. The values may be added for all the appliances served by a particular length of pipe, then by using Fig. 11.12 it is possible to arrive at a suitable value for design of the pipework. Sanitary appliances running continuously are catered for by adding the actual flow to the design flow for intermittently used appliances read from Fig. 11.12. Table 11.6 gives a schedule of values for water supply to sanitary appliances. The values are called simultaneous demand units, or simply demand units.

Figure 11.12 can only be used for pipes serving groups of appliances having a total of over 30 demand units. The pipes serving appliances whose total of demand units is less than 30 must be sized on the basis of full simultaneous demand. This creates an anomaly where, following the direction of flow, the pipe size required increases suddenly at the point of change in the calculation procedure. The pipe sizes would normally be expected to reduce progressively in size as the flow in each section becomes less. The problem arises because an abrupt change is made from one

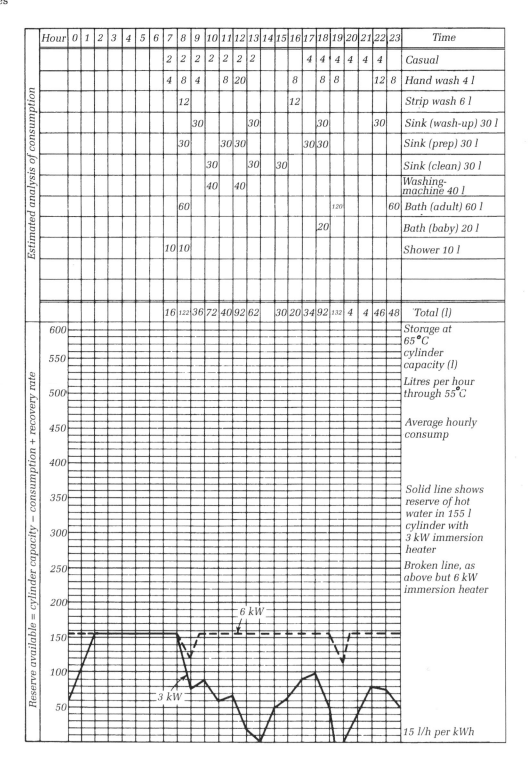

Figure 11.11 Sheet for estimating storage capacity and heat input (recovery rate) for hot-water supply installations.

calculation method to the other without taking into account a transitional zone. There is no convenient method available to estimate sizes in the transitional sections. The normal procedure is to continue the last size determined by the demand unit method until it corresponds with the size calculated by full simultaneous demand.

Pipes are sized so that the friction loss in the pipe balances the head available from cistern or main. Appliances such as showers require some head to give the spray. This is described as maintained head and, for calculation purposes, must be deducted from the head available in the system. Figure 11.13 is a nomogram for sizing pipes in relation to head available, flow required, diameter and length. The following procedure is a step-by-step guide to pipe sizing. Table 11.7 and Figs 11.14 and 11.15 provide an example of the method.

Step-by-step method for water supply

1 *Diagram*

Prepare a line diagram of the installation showing draw-off points, lengths of pipe, values, cisterns and cylinders and heights between cistern centre and draw-off points. Annotate draw-off points and pipe junctions by letters to identify pipe sections. A three-dimensional diagram will often be helpful for sizing and general use. Axonometric projections can be quickly made from ordinary plans and they enable pipe length to be scaled off (in horizontal and vertical planes).

2 *Peak loading*

Decide which sanitary appliances will contribute to peak flow and what frequency of use should be catered for in design (see Table 11.6). It may be necessary to consider more than one case if the peak flow conditions is not clear from initial inspection.

3 *Flows required at appliances*

Using Table 11.6, mark the diagram with the following information for each draw-off point:
- flows required (l/s)
- demand units
- maintained head required at appliance, if any

4 *Flows required in pipework*
- On each section of pipework mark the total of demand units for all appliances supplied by that section.
- Using design flow and the figure for demand units calculated above, mark each section of pipe with the appropriate design flow. Steady flows for showers or any other fitting should be added.
- For pipe serving single appliances and in other cases where the total of demand units is not large enough for Fig. 11.12, use the sum of the actual flows required at the appliances.

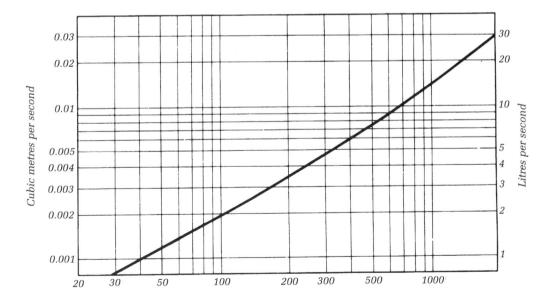

Figure 11.12 Design flow chart for sizing pipes in hot and cold water supply installations.

5   Equivalent pipe lengths

Allowance must be made for the loss of head through fittings and taps. The most convenient method is by adding an additional length to the pipe for calculation purposes. A 25% addition to the length for elbows, tees, etc., plus the appropriate value for the guessed size of valves and known size of taps from Table 11.8 should be added to the pipe length and the total marked on the diagram.

6   Establish critical run of pipework

The draw-off point having the most critical supply must be established. Using Fig. 11.13 link the head available and pipe run to each appliance and project the line to cut the rate of loss of head scale. The lowest value found establishes the critical run of pipework which extends from the supply point, usually the cistern to the draw-off point. In practice it will not be necessary to test every point since the majority can be eliminated by inspection.

7   Establish pipe diameters for critical run

Using the point found on the rate of loss of head scale (Fig. 11.13) for the critical run, draw lines through the various flows required along the sections of the run to cut the pipe diameter scale at the appropriate point. The next largest pipe diameter will normally be selected unless a reduced flow would be more acceptable than a larger pipe.

8   Check

Check that the diameter found does not vary from the assumption of valve size made in step 5. If it does, recalculate on the basis of the new valve size.

9   Branches

Branches may be dealt with as described above, starting from the junction with the critical run and using as available head the total head to the sanitary appliances less the head loss in friction in the critical run up to the point of junction. The head lost in the section of the critical run up to the branch is determined by multiplying the rate of loss of head (established during calculation of the critical run) by the length up to the branch, including allowance for fittings.

This system for pipe sizing enables various frequencies of use to be taken into account, a facility not provided by the example in BS 6700.

Table 11.6   Simultaneous demand units for sizing hot- and cold-water supply pipes

Sanitary appliance	Tap size (mm)	Frequency of use (min)	Simultaneous demand units
WC			
(flushing cistern)	13	20[a]	1.5
9 l		10	3.0
		5	6.0
WC			
(flushing cistern)	13	20[a]	0.9
4.5 l		10	1.8
		5	3.6
Lavatory basin	13	20[a]	0.3
		10	0.8
		5	1.5

Spray tap: add 0.015 l/s per tap to hot and cold

Bath	19	75[b]	1.0
		30	3.3
	25	75	2.0
		30	6.6
Sink	13	20[a]	0.7
		10	1.8
		5	3.5
	19	20	1.0
		10	2.5
		5	5.0
	25	20	2.0
		10	5.0
		5	10.0

Showers	Add 0.6 l/s to both hot and cold
Urinals	Will fill during slack periods

[a] Usage intervals

	Every
Peak domestic usage	20 min
Peak commercial usage (offices, etc.)	10 min
Use with queues forming, etc.	5 min

[b] Peak bath usage does not normally coincide with domestic peak demand.

Example sizing calculation

1   Diagram

Figure 11.14 shows a hot-water supply distribution layout, including diagrams of the sanitary appliances. Figure 11.15 shows the related cold-water supply layout containing a minimum of detail and appropriate to this type of calculation. In this very simple case elevational diagrams are appropriate. In more complex installations an axonometric diagram is likely to be helpful. Provided the diagrams are drawn to scale dimensions may be scaled off with sufficient accuracy for sizing. All pipe junctions are labelled (e.g. Ⓐ,Ⓑ, etc.) so that all runs can be identified.

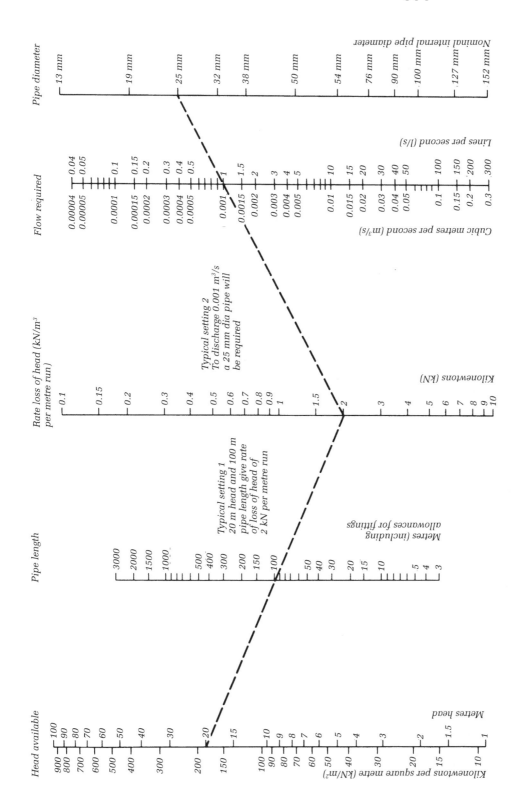

Figure 11.13 Nomogram for sizing water supply pipes.

Table 11.7   Allowing for frictional losses in taps and valves: equivalent pipe lengths

Type	Equivalent length (m) for nominal diameters									
	13 mm	19 mm	25 mm	32 mm	38 mm	50 mm	62 mm	75 mm	87 mm	100 mm
Taps and globe-type isolating valves	5	6	9	11	14	18	21	25	30	36
High pressure ball-valves	75	40	40	35	21	20				
Low pressure ball-valves	8									

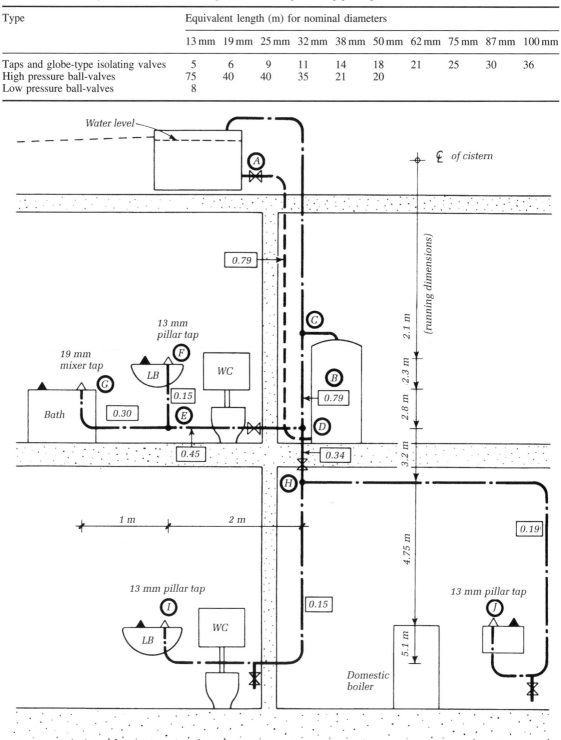

Figure 11.14   Pipe sizing: an example.

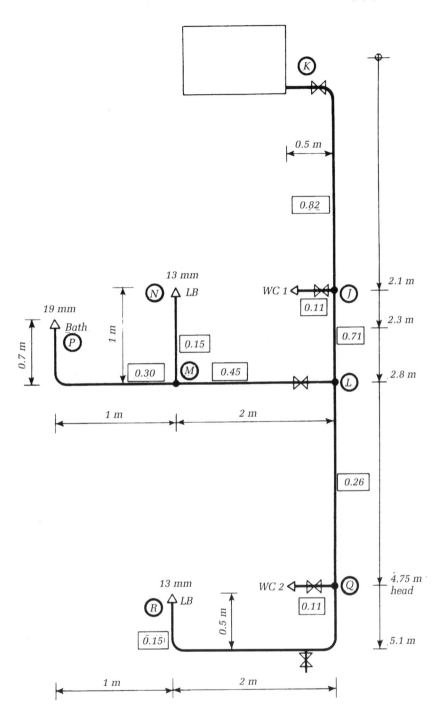

Figure 11.15   Pipe sizing: an example.

2 Peak loading

In this case no reduction in flow for diversity is appropriate and the flows marked in boxes (e.g. 0.45) for each section of pipe represent the sum of the actual flows. In larger installations the diagram would be marked with the demand unit value for each appliance and for each pipe length the total demand units for all appliances served. In this small installation all the sanitary appliances may be in use at peak times.

3 Flows required at appliance

The flow in litres per second for each draw-off point is marked on the pipe branch leading to it (e.g. 0.30); see Table 11.5 for typical rates.

4 Flow required in pipework

In this case all appliances may be used simultaneously and no diversity allowance is appropriate. Each pipe length is therefore marked with the sum of the flows of all the fittings it serves. In larger installations the diagram would be marked with the demand unit value for each sanitary appliance (Table 11.6) and each pipe length marked with the total of demand units for all sanitary appliances served. The appropriate design flow can then be determined from Fig. 11.12 and marked on the pipe length.

5 Equivalent pipe lengths

Scale the length of each pipe run between junctions, add allowance for elbows, tees (25% assumed) and allowance for valves and taps based on assumed diameter (see Table 11.7).

Pipe	Length	Allowance for elbows and tees	Allowance for valves and taps	Total
Figure 11.14				
AB	4	1.0	11	16
BC	0.5	0	0	0.5
CD	1.0	0.25	0	1.25
DE	2	0.5	9	11.5
EF	1	0.25	5	6.25
EG	1.7	0.5	5	7.2
DH	0.75	0	9	9.8
HI	4.5	1	5	10.5
HJ	5.5	1.5	5	12
Figure 11.15				
KJ	3	0.7	11	14.7
JL	0.7	0.2	0	0.9
J to WC	0.5	0	13	13.5
LM	2	0.5	9	11.5
MN	1	0.25	5	6.25
MP	1.7	0.4	6	8.1
LQ	2.0	0.5	0	2.5
Q to WC	0.5	0	13	13.6
QR	3	0.7	5	8.7

6 Critical run

Consider the following draw-off point in accordance with the method described in the previous list.

Pipe run	Total length	Head available (scaled)[a]	Rate of loss of head[b]
Figure 11.14			
AF	35.5	2.0	0.6
AG	36.5	2.3	0.6
AJ	39.3	4.75	1.2
AI	37.8	4.75	1.3
Figure 11.15			
KN	35.2	2	0.61
KP	37	2.3	0.62
KR	26.8	4.75	1.8
K to WC	28.2	2.1	0.7
K to WC	31.6	4.75	1.5

[a] Head available is the vertical height from the centre line of the cistern to the tap on draw-off point.
[b] The lowest value found for Fig. 11.14 is 0.52, therefore AF is the critical run. For Fig. 11.15, KN is the critical run.

7 Pipe diameters

Using the nomogram (Fig. 11.13) and the data above, determine the diameter appropriate to the various sections of AF and KN.

Section	Flow required	Diameter (mm)
Pipe run AF		
AB		
BC	0.79 l/s	32
CD		
DE	0.45	25
EF	0.15	18
Pipe run KN		
KU	0.82	32
UL	0.71	32
LM	0.45	25
MN	0.15	18

8 Branches

Figure 11.14: consider pipe from D onwards. The critical run must be established. Since I and J are at the same level J is more critical since pipe length is greater. Establish head available at J and size pipe run DJ. Head available is 4.75 m less the friction loss up to D. The rate of loss of head in AF is 0.6 and the length AD is 18 m. Therefore the loss of head due to friction in AD (from Fig. 11.13) is 1.1 m. Head available for flow in DJ is $4.75 - 1.4 = 3.4$ m. The length of DJ is 21.5 m and the rate of loss of head is 1.5. The appropriate size for DH is therefore 18 mm and for HJ 12 mm. The other branches may be similarly sized.

9 *Note*

In practice it will not be essential to set down all the figures recorded in this example. Critical runs can often be determined by inspection and pipe sections can be sized from the nomogram without tabulation. In Fig. 11.15 the WCs have not been considered in the critical runs on the assumption that some restriction of the cistern filling rate will not be significant.

Pipe sizes

Before metrication UK pipe sizes were defined, even if not always exactly, by internal diameter. In the metric system the outside diameter is used. However, manufacturing and, in some cases, Standards have not yet changed. Pipe flow capacity is determined by inside rather than outside diameter. Thus the present situation in relation to pipe sizes is confused. Designers will need to be sure which dimension is

used. In this book, where the main question is normally discharge capacity, inside diameters are used in all cases.

Table 11.8 illustrates the problem and gives examples of typical sizes in millimetres.

Table 11.8 Pipe sizes: internal diameter in millimetres

Imperial size (in)	Metric equivalent (nearest mm)	Copper, BS 4127		Steel, BS 1387		uPVC	
		ID	OD	ID	OD	ID	OD
$\frac{1}{2}$	13	14	15	15	21	15	17
$\frac{3}{4}$	19	20	22	20	27	22	27
1	25	26	28	25	34	28	34
$1\frac{1}{4}$	31	32	35	32	43	35	42
$1\frac{1}{2}$	38	40	42	40	48	–	–

12 Drainage installations

12.1 General principles

A long process of development by trial and error during the late nineteenth century resulted in the establishment of a set of rules for pipe sizes and layout of sanitary installations which, when followed, ensured satisfactory performance. Drainage bye-laws were framed as an expression of these rules and for nearly half a century drainage design consisted almost solely of applying these rules. It was not necessary for designers to understand the basic principles involved. Departures from established solutions, even if justified by satisfactory performance and based on sound principles, were not acceptable.

Performance standards

In recent years problems of size of installation, and particularly height of building, unprecedented in this country, have had to be solved and completely internal plumbing achieved. Economy of established methods has been questioned. Fundamental studies of the hydraulic behaviour of drainage installations have been made, principally by the American National Bureau of Standards and the Building Research Establishment. New methods have been developed and most significantly the legislation affecting drainage has been revised so that it now requires the achievement of standards of performance rather than detailed compliance with specific precepts. New problems and methods and the freedom to design for a standard of performance in fact impose considerably more responsibility, and require from designers, if improved installations are to result, much more fundamental knowledge than was previously the case. A comparison of past and current regulations (Table 12.1) shows clearly the change from insistence on

fixed solutions to a specification of performance requirements.

Hydraulic design

The design implications of these changes in the form of legal requirements and of research into economical systems relate mainly to the hydraulic performance of installations. It is worthwhile and not difficult to set down the hydraulic flow problem posed by removal of waste and soil flows from buildings.

Pipe systems for drainage must be able immediately to accept the flow from any sanitary appliance connected to the system. These flows must be discharged completely and without delay to sewer or treatment plant. Entry and exit points cannot be protected by valves but no water must emerge under any condition of flow and in spite of displacement by flows of water, air must at no time be allowed to escape from the pipes into the building. Solid matter is present in some of the flows and must be carried by the flow of water without becoming stranded or lodged.

Posed with precedent, the solution to this might appear to be a formidable problem. However, the principles of solution have already been developed and are implicit in the form of established drainage systems involving pipes constantly falling towards discharge points, empty except when flow is taking place, with entry points closed by water seals and within which air pressure fluctuations are limited. If it can be taken for granted that watertight, properly aligned, smoothbore pipes laid to suitable and continuous falls are part of all drainage installations then two major hydraulic flow design problems remain:

- How to estimate flows so that satisfactory but economical pipe sizes may be employed.

Table 12.1   Past and present regulations affecting drainage installations

	Past	Present

London	1 If the soil pipe or waste pipe of any soil fitment shall be in connection with the waste pipe of any waste water fitment, the trap of every such soil fitment or waste water fitment shall be ventilated in the following manner:
	
	A trap ventilating pipe shall:
	(a) be connected with the trap or the branch soil or waste pipe;
	   (i) at a point not less than 75 mm nor more than 300 mm from the highest part of the trap;
	   (ii) on that side of the water seal which is nearest to the soil pipe or waste pipe;
	   (iii) in the direction of the flow
	
	2 The branch and main trap ventilating pipes respectively shall have in all parts an internal diameter of not less than:
	
	(a) 50 mm where connected with a soil pipe, or a waste pipe 75 mm more in internal diameter.
	
	(LCC *Drainage bye-laws* 1934, clauses 9(1)A and 9(2)A)
		In order to prevent destruction of the water seal of the trap of any sanitary appliance, the trap shall be ventilated whenever necessary by a ventilating pipe positioned so as to prevent any nuisance or injury or danger to health arising from the emission of foul air from such a pipe and, where necessary, have the open end fitted with a suitable grating or other cover constructed in the manner prescribed in by-law 6 paragraph (15)(*c*)(*iv*) for gratings to drain ventilating pipes, or connected to the soil or the waste ventilating pipe above the highest appliance.
		
		(LCC *Drainage bye-laws* 1962, clause 11)

England and Wales	The drains intended for conveying foul water from a building shall be provided with at least one ventilating pipe, situated as near as practicable to the building and as far as practicable from the point at which the drain empties into the sewer or other means of disposal: Provided that a soil pipe from the water closet, or a waste pipe from a slop sink, constructed in accordance with these by-laws, may serve for the ventilating pipe of the drain, if its situation is in accordance with this by-law. (*Model Bye-laws*, Series IV, Buildings 1939, clause 102)	Water seals in traps. Such provision shall be made in the drainage system of a building, whether above or below the ground, as may be necessary to prevent the destruction under working conditions of the water seal in any trap in the system or in any appliance which discharges into the system. (*Building Regulations* 1965, clause N3)

● How to ensure, with economy of pipework, that excessive pressure fluctuations do not occur.

Estimation of flows

There are two cases to be considered: where the pipe carries the flow from only one sanitary appliance, and where two or more appliances can contribute to the flow.

In sanitary appliances subject to a fixed rate of flush, such as WCs, urinals and showers, the outgo from the appliance must be large enough to discharge the full rate of input. In other cases, however, such as baths, lavatory basins and sinks, the size of the waste outgo and associated pipework governs the rates of outflow and is itself governed only by the time considered acceptable to empty the sanitary appliance. Existing precedents for sizing are valid and well established and this aspect of sizing presents little problem.

When several sanitary appliances can contribute to flow in a length of drainage installation a problem does arise. It would be simple to add the combined flows of all the appliances but it is uneconomic to do so since the changes of all the appliances contributing to flow at the same time may be so small as to be ignored. The probability of finding a 91 WC in the process of flushing at a time of peak domestic use is 0.004 (where 1 represents certainty of finding the WC flushing and 0 certainty of finding the WC not flushing). Even in congested use this probability will only increase to 0.02. The probability of finding two WCs flushing at the same time is the product of their individual probabilities. Therefore the probability of coincidence of flush from two domestic 91 WCs is 0.000 16. From three it would be 0.000 000 064. This very low probability of finding WCs flushing together demonstrates the need, if economy of pipework is to be obtained, for taking this into account in estimating flows and consequent pipe sizes. Data for estimating flows given in *Architects' Journal* Information Sheet 1417 take this factor into account. A detailed description of the theory and method by which the design data were established can be found in an article by Burberry and Griffiths (Service engineering:

hydraulics, *Architects' Journal*, 21 November 1962, 1185–91). The use of these data will in large installations usually enable economical pipe sizes to be used. The same benefit does not arise in very small installations since the minimum size of pipe permissible from considerations of blockage (100 mm diameter in the case of foul drains or soil pipes carrying flows from more than one WC) is capable of dealing with the flows from many appliances. But even in these installations, unnecessary increases of drain diameter may be avoided.

Avoidance of pressure fluctuations

Modern opinion on hygiene does not regard the escape of drain air into a building with the same degree of horror that was the case in the past. No one, however, would advocate systems which would permit this – on grounds of amenity if not of health. Consequently, air pressure fluctuation in the pipework must not be allowed to reach intensities which would affect the water seals of traps. The classical method of doing this is by vent pipes opening into waste or soil pipes near each trap outgo and leading to the external air. By this means the air pressure at each side of the trap can be kept stable at atmospheric pressure. The arrangement is expensive and unsightly, and in an increasing number of suitable instances is being replaced by neater and cheaper arrangements which avoid the development of pressure fluctuations by using more sophisticated hydraulic design rather than ugly and unnecessary trap venting.

In principle the approach depends on sizing and arranging the pipework so that full-bore flow and the consequent pressure fluctuation are avoided. Even a momentary bridging of the pipe can lead to gross pressure fluctuations. An understanding of the way in which flow in partially filled pipes behaves is vital if design is to be carried out. Figures 12.1 to 12.7 show some of the important cases.

Horizontal and sloping pipes flowing part full   Flow occurs in the bottom of the pipe (Fig. 12.1(a)). As the

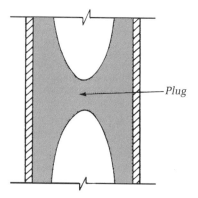

Figure 12.2   Flow in a vertical pipe showing the formation of plugs.

water level rises, air is displaced from the pipework; and as flow of water takes place, friction with the air in the pipe induces movement of air in the direction of flow. (It is interesting to compare this with the thermosiphon theory of drain venting popular at one time.) The induced pressures are small.

Vertical pipes flowing part full   Flow from a branch entering a vertical stack accelerates downwards because of gravity and quickly assumes the form of a sheet on the wall of the stack (Fig. 12.1(b)). Velocity of flow increases until friction balances the pull of gravity. Airflow is induced down the stack and positive pressures are likely to develop at the foot of the stack due to restriction in area available for airflow as the rate of water flow is reduced on entering the horizontal drain. Paper and solid matter tend to fall in the centre of the pipe. They are discharged very rapidly.

The pattern of flow described does not continue progressively with increasing flow until the pipe is flowing full-bore. When one-quarter to one-third of the full flow capacity of the pipe is reached the whole section of the pipe is bridged at intervals by 'plugs' of water (Fig. 12.2) which move along the pipe at the velocity of flow. Gross pressure fluctuations occur as a result of the movement of the plugs, and complete trap venting would be required in any system where the concentration of flows could give rise to the formation of plugs. Essential if vent pipes are to be avoided, complete trap venting is in any case desirable to avoid plug formation; this is one of the factors governing the number of appliances that can be connected to a stack of given size. For economical but satisfactory pipe sizing it is necessary to be able to estimate the peak flow resulting from the operation of numbers of sanitary appliances.

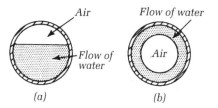

Figure 12.1   Water flow in a pipe: (a) horizontal pipe; (b) vertical pipe.

Transition from vertical stack to horizontal drain   At the foot of a stack the flow leaves the walls and runs in the bottom of the pipe; the reduced velocity of the horizontal flow will cause a greater cross-sectional area of the pipe to be occupied, with a corresponding reduction in air space (Fig. 12.3(a)). Positive pressures will develop at the foot of the stack and may be great enough to blow traps. Sharp bends at the foot make the problem worse.

It is quite possible in building drainage installations to have a combination of flow rate, pipe diameter and drain fall which produce a total bridging of the pipe in the drain; this is known as a hydraulic jump (Fig. 12.3(b)). If this situation arises gross pressure fluctuations will occur. The same result can come from a drain not subject to hydraulic jump but which has no outlet for the induced airflow, as in the case of connections to sewers with no interceptor or fresh air inlet when the sewer is flowing full.

Discharge of branches into the stack   The discharge of a branch into a stack will at least partially block the cross-section, and if gross pressure fluctuations are to be avoided, the rate at which flow can be introduced from any one branch and the grouping of branches must be controlled. Figure 12.4 shows the plan and

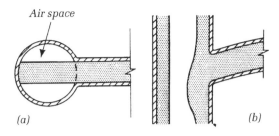

Figure 12.4   Flow from waste branch entering stack: (a) plan; (b) section.

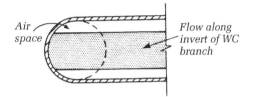

Figure 12.5   Plan showing flow from WC entering stack.

section of a waste inlet to a stack, indicating how the airflow in the stack is restricted by waste discharges. Figure 12.5 shows the flow from a WC entering the stack. Since only the bottom part of the pipe, not the full diameter, is occupied by the flow, some space for airflow is left.

Pipes for individual waste appliances   In most waste appliances the theoretical free discharge capacity of the waste pipes is greater than the flow that can occur through the waste, which is partially restricted by a grating. In practice the direct connection of sanitary appliance, trap and pipe, introducing water into the waste pipe at a sharp angle, results in full-bore flow at reduced velocity (Fig. 12.6(a)). The system operates as a siphon.

Towards the end of the discharge air will pass through the trap and the full section of pipe will become a diminishing plug as the flow of water ceases (Fig. 12.6(b)). Provided the plug is near enough to the trap, water will flow back and reseal the trap as the plug decays. If the pipework is too long or too steep this resealing flow will not be possible. Sanitary appliances with flat bottoms have a long-drawn-out tailing off of flow after the siphoning has occurred, which tends to reseal the trap.

The limits of slope and length of wastes in non-vented systems and the differences between lavatory basins and baths are the result of the phenomena described.

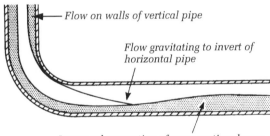

(a)

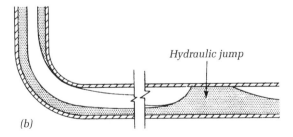

(b)

Figure 12.3   Vertical section through pipe, flow at foot of stack: (a) change in relative position of flow from vertical to horizontal section of pipe; (b) hydraulic jump.

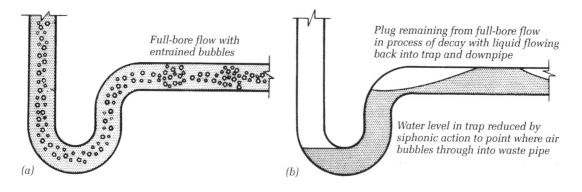

Figure 12.6 Vertical section showing flow in trap and waste: (a) full-bore flow; (b) how the trap reseals when the waste has appropriate fall and length.

Pipes from WCs and gullies connecting to stacks or drains Full-bore flow in the outgoes of WCs and gullies is not likely to occur. The flow rate is so small in relation to the capacity of the pipe that only a small fraction of the total cross-sectional area will be occupied by the flow. Figure 12.7 gives an idea of the flow area in relation to the whole by comparing the cross-sectional area of a 32 mm diameter flush pipe with that of a 100 mm diameter soil pipe. Flows into gullies depend on the number and type of appliances but clearly in normal use the situation is similar to that of WCs.

Groundfloor WCs and gullies are conventionally connected directly into manholes, without any trap venting, using branches which are sometimes long and steeply sloping and may have bends at the gully outgo and the manhole entry. No pressure fluctuations should develop and in normal circumstances there is little risk that any will arise.

Design considerations

Venting

Where pressure fluctuations cannot be kept within acceptable limits (normally ±25 mm water gauge) by means of controlling size, length and fall of pipes with only terminal opening to atmosphere, separate vent pipes must be provided that can discharge or supply air to match the rate of air movement in the installation caused by flows of liquid. In below-ground drainage where velocities are low and pipes very rarely if ever flow full, very simple venting arrangements will serve a whole installation. In most cases the soil stacks coupled with the fresh air inlet or direct connection to the sewer will be more than enough. Inside the building small diameter pipes present a more critical problem which may be solved by venting the pipe serving each trap or, if the design of these branches themselves avoids pressure problems arising within them, by venting the main pipe runs at suitable intervals. The *National Plumbing Code Handbook*, by V.T. Manus (McGraw-Hill, New York, 1957) gives data on the required airflows and the capacity of vents to deliver them. To maintain their efficiency vent pipes must be designed so that they do not become flooded in use and so that any condensation will drain to the waste or soil pipes.

Oversize pipes

It will be clear from the foregoing that the use of large-diameter pipes could avoid any problems of pressure fluctuations in a drainage pipework system. If it is proposed to adopt this approach with horizontal wastes, trouble may be experienced from a build-up of solid deposits in the unflushed section of the pipe, reducing the effective diameter.

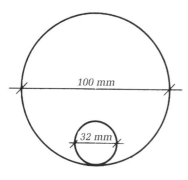

Figure 12.7 Comparison of cross-sectional areas of pipes with diameter 100 mm and 23 mm.

Detergents

The main effect of detergents in drainage pipework is to keep the interior very much cleaner and freer from deposits than was the case when soap was used exclusively. But excessive detergent foam may cause problems where waste flows are subject to pipe arrangements that disturb the flow and produce turbulence and foam. This may arise at sharp elbows at the foot of stacks, at gullies and at hopper heads. To avoid this, waste stacks should be connected to the drain by easy bends. Use neither hopper heads nor the so-called *two-pipe system*, in which gullies trap waste stacks from the drain.

Common misconceptions

Velocity in vertical stacks

Architects have expressed concern about the potential flow velocities in long vertical pipes for tall buildings. Some tall drainage stacks have offsets to interrupt the flow. In fact, the forces of gravity and friction acting on the flow in a vertical stack soon balance and a *terminal velocity* is reached. The vertical distance required for flow to reach terminal velocity varies with flow and pipe diameter. It is very likely to be no more than a single storey, so no precautions are necessary to reduce velocity in tall stacks. Offsets should be avoided since velocity of flow is reduced and plugs may be formed, creating gross pressure fluctuation.

Minimum falls

A velocity of flow of about 0.6 m/s is required to avoid stranding solid matter; the critical case is for a pipe carrying the flow from one WC (2.3 l/s). For a 100 mm or 150 mm diameter drain in perfect condition a fall of 1 in 200 would give this velocity. In practice irregularities of laying, ground movement and distortion of pipes mean that sharper falls are needed. Local experience is likely to be the best guide. Local authority recommendations for minimum falls vary greatly; for 100 mm diameter pipes they go from 1 in 40 to 1 in 110. One distressingly common misconception about minimum falls arises from the misuse of the old rule of thumb saying 100 mm drains, 1 in 40; 150 mm drains, 1 in 60; and so on. If the fall available is only 1 in 60 then a 150 mm diameter pipe may be mistakenly selected, even though a 100 mm diameter pipe would have adequate capacity. This is very wrong. It is quite clear that for a given fall the smallest pipe capable of discharging the flow will be the best pipe to use since the depth of flow will be greater.

Design of drainage installations

The previous sections have described some of the phenomena involved in drainage installations. They have not given quantitative data for design. There are three ways to approach the collection of data.

Following established precedents

This is rightly the most popular method for drainage designers. The main limitation is that an existing installation which demonstrably functions satisfactorily is not necessarily an economical one. If there are doubts on this score recourse must be had to the other two approaches to check the performance. The main pitfall is that trouble can occur if the new situation, although apparently similar to the model, does in fact differ in some significant respect. The analysis of the other two approaches may bring to light differences which could otherwise be missed on more superficial inspection.

Use of published hydraulic design data

In a circumstance where direct precedent is not available and where the mode of performance of the installation is understood, the necessity for venting and the size of waste, soil, drain and vent pipes may be established from published data covering flows of water and air and related pipe sizes. The references give a wide range of values but the data are not comprehensive at present.

Practical experiment

Since standards for drain installations are now based on performance, observation of a specially made prototype can be of fundamental value in design. In cases where designers suspect that traditional practice is uneconomic or wish to employ some arrangement which seems theoretically sound but is as yet untried, conclusions can be reached by constructing and testing a prototype, particularly of work above ground.

A building contract provides opportunities for this and in cases where there is frequent repetition of a particular plumbing detail the procedure can be well justified. It is also possible where the designer feels that simpler than normal venting arrangements would be satisfactory to delay this part of the installation so as to give opportunity to observe the performance of the rest.

In addition to hydraulic considerations there are a number of other conditions which must be satisfied by drainage installations. Table 12.2 shows a summary of the general criteria for performance together with the means available to achieve them.

Table 12.2 Performance criteria for drainage installations

Requirement	Achieved by
Watertight: in the case of underground drains entry of water as well as loss must be avoided and either case would encourage root penetration	Appropriate selection of pipe material and jointing system and by satisfactory work standards
Non-blocking	Adequate minimum diameter Smooth internal surfaces Properly aligned joints No reduction in diameter in direction of flow
Adequate discharge capacity	Flows in pipe depend on diameter, gradient, internal surface and alignment of pipes
No nuisance: overflowing should not occur in free-flowing installations of adequate size but nuisance may be caused by the discharge of drain air into the building	Traps at all entries to drainage pipework (surface water systems outside the building are not subject to this) Pipe installation designed to avoid air pressure fluctuations which might unseal the traps
Durable: durability of drain installations is the problem; very durable materials may not necessarily provide a long-lived drain; factors include ● Normal life of drain materials and jointing materials ● Effects of moisture, building settlement and ground movement on drain ● Deterioration of pipes and joints due to chemical nature of flow ● Deterioration of pipes or joints due to surroundings (e.g. soil chemicals) . ● Mechanical damage or crushing ● Rust penetration	Selection of pipe material Selection of jointing system and material Support or bedding Work standards for the above
Traceable and accessible for maintenance	Pipework arranged in straight lines between access points Access point provided so that all sections may be inspected and if necessary cleared by flexible lines and rods
Economic	Satisfactory siting of building in relation to disposal point for drainage Grouping of sanitary appliances to give short pipe runs

12.2 Drainage above ground

Types of drainage system

Drainage installations for buildings have to deal not only with flows from sanitary appliances but also with rainwater from roofs and paving. The established techniques for these installations are divided into work above ground (including work inside basements) described as waste and soil systems for flows from sanitary appliances and as rainwater pipes for surface water. Below ground all the work is referred to as drainage. Installations taking flows from sanitary appliances and those taking surface water are called *foul drains* and *surface water drains* respectively. In some cases the same pipes carries the two flows and is termed a *combined drain*. As long as the pipes remain within the boundary of the site of the particular building which they serve, the term *drain* is applied

irrespective of the size of the pipes. But when they pass from the site they will be termed *sewers* and some different considerations of design and construction apply.

For all waste and soil and drainage systems a satisfactory standard of materials and work is required. This will include watertight pipes with smooth internal bores, satisfactory junctions that avoid the risk of blockage, and properly aligned joints with no ridges to catch solid matter. Pipe materials and joints have been dealt with earlier and the use of proper material and work standards is assumed in the following notes, which deal with the design of drainage systems.

It is important to remember that all waste and soil pipes and drains must discharge the flows they receive immediately and that consequently all pipes must fall continuously to ensure a proper flow. Arches and dips to avoid obstructions are not acceptable (unlike water supply or other pipe systems under pressure). This point has considerable implications for planning and economy in installations both inside and outside buildings.

Considerable developments in theory and practice have taken place in recent years and many existing buildings have installations which would not now be repeated in new work, so it is important to consider past as well as present methods.

Waste and soil systems

Waste and soil systems have to convey the flows from sanitary appliances to the drains without causing any nuisance from leakage, blockage or smell and in a neat and economical manner. Soil pipes carry flows containing excrement, whereas waste pipes carry flows from baths, sinks, lavatory basins, etc. The hydraulic problems involved have been described. In the early days of plumbing in buildings these phenomena were not so well understood and it was also thought necessary to exclude drain air from the building, not only because smell could constitute a nuisance but because it was thought that smell was a direct menace to health. Drainage pipes were therefore taken outside the building, above or below ground, at the first opportunity and sanitary appliances were usually sited against external walls. Each sanitary appliance was provided with a trap (Fig. 12.12) to seal the air in the pipework from that in the room. In the case of WCs the single trap was accepted and the outgo pipe was connected directly to the drain at a manhole either by a pipe underground, in the case of groundfloor appliances, or by a vertical stack for those on upper floors. But in the case of waste flows where, unlike soil systems, nuisance would not result from the content of a gully, an additional safeguard against the penetration of drain air into the building was provided by disconnecting the waste pipes from the drain by discharging them over a gully (Fig. 12.18). In addition to the air gap at the termination of the waste pipe the gully itself contains a trap to seal off the drain air. Waste appliances on first floors were discharged to a hopper at the head of a vertical stack, itself discharging as before over a gully. Buildings over two storeys in height, or with ranges of sanitary appliances discharging into common waste or soil pipes presented special difficulties in that the flows in the pipes could cause pressure fluctuations liable to affect the water seals in the traps. This was overcome by providing each trap with a vent pipe (full name trap ventilating pipe, sometimes called antisiphon pipe) connected near the trap outgo and leading to the open air. This

vent pipe allowed air to escape when pressure developed, or fresh air to enter to overcome suction, thereby preventing pressure fluctuation and consequent trap unsealing. The arrangement described, where waste flows are kept separate, is known as the *two-pipe system*. Although many of its features are still in use, the complexity and expense of pipework, and the serious disadvantages of the waste arrangements, have led to improvements in design.

The main disadvantages are threefold:

- The discharge of waste over a gully grating was an unpleasant feature and caused considerable nuisance when the gully grating became blocked by leaves. This can be overcome by the use of the back inlet type of gully (Fig. 12.18), which introduces the waste flow above the water level but below the grating. This is a considerable improvement but an open waste gully is still a possible source of smell and is by no means an advantage close to a building, except when it is used for pouring away liquids that cannot be disposed of through the normal sanitary appliances.
- The hopper used at first-floor level in two-storey work (particularly housing) to collect the flows from waste pipes at this level, constitutes a potential nuisance from smell, blockage and overflowing. The problem can only be avoided by eliminating the open hopper and connecting the wastes directly into the waste stack, which would then have to be carried up above the eaves as a vent, and if more than one sanitary appliance contributed to the stack then a vent pipe to protect the trap seals would also be needed.
- The widespread use of detergents has produced a very critical problem, particularly in multistorey buildings, where the turbulence arising as the waste is discharged into the gully causes foaming of the detergent, which can rise and spill over the gully grating, causing considerable nuisance.

These disadvantages, coupled with the extent of the pipework required, led to a desire for economy and improvement. More confidence in plumbing design, observation of American practice and less fear of drain air as a health risk, led to the use of the *one-pipe system*. In this system waste flows are connected into soil stacks and thereby to the drain, without the trapped gullies usually provided. By this means the gully and hopper are eliminated and the duplication of pipework substantially reduced. But in this system each trap must be provided with a vent pipe. In domestic work, where hoppers may be used to avoid common wastes that require venting, this means the savings in waste stacks and pipes might well be more than offset by the additional vents needed, so the two-pipe system has continued to be employed.

The desire for economy and efficiency in the extensive postwar building programmes directed the attention of the Building Research Establishment to this problem and, as a result of studies and laboratory work carried out there, it became apparent that undesirable air pressure fluctuations in drainage pipework could, in a certain range of circumstances, be eliminated by the observation of a number of simple rules without the necessity for trap ventilating pipes. The method of design was termed the *single-stack system*. Initially it was limited to five storeys in its application; after some years modifications were proposed which enabled buildings up to about 10 floors to be served, and installations up to 20 floors high, using 150 mm diameter stack, have recently been considered feasible. The single-stack system has two principal functions:

- Preventing pressure fluctuations arising from the operation of other sanitary appliances by
  - connecting each appliance separately to the stack
  - limiting the flows in the stack so that plugs cannot form (see page 260)
  - maintaining the stack straight to avoid plug formation (see page 260).
- Preventing self-siphonage arising from the operation of sanitary appliances. This is achieved by careful control of the diameter, length, fall and connection to the stack of the waste pipe servicing the appliance (see page 261 and Figs 12.6(b) and 12.8).

Another important development in waste and soil installations, intended to avoid freezing, was the requirement in the Building Regulations 1985 for all waste and soil pipes (except wastes to back inlet gullies) to be inside the external walls. This provision was a direct result of the extremely cold winter of 1963. More favourable winters since then have encouraged relaxation of the regulations in 1976 to permit outside soil and waste pipes for buildings up to three storeys in height.

The single-stack system giving neat and economical pipe installations is clearly the most desirable system to employ wherever possible. It lends itself mainly to dwellings. The need for a straight stack and limited length and fall to the individual wastes means that, if advantage is to be taken of this system, it is important to bear its requirements in mind at the planning stage of the building design and in the layout of sanitary appliances. Figure 12.8 shows the requirements for

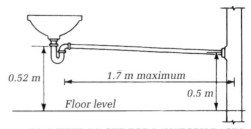

32 mm DIAMETER WASTE FOR LAVATORY BASIN OR SMALL SINK

If L is less than 0.69 m slope may be not more than 5° (1 in 12)
For lengths of 0.69 - 2.3 m the (maximum) slope should be not more than 2½° (1 in 24)

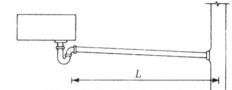

38 mm DIAMETER WASTE FOR LARGE SINK

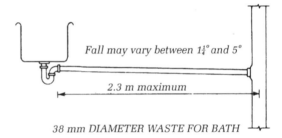

38 mm DIAMETER WASTE FOR BATH

Figure 12.8 Limits of fall and length for single-stack wastes from lavatory basins, sinks and baths. In all cases slow bends (but not elbows) may be used in the horizontal plane.

single-stack wastes and Fig. 12.9 shows some problems which may have to be overcome when branch connections from WC and bath meet the stack itself. The difficulties arise because the levels at which the bath waste and WC outgo meet the stack are inevitably very close. If the bath waste centre is not lower than the centre of the WC outgo a correction can be made, as shown in Fig. 12.9. At points lower than this, however, connection is not satisfactory because of the possibility of the flow from the WC affecting the bath outgo. Connection cannot therefore be made until 200 mm below the WC junction. A suitable detail for achieving a bath waste connection below the unacceptable zone is shown in Fig. 12.9.

The single-stack system is now very widely used and has had a major effect in improving the efficiency,

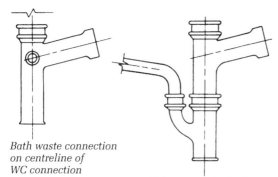

Bath waste connection
on centreline of
WC connection

*Parallel junction for bath
waste where otherwise
connection would enter stack
immediately below the WC
connection*

Figure 12.9 Acceptable arrangements for juxtaposition of
bath waste and WC outgoes in single-stack systems.

economy and appearance of waste and soil
installations. But it has three major disadvantages.
First the limitation of length of wastes from sinks and
lavatory basins; second the very flat gradient
appropriate for these wastes, which is likely to conflict
with other appliances; and third the problem of bath
waste connections described above. Development
work at Marley Plumbing has produced the Marley
Collar Boss fitting (for the Marley PVC system). This
fitting consists of a gallery with bosses for pipe
connections running round the 100 mm soil stack.
Flows from the branches are introduced into the stack
without risk of interference, and small vents can also
be connected to the gallery. This fitting enables the
layout shown in Fig. 12.10 to be achieved. All three
disadvantages of the single-stack system are overcome.
The bath connection can be made in the previously
unacceptable zone below the WC connection. The sink
and lavatory basin wastes can drop vertically since
they are provided with 13 mm diameter vents to
overcome siphonage and the same vent makes it
possible to connect the lavatory basin waste into the
bath waste, instead of having a separate connection to
the stack.

Where it is not possible to plan the location of
sanitary appliances within the limits described for the
single-stack system, or the stack cannot be kept
straight, trap ventilation pipes will have to be used to
ensure that trap seals are not destroyed. Figure 12.11
shows arrangements which may have to be made when
a lavatory basin is so situated that normal single-stack
principles cannot be applied and the same layouts

would be appropriate for other sanitary appliances
similarly situated. Where the stack cannot be kept
straight, pressure fluctuations due to the bends may
occur and it may be necessary to vent all traps.

BS 5572 *Sanitary pipework above ground* gives
very comprehensive details of the principles, details
and sizing of waste, soil and vent pipes. The most
significant additional material in the code deals with
extensions to the range of conditions under which the
single-stack principles can be used.

The maximum lengths of unvented wastes from
lavatory basins can be increased to 3 m, provided the
diameter of the waste pipe is increased from the
normal 32 mm to 40 mm. The fall can be $1-2\frac{1}{2}^{\circ}$. The
trap remains at 32 mm diameter. Similarly the
maximum length of unvented wastes to individual
sinks and baths can be increased to 3 m provided a
40 mm diameter pipe is used. In these cases falls of 1–
5° are acceptable.

New standards are also introduced by the code for
ranges of lavatory basins and WCs. A range of four
lavatory basins, the traps from which discharge into a
straight run of 50 mm diameter waste not more than
4 m long and with a fall of $1-2\frac{1}{2}^{\circ}$, will not give rise to a
need for venting. A single 25 mm diameter vent pipe
connected to the waste at the end remote from the
stack will enable the number of lavatory basins to be
increased to five, the length to 7 m and for two bends
to be accommodated.

Branch discharge pipes servicing ranges of WCs
will not require any venting if the pipes are straight,
the length does not exceed 15 m, the fall is 0.5–5.0°
and not more than eight WCs are connected. A single
ventilating pipe, not less than 25 mm diameter,
connected to the branch at the end remote from the
stack will enable larger numbers of WCs or bends in
the branch to be accommodated.

It is necessary to test new installations to ensure
they perform adequately and that trap seals are not
destroyed in use. Flushing all appliances
simultaneously would be an unreasonable test since
this situation would not occur in reality. BS 5572
Sanitary pipework above ground gives tables which
enable appropriate numbers of appliances to be flushed
for testing purposes.

Traps

In the case of WCs the trap is an integral part of the
appliance (Fig. 12.19); in other sanitary appliances it is
fixed to the outgo. A wide variety of designs, materials
and depths of seal are available. Appearance may be an
important feature, or ease of access for cleaning,

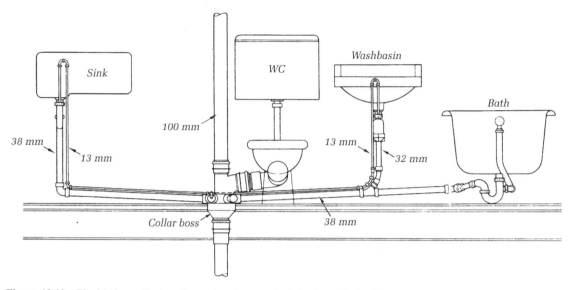

Figure 12.10   The Marley collar boss for neat and economical single-stack plumbing.

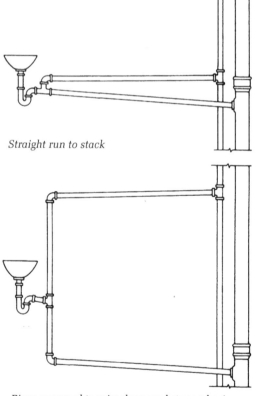

Straight run to stack

Pipes arranged to miss doorway between basin and stack

Figure 12.11   Waste and trap ventilating pipe configurations for a lavatory basin too far from the stack for use of single-stack arrangements.

flexibility of outgo position or a variety of other features. Figure 12.12 shows basic trap types; Fig. 12.13 a bottle trap, which can be chromium plated for neat appearance; Fig. 12.14 a three-jointed trap which can give a wide range of outgo positions and the bend of which can be removed bodily in the case of blockage (this makes clearing easier than the conventional clearing eye and also avoids the internal ridges associated with the clearing eye, which can create blockages themselves); and (Fig. 12.14) a trap, the body of which is formed by a glass bottle which can retain kitchen sediment and give visual indication of the need for cleaning. These considerable variations mean that, as with sanitary appliances, it is usual to select and specify a particular type from a particular maker for each application.

In addition to the features described above the depth of seal must be specified. It is normal at present for a minimum of 25 mm of seal to be retained under all conditions of operation. Traps were used with seals of 75 mm, 38 mm and of nominal depth of a few millimetres. WC traps have 50 mm seals. The nominal depth traps were used in situations where in past regulations very short lengths of waste pipe discharging into the fresh air were not required to be trapped but were provided with traps to avoid cold draughts and the entry of insects. This arrangement was particularly popular for bath traps, since space under the bath was restricted and the waste was often very short, discharging into a hopper on the wall immediately outside the bathroom. Thirty-eight millimetre seal traps normally give satisfactory

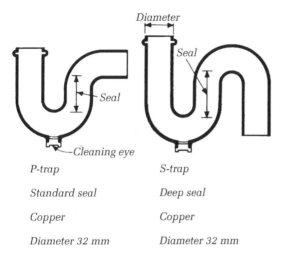

Figure 12.12   Standard seal P-trap and deep seal S-trap.

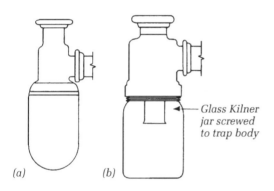

Figure 12.13   Two 32 mm diameter traps: (a) chromium-plated resealing bottle trap; (b) Econa trap with Kilner jar.

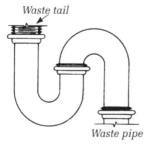

Figure 12.14   The third joint in this two-piece copper S-trap allows part of the trap to be removed for cleaning, obviating the need for a cleaning eye. The extra joint also allows the waste pipe to be swivelled into many orientations.

performance when provided with trap ventilating pipes, although when the one-pipe system was introduced 75 mm seals were used because of the fear of health risk from soil pipe air. In the single-stack system 75 mm seals are used, partly to overcome the slight seal losses occurring in normal operation of this system and partly to give an additional safety factor. The Building Regulations 1985 require 50 mm minimum depth of seal for traps discharging to drains, and the use of 75 mm seal traps is becoming usual.

The diameters of traps given in Chapter 10 have no special theoretical basis. They are sizes that have been found to give accepted times of discharge for normal use.

A quite different approach to the problem of maintaining trap seals, different from the approach in the paragraphs on waste and soil installations, lies in the design of special traps which will allow air to pass through while retaining sufficient water to re-establish the seal. Figures 12.15 and 12.16 show several traps of this type where a special reservoir of water, coupled in one case with an air bypass, enables performance of this sort. Opinions differ on traps of this sort and they are not universally accepted for use in buildings. It is possible by their use to make substantial pipework economies but many people take the view that deposits inside the trap over a period could render its operation less effective. When called upon to function they have a serious disadvantage, particularly in domestic applications; this is the gurgling noise of air passing through the water.

Air bubbling through trap when inlet level falls

Sufficient remains in reservoir to reseal the trap

Air passes through small air pipe when water level in inlet falls due to siphonic action

Sufficient remains in reservoir to reseal the trap

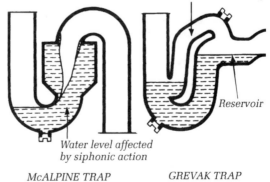

Water level affected by siphonic action

Reservoir

McALPINE TRAP   *GREVAK TRAP*

Figure 12.15   Sections through typical resealing traps.

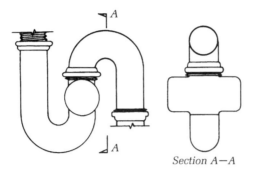

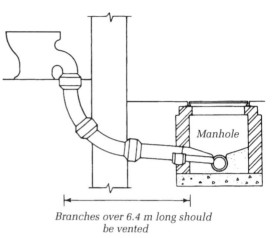

Figure 12.16   The Econa resealing trap.

*Branches over 6.4 m long should
be vented*

Figure 12.17   This groundfloor WC is connected directly to the drain.

Practical installation details

Entry to the drain

Flows from waste and soil installations have to be introduced into the drain without any discharge of drain air. This is accomplished in two ways.

Stack   Wastes and WC outgoes are discharged into a vertical stack, rising to an open vent at above eaves level and leading down to a bend below ground level, which delivers the flow into the drain itself. The junction is almost invariably made at a manhole.

Groundfloor WC   It is well-established practice to connect groundfloor WCs, usually of S-trap types, directly to the drain via an underground bend. The junction is usually made in a manhole. The problem of self-siphonage of the trap does not arise because the rate of flow from the flush only fills a small part of the pipe (Fig. 12.7). Induced siphonage does not arise since the drain is itself ventilated and usually not subject to marked pressure fluctuations. Figure 12.17 shows a typical arrangement.

Groundfloor gully   The connection is similar to that of the groundfloor WC and the traps of gullies are not subject to siphonage for the same reasons as WC traps. Gullies are therefore used to introduce waste flows into the drain. Figure 12.18 shows a typical detail. The waste connection to this type of gully is the only waste or soil pipe which is now permitted outside external walls. (A drop inside the wall, leading to a horizontal back inlet, gives additional frost protection but is more complicated to install and maintain.) Figure 12.19 shows caulking ferrules which can be used to join the small-diameter waste pipe to the 200 mm gully inlet.

These limits on drain entry must be borne in mind when planning buildings. Sanitary appliances on the

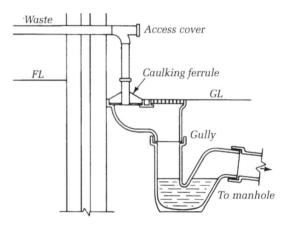

Figure 12.18   This waste is connected to the drain via a gully.

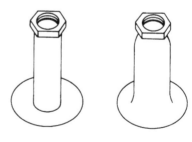

Figure 12.19   Caulking ferrules for joining small-diameter (32 mm and 38 mm) waste pipes to gully inlets or to 100 mm diameter pipes.

ground floor may be freely disposed; the only limit is the need to construct drains to lead to them. It is recommended that branches to WCs and gullies should not be longer than 6 m unless they themselves are provided with a vent pipe. On upper floors, however, unless the sanitary appliance is within reach of a stack, a special one will have to be provided, running down to the ground to the drain and up to the roof as vent. This means that isolated sanitary appliances on upper floors should be avoided as much as possible; appliances should be grouped on each floor and groups on different floors should be vertically over one another.

Groups of sanitary appliances

Several appliances discharging into a common waste or soil branch is an arrangement frequently required in practice, particularly in non-domestic buildings. The conditions for single-stack design do not apply and it has in the past been thought necessary for every trap in such a layout to be individually vented. Figure 12.20 shows such an arrangement for a group of WCs. But it is now becoming standard practice for ranges of up to five WCs to be connected to a common branch without individual trap ventilation. Larger numbers of bends in the common branch would necessitate ventilation but vent pipes leading to selected traps or to the common branch are now more usual than venting every trap and adequate performance is achieved.

Ranges of lavatory basins have often been individually vented as shown in Fig. 12.21(a). More

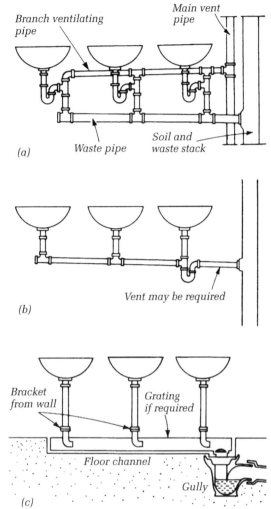

(a)

(b)

(c)

Figure 12.21 Groups of lavatory basins: (a) with common waste and vented traps: (b) with common waste and vented trap (maximum length of waste 5 m); (c) discharging into a floor channel.

economical and neater possibilities are shown in Figs 12.21(b) and 12.21(c). The channel and gully make hosing down and mopping the floor easier. The arrangement is a particularly convenient one on the ground floor since there is no need for a vent stack.

Where rows of WCs are required in adjacent male and female lavatories the plumbing arrangements can be made very much simpler and neater if a duct is used. A space approximately 1 m wide can accommodate and give access to all the plumbing for the WCs. Flushing cisterns can also be fixed inside the duct operated by a lever fitting that passes

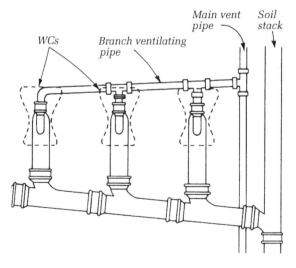

Figure 12.20 A group of WCs with vented traps and discharging to a common branch.

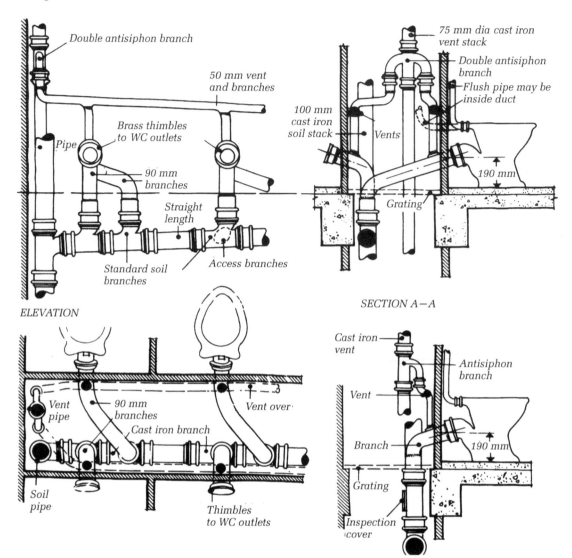

Figure 12.22   Plumbing duct between two lavatories.

through the wall. Figure 12.22 shows a duct of this type. Although the duct illustrated runs through several floors it is possible to have one on a single floor. But note that the floor level of the duct in this case will have to be somewhat lower than the floors of the lavatories.

Sizes for waste and soil pipes

Chapter 10 gives the diameters of the waste or soil outgoes of the most common sanitary appliances. The first section of pipe is normally of this diameter, and if

a single-stack system is used, all branches will be sized in this way. The discharge capacity of stacks can be determined using the discharge unit method described earlier, and the units given in Table 12.8 may be used in conjunction with Table 12.3 to determine appropriate sizes for vertical stacks. BS 5572 *Sanitary pipework above ground* gives details of sizing for more complex installations and for large trap venting pipes. The step-by-step method for foul drains and the associated data, together with Table 12.3 for capacity of vertical pipes may be used for pipe sizing in large installations.

Table 12.3 Discharge capacity of vertical waste and soil stacks

Nominal stack diameter (mm)	Maximum number of discharge units
50	20
76	200[a]
100	850
125	2 700
150	6 500

[a] Not more than one WC.

12.3 Rainwater pipes and gutters

Materials for gutters and rainwater pipes

Cast iron was for many years the most usual material for gutters and downpipes, but it was heavy, liable to fracture and required maintenance. Plastics are now replacing cast iron in popularity. Pipes and gutters made of plastics (usually uPVC) are light, easily fixed and require no painting. In general the plastic systems are similar to the cast iron they replace, but they use special jointing methods and support and gutter sections designed for better flow; manufacturers' catalogues should be consulted for specific details.

Other materials are also used. Asbestos cement will not corrode but is not as strong as cast iron. Cast aluminium is usually in the shapes of cast iron. It is quite strong, rustless and very light. Long lengths of pipe are available, thus reducing the number of joints. Extruded aluminium is used for gutters. BS 1430 covers both types of aluminium rainwater goods. Galvanized mild steel, 1.3 mm for gutters, 1.0 mm for pipes, is made to BS 1091. Galvanized steel is light and cheap but subject to corrosion if the galvanizing becomes damaged. Shapes are somewhat similar to cast iron.

Wrought copper and zinc pipes and gutters are covered by BS 1431. They resist corrosion, but are not as robust as cast iron or aluminium, and therefore have a different shape; the edges of the gutters are beaded. The ogee gutter is fixed through tubular stays. The sockets on the pipes are slimmer and neater than with cast iron.

Outlets for flat roofs are made of cast iron as shown in Fig. 11.4. The outlets in Fig. 11.4 are for use with asphalt, which is turned into a special flange. The outlet with junction is useful for balconies in flats.

General layout

The nature of rainfall and design standards for its intensity together with the vital need for overflow provision have been covered in Chapter 2. Removal of the rain falling on the roof is usually by gutters that collect the water and direct the flow to downpipes, which discharge it to the drains. 'Flat' roofs are provided with falls that direct the water to the gutter. Large flat roofs are sometimes provided with roof outlets, leading directly to downpipes, and falls are arranged to direct the flows. Shallow falls are unlikely to control any flows of water across the roof during high winds, but small upstands at or near the edge of the roof can prevent this undesirable run-off. In tall buildings, particularly if the wall surfaces are fully glazed or impervious, it may be necessary to provide gutters running across the wall. These gutters may be combined with sill details.

Downpipe locations will be constrained by planning, structural, aesthetic and cost considerations. It is important to include consideration of the surface water drainage layout in decisions about the roof drainage. Drains are very much more expensive than gutters and careful consideration of the roof layout may be able to reduce drain runs and even manholes. In the case of industrial buildings it has been conventional to provide a downpipe at each bay on each side of the building, creating long runs of surface water drain with many junctions. In this type of building it should be possible to run horizontal pipes at high level and reduce the number of downpipes and drain runs. Rainwater pipes of this sort should be provided with falls and calculated in the same way as surface water drains.

Rainwater pipes may run externally or internally; in both cases the joints should be watertight. The nature of flow in rainwater pipes with substantial amounts of entrained air gives rise to noise. The location and enclosure of internal rainwater pipes should take this into consideration. Where pipes pass through habitable rooms the casings should have adequate weight to prevent noise penetration.

Maintenance

Particularly when trees are nearby, gutters rapidly accumulate deposits and should be cleaned regularly. Wire balloons are often used at the entry to downpipes to prevent solid matter from the gutter entering the pipe. The result is to encourage deposition of sediment and progressive obstruction of the gutter. Rainwater pipes should be accessible to rodding.

Need for overflow provision

Although the standard design rate of 75 mm/h gives

satisfactory performance for the great majority of the time, this standard will inevitably be exceeded. This is not only the result of intense rainfall rates described in Chapter 2, gutters and outlets are subject to progressive build-up of deposits, to obstruction by objects left on the roof, and by windblown debris such as leaves and plastic bags. It is essential, therefore, that provision is made for overflow to take place without damage to the building or its contents. North-light roofs where beams occur on the line of the external wall present a particular problem. The opening to allow rainwater to pass may become blocked. Any resulting overflow should discharge outside the building, not into the rooflights.

Hydraulic considerations

Ideally gutters should be provided with a very slight fall, 1 in 360 is recommended. The object is not to provide a hydraulic gradient but to ensure there is no standing water in the gutter. Falls may not always be practical where gutters are fixed to fascias. Figure 12.23 shows a typical eaves gutter detail.

The flow of water from gutter or roof to downpipe is the critical element in pipe sizing. Figure 12.24 shows how the flow accelerates as it enters the downpipe, leaving much of the downpipe empty. The depth of flow in a downpipe or across a flat roof will never be great enough to give full-bore flow. A variety of devices attempt to overcome this problem. Figure 12.25 shows a tapering roof outlet where the diameter of entry is greater than the diameter of the downpipe. Sumps with outlets at the bottom attempt to give a greater depth over the outlet. Precise hydraulic calculations cannot be applied to the situations described, but simple procedures give satisfactory results in most buildings. Using the methods described earlier, moisture rates of rainfall on roofs can be estimated and the run-off to each gutter and downpipe calculated from the roof areas involved. Appropriate sizes for gutters and downpipes can be obtained from Tables 12.4 and 12.5.

Bends reduce the flow in gutters so that reduction should be made as follows. Within 2 m of outlet:

Sharp-cornered bends	20%
Round-cornered bends	10%

More than 2 m from the outlet these percentages should be halved.

For valley and parapet gutters use this formula:

$$\text{Flow capacity (l/s)} = \text{area of cross-section (mm}^2) \times 0.000\,156$$

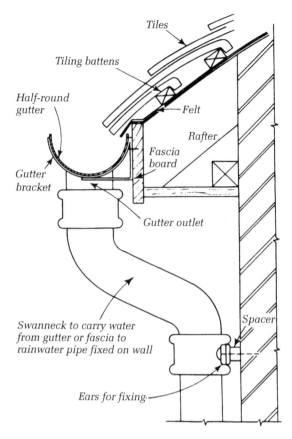

Figure 12.23   Detail of typical eaves showing half-round gutter, gutter outlet, swan neck and rainwater pipe.

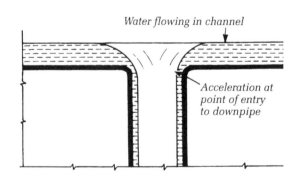

Figure 12.24   Demonstrates why the water flowing in a gutter or a roof channel is unlikely to give full bore flow in a rainwater downpipe (wire balloon to prevent leaves entering pipe not shown).

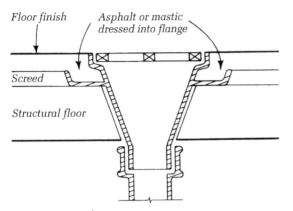

Figure 12.25 This roof outlet for surface water tapers from the diameter of the outlet to the smallest diameter of rainwater pipe connection.

Table 12.4 Flow capacities in level half-round gutters

Gutter size (mm)	Flow in true half-round gutters (l/s)[a]	Flow in nominal half-round gutters (l/s)[b]
75	0.38	0.27
100	0.78	0.55
115	1.11	0.78
125	1.37	0.96
150	2.16	1.52

[a] Pressed steel to BS 1091; asbestos cement to BS 569.
[b] Aluminium to BS 2997; cast iron to BS 460.

Table 12.5 Recommended downpipe sizes

Half-round gutter size, diameter (mm)	Diameter (mm)	
	Outlet at one end of gutter	Outlet not at one end of gutter
Sharp-cornered outlet		
75	50	50
100	62	62
112	62	75
125	75	87
150[a]	87	100
Round-cornered outlet		
75	50	50
100	50	50
112	50	62
125	62	75
150[a]	75	100

[a] Results for 150 mm gutters, sharp-cornered and round-cornered, are currently provisional.

It may give some guidance as to suitable sizes, provided the proportions of the cross-section of the gutter remain somewhat similar to the half-round or ogee section (ratio of width to depth is 2 : 1).

Outlets with sharp corners will restrict the flow in the gutter and downpipe and should be avoided if possible. BS 6367 and BRE Digest 34 give further data on sizing.

Siphonic rainwater systems

For large roof areas the sizes and number of downpipes estimated by conventional methods are very large. If rainwater pipes could be made to flow full-bore the discharge capacity of the pipes would be increased dramatically. This is partly the result of the whole area of the pipe being used but more especially it is the result of the head driving the flow being the whole height of the full pipe. In conventional design the head available to supply the pipe is only the few centimetres of depth of flow in the gutter.

Systems to achieve full-bore flow are now commercially available. They avoid entrainment of air by using specially designed roof outlets; Fig. 12.26 shows the principles. When the water level in the outgo submerges the baffle plate, air entrainment is prevented and very large rates of flow can take place. In practice the flow pattern will oscillate between conventional and siphonic flow since, unless the rainfall rate is very high, the very rapid rate of flow will reduce the water level in the roof outgo to below the baffle plate. More than one outlet can be connected to a downpipe.

There are some special problems associated with this type of system. Pipes must be strong enough to

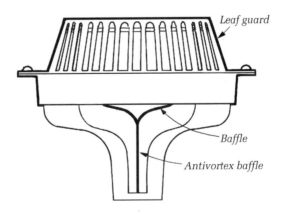

Figure 12.26 The Fullflow Systems siphonic roof outlet.

withstand the considerable pressure fluctuations, both negative and positive, which can result from the flows in the pipe. The discharge rates from rainwater pipes to the drain will, for short periods be very high. A 100 mm downpipe operating with a head equivalent to the height of the building will discharge at a much greater rate than the capacity of a 100 mm, nearly horizontal drain. The drains and manholes must therefore be capable of accepting not merely the rainfall rate, as in conventional systems, but the very high though intermittent flows from the siphonic system. The design of this type of system is a specialist skill. Companies that supply equipment for the system normally offer a design service.

Large gutters

Large gutters for industrial buildings are available in asbestos cement, in rectangular sections, BS 569; widths up to 600 mm; depths usually 150 mm; joints, spigot; and socket, external. These are very strong with all thicknesses of 9 mm or 12 mm. Similar profiles are used for heavy pressed steel gutters in a range of sizes from 300 mm wide up to nearly 1 m, the thickness being in 12 BG for the smallest up to 8 mm plate for the largest. Standard length, excluding socket, is 3 m. Standard fittings for both asbestos cement and steel are stopped ends and outlets; bends are not normally needed. These gutters are normally used with the roof sheeting, asbestos cement corrugated, metal deck, etc., or roof glazing clipped to both edges. For this reason a fall is difficult to arrange and they are usually fixed level. This necessitates rainwater pipes at frequent intervals. Details of factory roofs using these gutters are shown in the MBS title *Structure and Fabric Part 1*. Joints are made in red lead, or more often bituminous mastic, and bolted. Support is best at 1.5 m intervals, usually on steel brackets as part of the roof framing, but sometimes on timber framing, which is easier for site adjustment.

Connections from rainwater pipes to drains

There are two distinct cases to consider in connecting a rainwater pipe (RWP) to the drains. If the drainage system is separate and the RWP is delivering to a surface water drain no trap is necessary and a direct connection between pipe and drain may be made. Figure 12.27 shows a typical arrangement. Where the drainage is combined, however, it is necessary to provide a trap between the surface water pipework and the combined system. This may be done by collecting several RWPs into a drain and making a trapped

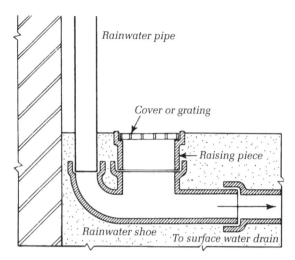

Figure 12.27 Rainwater downpipe delivery to untrapped rainwater shoe: only possible with separate drainage systems.

connection to the combined drain, or by providing a trap at the foot of each RWP as shown in Fig. 12.28. In most new buildings with combined drainage systems both systems will be used to achieve economy of pipework, and although some RWPs will be trapped and deliver to the combined drain, others may contribute to areas of separate surface water drains only trapped at the point of junction with the combined drains.

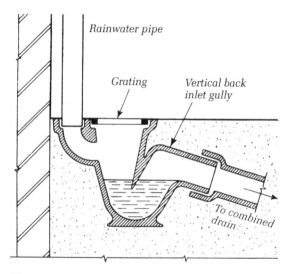

Figure 12.28 Rainwater downpipe delivery to trapped gully: essential if the rainwater pipe is delivering to a combined drain.

Drainage of paved areas

Paved areas must be laid to falls towards channels or gullies in the same way as flat roofs. This is very frequently overlooked in modern buildings which, in wet weather, force everyone approaching the entrance to splash through puddles formed on the paving. As a general rule all paved surfaces over 6 m² in area should be drained and should be given a minimum fall of 1 in 60, which will discharge rainwater swiftly and avoid puddles. Falls should never run up to the external walls but preferably run away from the building. Paved areas may discharge to channels in site roads, or to specially formed channels, forming part of the paving. Channels, which are more carefully laid and levelled than paving and where some pounding would not matter, because of their defined narrow shape, can be laid to substantially lower gradients down to 1 in 200. Often the slope of the building site will give greater falls than this in any case. Gullies, either untrapped or trapped, depending on whether the drainage system is separate or combined, will discharge the surface water flows to the drains.

Snow

Snow is not a serious problem in buildings in most of the country. Drifting across openings or excessive weight on roofs rarely occur and the rate of run-off from melting snow is invariably lower than the 75 mm/h standard for gutter and downpipe design. Large valley gutters may be provided with snowboards to ensure a free passage for water underneath the snowboards, particularly if overflowing could be a problem. Where sharply pitched roofs could spill snow directly on to glazing below or outside an entrance, vertical snow guards are sometimes provided at roof eaves.

12.4 Drainage below ground

The general principles of good drainage described on page 258 and in Table 12.1 apply to underground drains. In addition, a number of special considerations arise from the planning, constructional, organizational and hydraulical points of view.

Planning

Drains carry foul and surface water away from buildings to points of disposal. In many cases disposal from the site will be by discharge to a sewer. Public sewers are maintained by local authorities and usually sited below roads. In some cases building owners may combine to have flows from several properties passing across the different sites in a private sewer leading to a public sewer or other means of disposal. Where discharge to a sewer is not possible, means of disposal on the site itself will have to be arranged. Whichever is the case, in designing a new building, thought must be devoted to the ways in which siting and planning may be influenced by the need for feasible and economic drain layout. In normal situations it will be desirable to site the building so it is at a level from which drains can be laid to fall towards the disposal point, allowing liquids discharged into the drain to flow by gravity. It will clearly be economical to keep the length of the drain involved to a minimum. Within the building itself the grouping together of sanitary appliances, and particularly their grouping vertically, reduces the number of entries to the drain, branch drains and manholes; the positioning of these groups and of rainwater pipes can reduce the length of drain running round the building.

Although it is possible by means of long runs, deep excavations and pumping to drain most conceivable building situations, it is desirable to avoid these features unless they make possible some important and otherwise unattainable planning requirement.

Construction

Table 12.1 sums up the main performance requirements which must be met by drain installations. Drains must be watertight; not affected by the flows they carry or the soil chemicals; able to resist root penetration; able to withstand earth pressures, the effects of earth movement and building settlement; and capable of withstanding thermal and moisture movement. The materials available for use as drains are in themselves watertight and durable. The main problems for drain construction arise because of the various types of movement to which drains are subjected as well as the effects of crushing. This situation has not been understood for very long. Its realization is giving rise to the use of new materials, jointing and bedding techniques. In the past drains were constructed to be as rigid as possible. Joints allowed no movement and, in cases where especially high performance was required, concrete beds were provided under the drains. These precautions were not able to resist the forces involved in moisture, thermal and ground movement, so drains laid in this way were liable to damage from fracture. The present tendency is to replace rigid constructions with flexible constructions. In fact, the construction of rigid drains with rigid joints must now be regarded as bad practice in almost every circumstance. Either materials which are themselves flexible or joints capable of accommodating elongation or contraction (draw) or angular movement (slew) are increasingly used in drain construction. Joints capable of adjusting to both types of movement are known as flexible and telescopic.

Materials

Vitrified clay (glazed or unglazed)

This is the traditional material for drains; until recently it was described as salt-glazed ware or stoneware. It is brittle but very durable. It is widely used and a considerable range of special junctions, bends and fittings is made and stocked by builders' merchants. The traditional method of jointing (Fig. 12.29(a)) is by placing the plain (spigot) end of one pipe into the socket forming the end of the next pipe. Tarred hemp is wrapped around the spigot and compressed into the space between spigot and socket, thereby centring the spigot in the socket and preventing the cement mortar, used to complete the filling of the socket, from penetrating into the bore of the pipe. The joint resulting is not only rigid, but is liable to fracture, even without ground movement, because of the differing thermal and moisture movements of the cement joint and vitrified clay pipe. To overcome this a number of flexible telescopic joints have been developed. Figure 12.29(b) shows one type employing plastic bedding rings fixed to socket and spigot along with a rubber sealing ring. Joints of this type can withstand up to 5° in bending and 18 mm extension or shortening. Using this type of joint with proper bedding, vitrified clay pipes make excellent and durable drains. They are available in a wide range of sizes (only some of which are stocked by merchants) between 75 mm and 750 mm diameter. Until new metric lengths become available standard lengths based on the imperial sizes are 610 and 914 mm.

Very recently a new type of joint for vitrified clay drains has been introduced. It uses plain pipes without sockets, an arrangement that has a number of advantages in terms of economy, ease of cutting and handling, linked by a plastic sleeve which will allow both elongation and angular movement. Figure 12.29(c) shows this type of joint. The relevant British Standard is BS 65 : 1971 *Clay drain and sewer pipes*.

Concrete

Concrete drainpipes are notable for the availability of very large diameters. Many manufacturers list

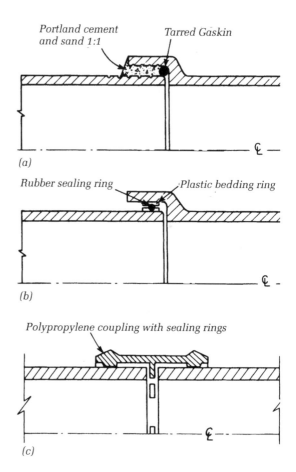

Figure 12.29 Jointing methods for vitrified clay drains: (a) cement joint; (b) flexible joint; (c) flexible sleeve joint for plain-ended pipes.

diameters from 100 mm to 1.83 m and special egg-shaped pipes are available to give improved depth of flow at times of slack discharge. In the smaller diameters concrete pipes are not usually considered competitive in price with vitrified clay, except where heavy crushing loads have to be met. Concrete pipes can be constructed to give special resistance to crushing. The methods of jointing are similar to those for stoneware. Flexible rubber ring joints have been available for concrete pipes for many years but their use has normally been for large civil engineering works. Several forms are available for building drains and their use is much to be preferred over the traditional cement joint. Figure 12.30 shows some of the jointing methods which are employed with concrete pipes. The relevant British Standard is BS 556 *Concrete cylindrical pipes and fittings*.

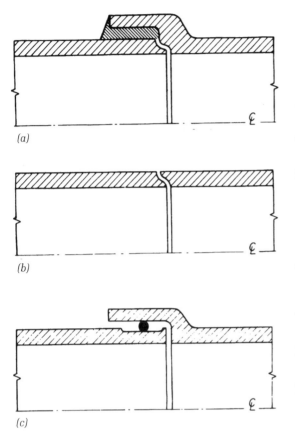

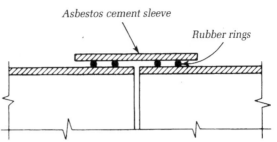

Figure 12.31 Rubber ring for asbestos cement drains.

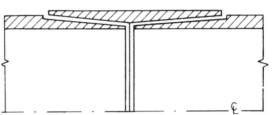

Figure 12.32 Tapered sleeve joint for pitch-fibre drains.

Figure 12.30 Jointing methods for concrete pipes: (a) cement spigot and socket; (b) ogee; (c) flexible rubber ring.

Asbestos cement

Asbestos cement pressure pipes have been available for many years and can be used for underground drainage. Recently, however, a special asbestos cement drainpipe has been developed. It is supplied in 4 m lengths, thereby reducing the number of joints to be made. The method of jointing, shown in Fig. 12.31, is inherently flexible and telescopic. Diameters from 100 mm to 600 mm are available.

The relevant British Standard is BS 3656 : 1963 *Asbestos cement pipe fittings for sewerage and drainage*. (Class 3 is appropriate for most building drainage application.)

Pitch-fibre

Pitch-fibre pipes are formed by impregnating wood fibre with pitch. The resulting pipe has adequate durability for normal drain use (although not suitable for continuous very hot discharges or some types of chemical flow). Pitch-fibre piping is available in 2.5 m lengths, is very light in weight in relation to other drain materials and has a very simple jointing technique. Figure 12.32 shows the type of joint. A taper is formed on the end of the pipe and this is driven into a tapered socket on a sleeve junction piece. A sledgehammer and a wooden dolly to prevent damage to the end of the pipe are the tools required for jointing. Although the formation of an individual joint is easy, care has to be taken to support the rest of the pipe, particularly where junctions are involved, to prevent displacement and care is required to ensure that successive joint drivings do not place excessive force on earlier joints. Long straight lengths of pipe are very easily and quickly laid. Indeed, it is possible to make up lengths of pipe at ground level then to lower them already jointed into the trench. This can be of great assistance in minimizing excavation or with difficult trench conditions. Junctions and fittings with appropriate joints are available for manufacturers. Bends are cut from lengths of pipe of appropriate radius and the tapered ends formed on site by the use of a special tool. Lengths of straight pipe are cut in the same way. Open channels for manholes are formed by cutting standard pipe. Although the jointing does not act telescopically, the pipes are flexible and will accommodate themselves to a considerable degree of ground movement.

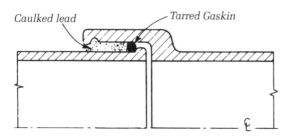

Figure 12.33 Caulked lead joint for cast iron drains.

Diameters from 75 mm to 200 mm are available and the range may be extended.

The relevant British Standard is BS 2760 : 1956 *Pitch-impregnated fibre drain and sewer pipes.*

Cast iron

Cast iron is traditionally used where ground movement might occur and where leakage must be prevented (e.g. drains under buildings), although modern thought favours short lengths of pipe with flexible joints. The spigot and socket joints of cast iron drains are either run with molten lead and caulked to take up contraction on cooling or filled with lead wool which is then caulked into a solid joint, as shown in Fig. 12.33. A wide range of junctions, bends and other fittings are available. Figure 12.34 shows some of them, including a typical cast iron inspection chamber which demonstrates the space saving obtainable. Three British Standards are relevant:

- BS 437 : 1933 *Cast iron spigot and socket drain pipes*
- BS 1130 *Schedule of cast iron drain fittings*
- BS 1211 : 1958 *Spun iron pressure pipes*

Plastic drains (made from unplasticized PVC)

The advantages of plastic drains are apparent: lightness, ease and speed of assembly and smooth internal surfaces. But designers have had doubts about the viability of plastic drains and about their cost. In 1968 an Agrément certificate was issued for 100 mm drains, and in 1971 there followed a British Standard for unplasticized PVC underground drainpipes and

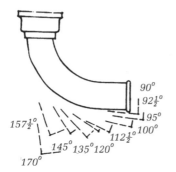

BENDS: SHORT AND LONG RADIUS

$92\frac{1}{2}^{o}$ only

TAPER BENDS:
SHORT AND LONG RADIUS
50–100 mm
75–100 mm
100–150 mm

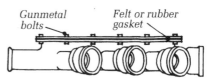

Access bends with round or rectangular bolted covers to heel or face of left- or right-hand bends in pipe diameters up to 300 mm

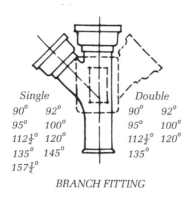

BRANCH FITTING

Single
90° 92°
95° 100°
$112\frac{1}{2}^{o}$ 120°
135° 145°
$157\frac{1}{2}^{o}$

Double
90° 92°
95° 100°
$112\frac{1}{2}^{o}$ 120°
135°

SPIGOT AND
SOCKET JOINT

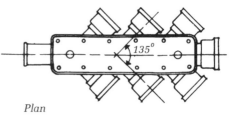

TYPICAL CAST IRON (DRAIN LEVEL)
INSPECTION CHAMBER

Figure 12.34 Cast iron drain fittings.

fittings. This standard also covers only 110 mm diameter pipes and fittings (the outside diameter is used to define plastic pipes rather than the inside diameter normally used for other pipes). Since 1968 there has been a rapidly increasing use of plastic drains and several Agrément certificates have been issued to a number of companies, including some certificates for 160 mm diameter pipes. The range of sizes available varies between different manufacturers. It is possible to obtain 82.4 mm (for surface water only), 110 mm, 160 mm, 200 mm and 250 mm diameter pipes. Imported pipes can be obtained up to 630 mm diameter. Pipes can be obtained up to 6 m in length. Although solvent welded joints may be used, they are difficult to make in the difficult conditions found in drain trenches, so push joints with rubber sealing rings are normally used. Figure 11.8 shows a joint of this type.

uPVC pipes are light in weight, relative to other drainage materials, easy to handle and joint and the drains may be tested as soon as they are laid, which enables trenches to be backfilled quickly, a significant advantage for progress of work on the site.

Although uPVC drains may be laid in conventional layouts with normal types of brick manholes, manufacturers working in this field have identified manholes as a source of some drainage troubles and as an important element in the cost of drainage systems. Most manufacturers include in their ranges of pipes and fittings some special manholes designed to minimize cost, or in one case a system which replaces a proportion of manholes with rodding points. Agrément certificates have been awarded to a number of systems of this sort. Designers wishing to use uPVC drain systems must therefore consult manufacturers' literature and should clearly satisfy themselves that Agrément certificates cover the system they propose to use.

Two systems are described on pages 246–7.

Drain construction

Proper bedding for drainpipes is essential for drain durability. Traditionally, and most existing drains will follow this pattern, trenches were dug and their bottoms levelled to the falls required. Pipes were laid by scooping grooves for the sockets or packing to support the barrels to the final level. Where a particularly sound job was required, as in the case of foul drains, a continuous strip of concrete was provided for pipe bedding. This is not now regarded as satisfactory. The concrete strip will not be capable of resisting the pressures of earth movement, fracture

will occur and serious drain failure is likely to result at the point of breakage. In trenches without concrete bedding local hard spots could cause pipe fracture.

At present the Building Research Establishment recommends the use of granular bedding, which is used under as well as at the sides and over the pipe and is laid and compacted in thin layers. This type of material provides a uniform bed for drainpipes, avoiding the problems of irregular support, and materially contributing to the crushing strength of drainpipes. The effects of ground movement are distributed rather than concentrated by granular bedding. Suitable graded sand, gravel and broken stone may be employed. Figure 12.35 shows how a pipe is bedded in this way. In cases where the pipe is not strong enough to resist the crushing forces extra strength can be obtained by providing a concrete base for the pipes in addition to the granular bedding. Figure 12.40 (right) shows this arrangement. It is important, however, that the base should not be continuous but should support each pipe separately. If laid as a continuous strip a gap should be cut out to correspond with each joint. Figure 12.36 shows a section through a typical joint and bedding. Note that the gap in the concrete and the gap in the socket of the drainpipe are filled with clay to prevent stone penetration and consequent loss of flexibility. Complete concrete encasement can be used where more strength is required than can be obtained using granular bedding, provided the casing has gaps at each joint.

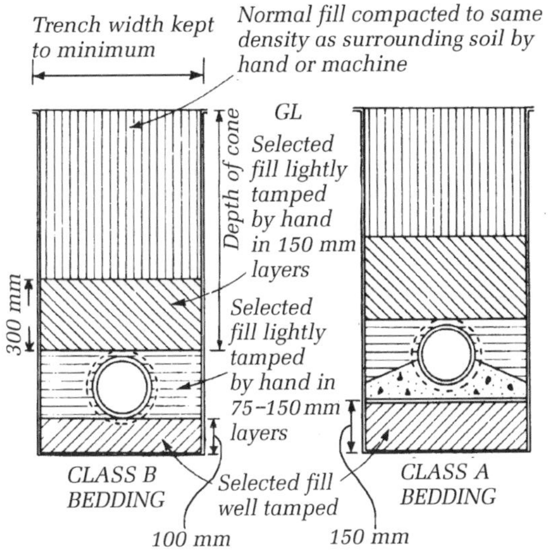

Figure 12.35 Drain beddings recommended by BRE. Class B is granular bedding whereas class A uses concrete to support the pipe against crushing (see Fig. 11.13 for joints).

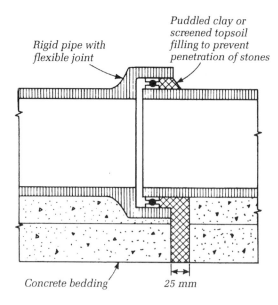

Figure 12.36 Concrete bedding for drainpipe: the gap cut into the bedding at the joint prevents fracture and allows flexibility.

BRE Digests 124 and 125 (first series) *Small underground drains and sewers* describe these principles and give details to enable pipe strengths and loads from the ground to be estimated, together with references to more detailed studies. It is worth noting that the load on drainpipes is determined by the width of the drain trench at the level of the crown of the pipe, above this the trench can be any convenient width. The width at the crown should clearly be kept to a minimum. BRE recommendations for width at crown are given in Table 12.6.

It is also recommended that the minimum cover over drains should be 1 m under fields and gardens, and 1.3 m under roads. Shallower drains than this may require a protective concrete slab placed over the pipe, but not in contact. Under roads this slab should be reinforced and wide enough to transmit the loads from above to the undisturbed ground beyond the sides of the trench.

Table 12.6 BRE recommendations for width at crown

Drain diameter (mm)	Width of trench (mm)
100	530
150	610
230	690
305	760

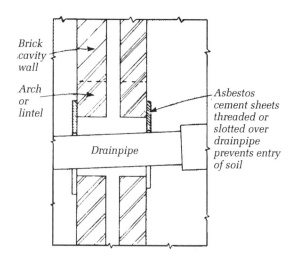

Figure 12.37 Detail of provisions for drain passing through a wall. Some uPVC systems have special fittings for this purpose.

Special precautions should be taken where pipes pass through walls to ensure that differential settlement does not cause drain failure by crushing. Figure 12.37 shows a suitable detail for a drain passing through a wall. Note that not only is the wall kept clear of the drain, but also the soil is kept out by asbestos cement sheets.

Inspection chambers

For ease of tracing and maintenance, drains are laid in straight lines between access points, which should allow drain rods to be introduced for clearing the drain if necessary. Suitable access points may be made by providing rodding eyes, access gullies or inspection chambers. Rodding eyes and access gullies can only be used at the heads of branch drains. At intermediate points and junctions access, if required, must be provided by inspection chambers. Rodding eyes are formed by bringing a drain bend up to ground level and sealing with a cover plate. Access gullies have a removable cover, usually on the outgo pipe, which allows access to the drain. Inspection chambers are pits formed below ground through which the drain runs in an open channel. They are required at all bends or changes of gradient and at junctions when either the main drain or the branches cannot be rodded from other points. Inspection chambers were traditionally provided at all junctions on soil drains. Present thought allows soil junctions without inspection chambers, provided that both main drain and branch can be completely rodded. A small number of manholes will enable main runs of drain to be rodded but branches from stacks or groundfloor WCs cannot easily be rodded except from manholes at the junction with the main run. Gullies can be provided with special access points and it is easy to connect branches from them to runs of drain, without manholes at the junction, while ensuring that each length can be rodded. Inspection chambers are also provided at intervals along straight lengths of drain without junctions. This is done to ensure that the whole length of drain can be satisfactorily rodded. Opinions vary about the satisfactory spacing of inspection chambers for this purpose. A survey of local authorities has shown that their requirements for manhole spacing varied between 23 m and 90 m. The Building Regulations 1985 require that no part of a drain shall be more than 45 m from an inspection chamber and that no junction, including junctions with private and public sewers, if not itself provided with an inspection chamber, shall be more than 12 m away from one.

Inspection chambers must be of adequate size to permit a person to work and also to allow the branches to enter sufficiently far apart not to impair the stability of the construction. Where manholes are very deep a working chamber of suitable dimensions is constructed, roofed by a concrete slab or brick arch, and this chamber is approached by an access shaft leading down from the surface. Table 12.7 shows minimum working sizes for rectangular manholes and gives details of sizes to allow for branches. Different considerations apply to manholes formed from circular precast concrete rings and manufacturers should be consulted.

Inspection chambers are usually constructed of either brick walls with concrete base and slab or of precast concrete rings. They must be able to withstand soil pressures and be watertight both to prevent leakage of liquid into the surrounding soil and to prevent seepage of groundwater into the drain. Table 12.7 gives the thickness of walls and slabs often employed in brick inspection chambers; Fig. 12.38 shows the general arrangements in a shallow manhole. Class B engineering bricks in English bond with joints fully filled with cement mortar treated with waterproofer form a suitable wall. Internal rendering should be avoided. Frost causes the rendering to break away from the walls near the top of the manhole, falling into the channel and perhaps blocking the drain. Step irons as shown in Fig. 12.38, must be provided if the inspection chamber is more than 610 mm deep. They should be set at about 300 mm centres and alternately placed to right and left to suit the movement of feet of workers climbing into the inspection chamber. They may be bedded directly into the brickwork. Where, as is usual, the cover is smaller than the manhole, a concrete slab should be used to support the cover. This gives a more permanent and satisfactory result than corbelled brickwork.

The main run of drain passes through the bottom of the manhole in an open half-round glazed vitrified clay (GVC) channel, either straight through the manhole or as a bend. Branch drains shown in Figs 12.38 and 12.39 are discharged into the main drain by open curved members known as branch bends, which rest on the edge of the channel and turn the flow so that it joins the main flow in the direction of its travel, irrespective of the angle at which the branch enters the

Table 12.7 Minimum sizes for rectangular manholes[a]

Manhole type	Depth to invert (m)	Minimum internal dimensions (m)			Thickness of walls (mm)	Thickness of slab (mm)
		Length	Width	Height above benching		
Shallow (no access shaft)	<0.6	0.75	0.7	–	115	100
	0.6–0.9	1.2	0.75	–	230	150
	0.9–2.4	1.2	0.75	–	230	230
Deep (working chamber and access shaft)[b]	>3.3	1.2	0.75	2[c]	230	230–450

[a] Rules for sizing manholes with branches:
- *Length* Established by side with most branches on the basis of 300 mm per 100 mm diameter pipe; 380 mm per 150 mm diameter pipe plus an allowance for the angle of entry if this is not 90°.
- *Width* Branch bends are designed for a 300 mm gap between wall and channel.
[b] Access shafts for deep manholes should not be less than 0.75 m × 0.7 m (corner step) or 0.85 m × 0.7 m (ladder). Manholes over 6 m deep should have a shaft provided with a rest platform every 6 m.
[c] Minimum cover opening size 0.6 m × 0.6 m.

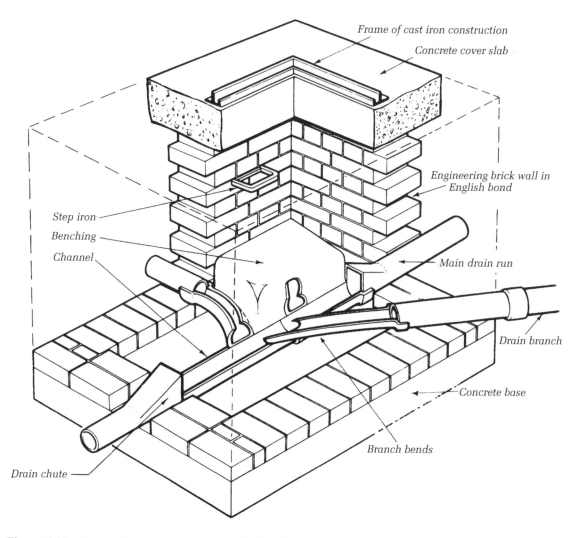

Figure 12.38   Cutaway showing construction of typical shallow brick manhole.

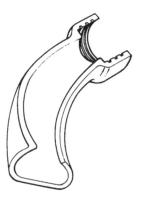

Figure 12.39   This branch bend is used in a manhole to create a junction between a branch and the main drain.

manhole. Both channels and branch bends are bedded in concrete *benching*, which rises vertically from the edges of the channel to above the level of the crown of the drain then slopes up to the walls of the manhole at a rise of 1 in 6. This allows a reasonable platform for workers to stand on but provides no lodgement for sewage in cases of surcharge.

Figure 12.40 shows a precast concrete inspection chamber base incorporating the channels and branches and one of the rings which would form the walls of the chamber. Shallow and deep types are available. The base is heavy and requires lifting tackle. Step irons are ready fixed into the rings. A special top ring gives support for a rectangular cover. One advantage of this type of inspection chamber is the facility to set the top

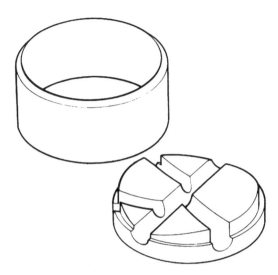

Figure 12.40   Base and typical ring for a precast concrete manhole.

Figure 12.41   Glass reinforced plastic (GRP) preformed inspection chamber for use with uPVC drains. There are five inlets and one outlet for 100 mm diameter drains. Any unused inlets are stoppered.

ring so that the cover is parallel to the pattern of paving. Delivery and lifting requirements mean that inspection chambers of this sort are most likely to be used on large contracts.

Figure 12.41 shows a glass reinforced plastic (GRP) inspection chamber and cover specially developed for use with uPVC drains. The inspection chamber is light and easily installed and the connection of drains is simple since the inlets and outlet of the chamber are

equipped with rubber ring joints similar to those in the run of drains. The chamber is supplied in stacking form so that delivery and storage are simplified and up to its maximum depth of 0.9 m, it can be cut to the right depth on the site with a carpenter's saw, thus obviating the need for different sizes. The cover is supplied especially for this chamber.

Covers for inspection chambers can be obtained in a wide range of sizes and patterns. The cheapest and most widely used is the chequer-pattern cast iron cover. Figure 12.38 shows the frame for such a cover. Light duty for pedestrian traffic, and medium duty for trucks and light vehicular traffic are available along with a wide range of special types, including recessed covers to take finishes to match surrounding paving, double-seal covers for use inside buildings, open-grating types for ventilating and heavy-duty types for roads. Heavy-duty types are often circular, but triangular or circular types with three-point suspension minimize the chance of rocking as traffic passes.

Connection to sewers

The final section of the drain leading to the sewer from the last manhole is almost always laid by the local authority, which will charge for this work. Connection is made towards the top of the sewer, since this is most economical and gives the minimum obstruction to flow. A hole is made in the sewer and a saddle junction cemented into place. Connections are made into the run of the sewer, not normally at inspection chambers.

In most of the areas served by earlier foul and combined sewerage systems a trap was required by the local authority as the outgo from the last manhole. This trap, known as a sewer gas interceptor, was intended not only to prevent sewer gases from entering the house drains but also to prevent the passage of rats. In fact, the interceptor was a major cause of drain blockage, since any heavy solid matter would remain there. Present practice in areas served by new sewers, taking account of this problem and recognizing that the flow of air is more frequently from drain to sewer than otherwise, is to make direct connections without interceptors. Some local authorities still require interceptors in new work and their use may also be desirable for new buildings in areas where interceptors were generally provided.

A cleaning eye to allow the connection to the sewer to be rodded is provided in each interceptor (Fig. 12.42 shows a typical example). It is desirable to use the type of interceptor which has a cleaning eye cover opened by a chain, the far end of which is secured near the cover. This arrangement enables the liquid to be drained easily from the inspection chamber rather than

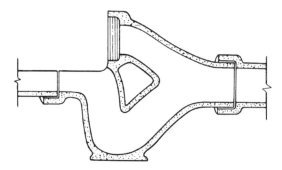

Figure 12.42   Sewer gas interceptor.

dipped out in the case of blockage. Where the interceptors are used, a special air vent, not less than 75 mm in diameter, is required to the final manhole.

Systems

It will normally be economical and desirable to discharge building drains to a sewer if one is available. Local authorities may require this if a sewer is within 30 m of the site or if they are prepared to bear the cost of any additional length of connection. In early sewage systems both foul and surface water drainage was discharged into the single 'combined' sewer. Purification plants capable of dealing with the foul flow cannot deal with the flow in wet weather, even if it is economic to do so. When six times the dry weather flow is reached, untreated sewage will be discharged direct. This is not a very satisfactory arrangement and more modern systems separate the foul flows, which are all treated, and surface water which is discharged untreated into watercourses. In country districts with suitable subsoil it may be convenient to use soakaways on the site to dispose of surface water and many local authorities will insist on this to keep down the load on existing combined sewers. The sewage system clearly affects the drainage layout. If it is a separate system then separate drains will be required for surface water and foul flows; otherwise the flows can be mingled in one set of combined drains. Local authorities will sometimes have proposals for providing a new surface water sewer to relieve an existing combined sewer. Once the new sewer is built the old combined sewer can be used for foul flows only, allowing the building drains to be separate. But until the new sewer is built, the local authority may require the drains to be constructed as separate foul and surface water systems with a common connection to the combined sewer.

Hydraulic considerations

It is necessary, in the case of foul drains, to be able to introduce flows into the drain without nuisance arising from the escape of drain air, and for flows to take place without causing air pressure fluctuations which would destroy gully seals. In both foul and surface water drains the discharge capacity of the drain must be adequate to deal with the flows likely to take place.

Entry to drain   Waste and soil stacks are connected to the drains normally at manholes directly by means of a bend below ground. The stack must be carried up to such a level that no nuisance can arise (usually above the eaves) from the escape of drain air.

Ground level WCs can be connected directly to the drain as shown in Fig. 12.17; there is no risk of siphonage of the WC trap.

Groundfloor gullies and urinal traps can also be connected directly to the drain since the same considerations apply as in the case of WCs. Figure 12.18 shows an external waste gully with waste pipe connection.

Rainwater pipes discharging to combined drains must deliver into a trapped gully. In conditions of heavy rain the flow may be substantially greater than the foul flows previously described. The restrictions and bends, however, will not normally allow sufficient water to pass to give full-bore flow in the 100 mm diameter outgo from a gully, so siphonage of the gully trap is unlikely. In any case rainfall flows do not end abruptly and it is certain there will be a final period of limited flow that will not produce siphonage. Where rainwater pipes discharge to surface water drains, direct connection, or connection to an untrapped gully and thence to the drain, is feasible and normal, since no nuisance problems arise from the escape of air.

Drain ventilation

Apart from the avoidance of air pressure fluctuations, drain ventilation is advocated as preventing the build-up of smells within the drain and as causing the drying and subsequent flaking off of deposits on the drain walls. It is questionable how critical the last two items are. In the case of air pressure fluctuations it is clear that the nearly horizontal drainpipes do not present such a critical problem as the vertical and horizontal pipes of waste and soil installations. Traditional practice has been to provide a low level vent at the last manhole (called the fresh air inlet) and a vent pipe running up the building at the head of the drain. A soil pipe could be employed to act as this vent. Branches longer than 6.4 m were often not separately ventilated but special ventilation for them was considered good

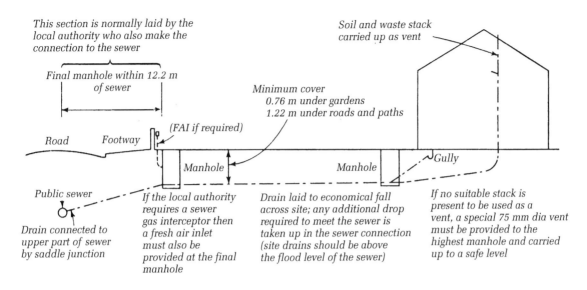

This section is normally laid by the local authority who also make the connection to the sewer

Final manhole within 12.2 m of sewer

Soil and waste stack carried up as vent

Minimum cover 0.76 m under gardens 1.22 m under roads and paths

(FAI if required)

Road   Footway

Manhole   *Manhole*   *Gully*

Public sewer

Drain connected to upper part of sewer by saddle junction

If the local authority requires a sewer gas interceptor then a fresh air inlet must also be provided at the final manhole

Drain laid to economical fall across site; any additional drop required to meet the sewer is taken up in the sewer connection (site drains should be above the flood level of the sewer)

If no suitable stack is present to be used as a vent, a special 75 mm dia vent must be provided to the highest manhole and carried up to a safe level

Figure 12.43   Principles of simple drain layout and drain ventilation.

practice. Figure 12.43 shows a simple system of this sort. The arrangement was based on the concept that warm drain air would rise up the high stack and be replaced by cold air entering the fresh air inlet. This would give a circulation of air through the drain. The fresh air inlet was provided with a mica flap intended to allow air to pass only in the desired direction. In operation this flap does not normally serve its purpose and the overriding influences on the movement of air are twofold:

- Filling of the drain, which forces air out at all vents.
- Entrainment of air in the direction of flow, which causes an opposite movement to that originally envisaged.

When interceptors are not employed the drain will be ventilated directly into the sewer where, under most conditions, a current of air is drawn in the direction of the flow. Although the number of points of ventilation of the drain are limited, groundfloor WC and gully traps are not affected by flows since building drains very seldom flow full so air pressure differences seldom arise. The Building Regulations from 1965 onwards, unlike the previous bye-laws, do not stipulate any specific provisions for drain ventilation. They require only 'such provision ... as may be necessary to prevent the destruction under working conditions of the water seal in any trap'. Figure 12.44 shows a typical drain layout for a housing scheme.

Transport of solids

One function which a drainage installation has to perform is the transport of solid excrement from WCs. Figure 12.45 shows the way in which solid matter is transported. Some of the flush from a WC will enter the drain ahead of the solid matter but, in a well-designed pan, much of the flush will follow the solids and provide the motive power causing movement. This water will leak past the solids, leading to stranding in the upper reaches of the drain until a further flush provides additional motive power.

Recent specialized developments in drain layout

The two systems described below were the result of extensive studies of the problems of drainage undertaken by manufacturers of uPVC pipes. Both systems are based on reducing the cost of manholes, a substantial proportion of the total cost of drains. In both cases the principles employed could be used with drain materials other than uPVC. Agrément certificates have, however, been issued for both systems using uPVC and, until equivalent support is available for the use of other materials, judicious designers are likely to restrict themselves to the certificated systems.

uPVC drain layouts designed in accordance with these systems are said to be economical in comparison with conventional layouts. An ultimate cost comparison can, however, hardly be made until traditional materials are used in comparably

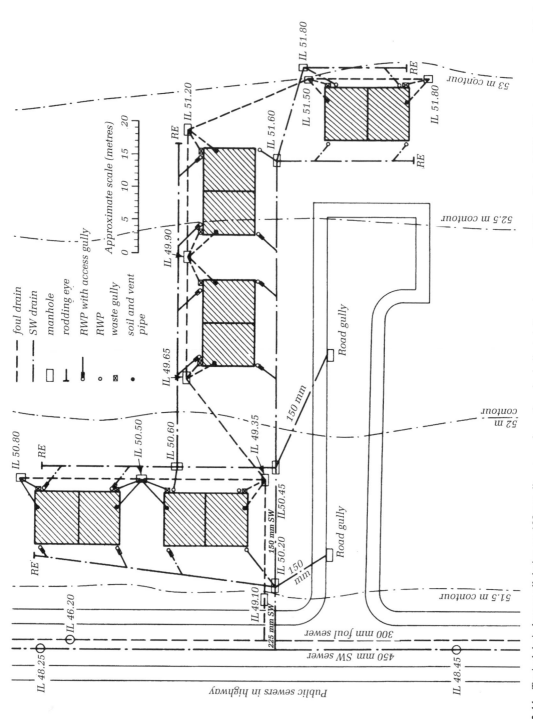

IL 51.80

IL 51.20

RE

RE

RE

53 m contour

IL 51.50

IL 51.60

IL 51.80

IL 51.80

RE

Approximate scale (metres)

0   5   10   15   20

foul drain

SW drain

manhole

rodding eye

RWP with access gully

RWP

waste gully

soil and vent pipe

52.5 m contour

RE   IL 51.20

RE

IL 49.90

IL 49.65

150 mm

IL 49.35

Road gully

52 m contour

IL 50.80

RE

IL 50.50

IL 50.60

IL 49.35

IL 50.45

IL50.20   IL50.45

150 mm SW

150 mm SW

150 mm SW

RE

Road gully

IL 49.10

225 mm SW

150 mm

Road gully

51.5 m contour

IL 46.20

300 mm foul sewer

450 mm SW sewer

IL 48.25

Public sewers in highway

IL 48.45

Figure 12.44 Typical drain layout: all drains are 100 mm diameter unless otherwise indicated. Paved areas round buildings (drained SW gullies) are not shown. Manholes would normally be accompanied by their number, their cover level and their invert level; for the sake of clarity only invert levels are shown here. Cover levels are at ground level, as indicated by contours.

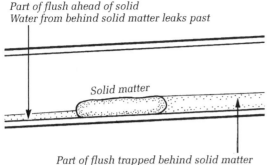

Part of flush ahead of solid
Water from behind solid matter leaks past

Solid matter

Part of flush trapped behind solid matter
This provides motive power for movement
of solids

Figure 12.45   Transport of solids in drains.

economical layouts. The manufacturers of these systems provide detailed literature on design and a technical consulting service.

Local authorities often have particular views on drain falls and manhole positions and spacings. They should be consulted at an early stage if specialized systems are to be used.

The Marscar system

In this system the great majority of manholes collecting branches from the building are replaced by a standardized bowl-shaped access pit. This pit, shown in Fig. 12.46, is shallow, requiring a minimum of excavation and has knockout holes so that a number of junctions can be made at convenient angles. All pits

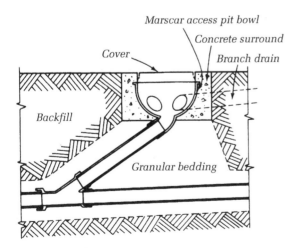

Marscar access pit bowl
Concrete surround
Cover
Branch drain
Backfill
Granular bedding

Figure 12.46   The Marscar drain system.

have a standard depth and the linkage with the varying depth of the main run of drain, which normally dictates the manhole depth, is made by an inclined branch coming from the access pit and connecting to the top of the main drain. Junctions between main runs of drain will require conventional manholes which, if suitably sited, will enable all lengths of drain to be rodded.

The Marley rodding point drainage system

This is an extremely well-detailed and documented approach to drain layout, replacing most inspection chambers, whose purpose is to give access to the drain, with rodding points which can be used to clear the drain of any obstruction even though rodding in straight lines is not possible. The method depends upon the use of flexible rodding equipment which can clear obstructions and negotiate slow bends. A 30 m length of 13 mm diameter low density polythene tube mounted in a reel has proved successful.

In preparing a scheme for rodding point drainage a conventional layout may be used but the number of branches coming from the building will be reduced as far as possible by connecting sanitary appliances to stacks rather than making individual connections to the drain. The collar boss fitting shown in Fig. 12.10 can aid in the layout. This has the effect of concentrating flows along a single length of drain, giving a better flush in the branch, and also reduces the number of junctions which would require rodding points. Having reduced the number of branches to a minimum, rodding points are provided at all junctions and bends in positions which would normally be served by inspection chambers. A minimum number of conventional inspection chambers should be provided, sited to allow visual inspection of flows, so that any debris dislodged by rodding can be removed from the drain. Very comprehensive details are available from the manufacturers.

Figure 12.47 shows the basin rodding point. Bends in the horizontal plane and junctions with other drains are made just downstream of the rodding point bend. Figure 12.48 shows how differences in drain level, such as would normally require a drop manhole, may be accomplished. Figure 12.49 shows the form of the rodding tube and the method of introducing this at a conventional inspection chamber. No difficulty of introduction exists at a rodding point.

Discharge capacity and fall

The discharge capacity of drains depends on the diameter of the pipe, the fall or gradient at which it is laid and the nature of the inner surface and accuracy of

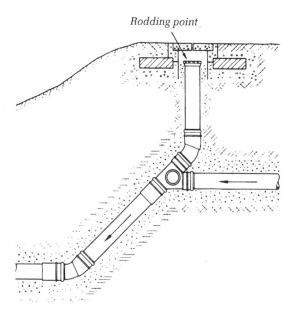

Figure 12.47   The Marley rodding point.

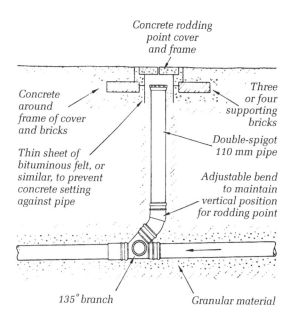

Concrete rodding point cover and frame

Concrete around frame of cover and bricks

Thin sheet of bituminous felt, or similar, to prevent concrete setting against pipe

Three or four supporting bricks

Double-spigot 110 mm pipe

Adjustable bend to maintain vertical position for rodding point

135° branch

Granular material

Figure 12.48   A ramp rodding point.

alignment of the joints. Formulae which give actual discharge capacities sufficiently accurately to be useful for drainage design have been available for some 200 years. Figure 12.50 is a nomogram, based on the Hydraulic Research Establishment's Research Paper No. 2 *Charts for the hydraulic design of channels and*

pipes. It relates fall, pipe material and diameter and flow and thus enables the size of pipe for a given fall and flow to be established, the information normally needed. Note that surface water and foul pipes are distinguished in the chart. This arises since the calculations for surface water drains assume that the pipes flow full, whereas in the case of foul and combined drains the design standard is only 0.75 × depth of flow, which allows some factor of safety and helps to prevent air pressure fluctuations. If the flows are known it is therefore a relatively simple matter to size the drains. The total fall available is usually established by the configuration of the site, and relative levels of building and disposal points as well as the appropriate diameter of drain may be read from the chart. The minimum velocity of flow in drains recommended by BRE is 610 mm/s. This means that a minimum gradient is required for each diameter of pipe. The guideline on the chart will indicate when gradients are hydraulically satisfactory for drains in good condition. Local authorities, however, have their own requirements for minimum gradient, which vary considerably from place to place. A survey showed that the variation is from a minimum fall of 1 in 40 to a minimum of 1 in 110 for 100 mm diameter drains. Larger diameters are laid to slacker gradients. This consideration may mean that the fall of a run of drain will vary as the diameter increases to take the flows from branch drains. Until recently it was thought that the maximum fall of drains should be limited to avoid stranding of solid matter and scouring of the pipe. This led to the provision of drop manholes on sloping sites. Views have changed on these points and, where possible, up-to-date practice is to allow drains to follow the fall of the land, thereby keeping the amount of excavation to a minimum. Drop manholes in these circumstances would only be employed to minimize drain excavation where branches have to be connected that are much shallower than the main drain. Figure 12.51 shows a typical drop manhole. Care must be taken at the foot of a slope that the flatter section of drain is capable of discharging the full flow likely to come down the steeper section.

Estimation of flow

For surface water drains this presents no difficulty. The Building Research Establishment recommend that design should be based on a rate of rainfall of 50 mm/h. To estimate the flow all that is required is to find the area of site and building which is contributing to the flow in the drain, multiply by 0.05 m to give the flow per hour in cubic metres and divide by 3600 to give the flow per second for use in

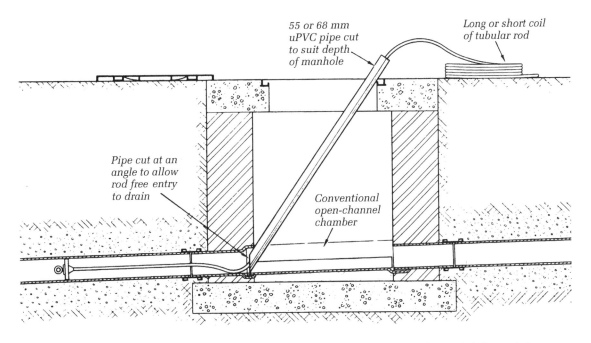

55 or 68 mm uPVC pipe cut to suit depth of manhole

Long or short coil of tubular rod

Pipe cut at an angle to allow rod free entry to drain

Conventional open-channel chamber

Figure 12.49   Using the Marley rodding point system: how to feed a rodding tube through a manhole into a drain.

Fig. 12.50. In sewer calculations more factors have to be taken into account. The intensity of rainfall fluctuates over a period and it is possible that by the time an area some way up the sewer is contributing at its full design rate, nearer areas are producing a very much reduced rate of run-off due to a decrease in the rainfall intensity. This problem is very rare on building sites; they are rarely large enough for this phenomenon to have any significance. Similarly, it is found that rain falling on grass and other non-paved areas does contribute a small proportion of its volume to the drain. The flow is unlikely to coincide with peak flows in building surface water drains and these can most satisfactorily be based on the assumption that there is 100% from roofs and paved areas.

Flows in foul drains present a more complex problem. Sanitary appliances operate intermittently and deliver varying quantities at varying rates of flow into the drainage system. It is difficult to decide what the peak flow will actually be. Table 12.8 and Fig. 12.52 provide a means of estimating flows based on the application of probability theory (see page 249). Each sanitary appliance has a number, called a discharge unit, which represents the rate of discharge, its capacity and its frequency of use. The flow in any section of the drain may be determined by summing all the discharge units of the sanitary appliances

contributing to the drain and consulting Fig. 12.52 to find the flow appropriate to the sum. Any continuous flows are added at their actual value. The sizing method given in BS 8301 is based upon the following method and the discharge units are given in Table 12.8.

1 Prepare drain layout plan for foul and surface water drains showing all points of entry (stacks, groundfloor WCs and gullies), drain runs, junctions, manholes and rodding eyes.

2 Mark cover levels at MHs and REs.

3 Mark invert levels at MHs and REs to give minimum cover. (Drain size assumed.) Use these values for minimum drain cover:

Under gardens and fields 0.9 m.
Under lightly loaded site roads 1.2 m.
Under heavily loaded roads special calculations will be needed.

Check the following items:
● Drain cover must be maintained between manholes. If cover is not adequate the depth of the drain must be increased. Resiting of manholes to coincide with changes of ground fall is usually economical.

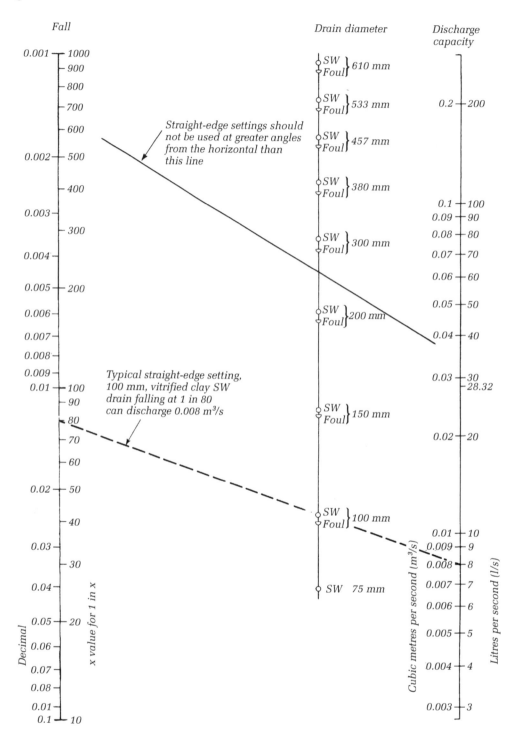

Fall

Drain diameter

Discharge capacity

0.001 ┬ 1000
      ├ 900
      ├ 800
      ├ 700
      ├ 600
0.002 ┼ 500
      ├ 400
0.003 ┤
      ├ 300
0.004 ┤
0.005 ┼ 200
0.006 ┤
0.007 ┤
0.008 ┤
0.009 ┤
0.01 ┼ 100
      ├ 90
      ├ 80
      ├ 70
      ├ 60
0.02 ┼ 50
      ├ 40
0.03 ┤
      ├ 30
0.04 ┤
0.05 ┼ 20
0.06 ┤
0.07 ┤
0.08 ┤
0.01 ┤
0.1   ┴ 10

Decimal

x value for 1 in x

Straight-edge settings should not be used at greater angles from the horizontal than this line

Typical straight-edge setting, 100 mm, vitrified clay SW drain falling at 1 in 80 can discharge 0.008 m³/s

SW } 610 mm
Foul}

SW } 533 mm
Foul}

SW } 457 mm
Foul}

SW } 380 mm
Foul}

SW } 300 mm
Foul}

SW } 200 mm
Foul}

SW } 150 mm
Foul}

SW } 100 mm
Foul}

SW 75 mm

0.2 ┬ 200

0.1 ┬ 100
0.09 ┼ 90
0.08 ┼ 80
0.07 ┼ 70
0.06 ┼ 60
0.05 ┼ 50
0.04 ┼ 40
0.03 ┼ 30
     ┴ 28.32

0.02 ┼ 20

0.01 ┬ 10
0.009 ┼ 9
0.008 ┼ 8
0.007 ┼ 7
0.006 ┼ 6
0.005 ┼ 5
0.004 ┼ 4
0.003 ┴ 3

Cubic metres per second (m³/s)

Litres per second (l/s)

Figure 12.50   Nomogram for estimating drain sizes.

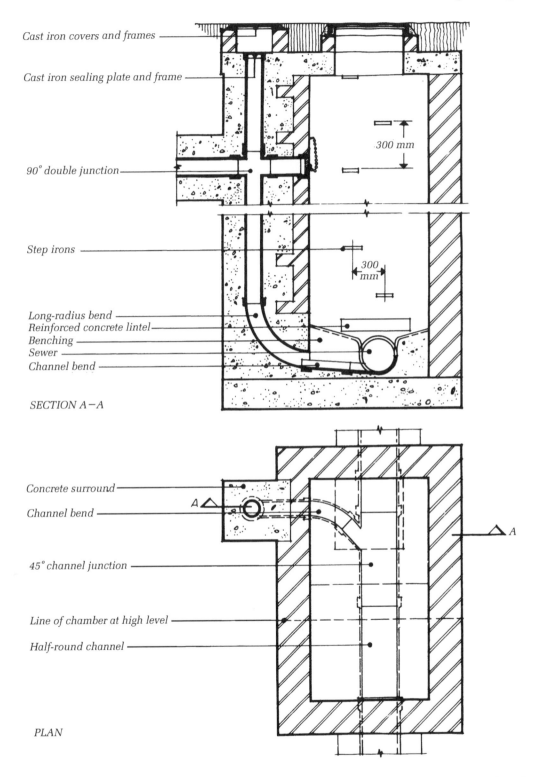

Cast iron covers and frames

Cast iron sealing plate and frame

90° double junction

300 mm

Step irons

300 mm

Long-radius bend
Reinforced concrete lintel
Benching
Sewer
Channel bend

SECTION A–A

Concrete surround

Channel bend

45° channel junction

Line of chamber at high level

Half-round channel

PLAN

Figure 12.51   Construction of a drop manhole to allow a shallow branch at high level to join the main drain at a lower level.

Table 12.8 Discharge units for estimating flows in foul drains

Appliance	Discharge units for given frequency of use				
	5 min[b]	10 min[c]	20 min[d]	30 min	75 min
Water closet					
14 l	40	20	10	–	–
9 l	28	14	7	–	–
4.5 l[a]	12	6	3	–	–
Lavatory basin (LB)					
With plug	6	3	1	–	–
With spray tap and no plug	0.5	0.25	0.15	–	–
Sink, 38 mm waste	27	14	6	–	–
Bath[e]	–	–	–	18	7
Shower[e]	Add 7 l/min per spray to the design flow				
Urinal, per stall	–	–	0.3	–	–

[a] Where dual flush is used (see page 232).
[b] Corresponds to congested use with queueing.
[c] Corresponds to normal commercial or office use.
[d] Represents the peak rate of domestic use (morning).
[e] Use does not contribute to morning domestic peak.

- Surface water and foul drains need to clear each other. In case of need reassess levels. Surface water drains are normally laid to minimum falls and foul drains dropped to clear.
- Drain falls must reach the minimum acceptable standards (hydraulically or to satisfaction of local authority). If not, reassess levels.

- Drains must be clear of service runs and any other underground features.

4 Size foul drains.
- Establish flows in each section:
  - Establish the type and number of sanitary appliances and the frequency of their use. Consider which appliances will contribute to the peak flow. In some cases, more than one combination of appliances may have to be tested to establish which is critical. (See Table 12.8 for discharge units and frequency of use in various building types.)
  - Mark against each point of entry the total of discharge units for sanitary appliances contributing to peak flow in the drain.
  - Mark cumulative totals of discharge units carried by each section of drain.
  - From Fig. 12.52 establish flows appropriate to each section of drain and mark on layout.
  - If there are any continuous flows (e.g. from manufacturing processes) add these to the flow in all sections of drain which will carry them.
- Size the drain using the nomogram in Fig. 12.50.

5 Size surface water drains.
- Establish flows in each section:
  - Measure areas of roof or paved areas contributing to each point of entry.
  - Calculate areas of roof or paved areas contributing to each point of entry. Normally at rainfall rate of 50 mm/h; mark on layout.

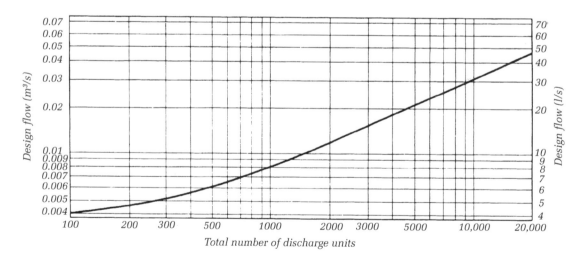

Figure 12.52 Design flow for foul drains.

– Mark cumulative totals of flow carried by each section of drain.

- Size drains using the nomogram in Fig. 12.50.

6 Size combined drains. Peak surface water and foul drain flows do not usually coincide and in most cases the combined peak may be assumed to be the same as the greatest individual flow rate. Care should be taken, however, to add any continuous foul flows to the surface water flow.

7 Check the following items:
- Adequate cover needs to be maintained.
- Adequate falls need to be maintained.
- Revised levels must not cause clashes between foul and surface water drains.

For housing schemes BS 8301 suggests that, to avoid inconvenience from blockages, not more than 20 houses should be connected to a 100 mm diameter drain (110 mm for uPVC) and not more than 150 houses to a 150 mm diameter pipe. The statistics which lead to these figures are not made clear but it would be injudicious for designers not to take them into account in case blockages occur and the design is blamed. This consideration will influence both layout and sizing.

Combined drains require separate calculation for surface water and foul flows. The drain sizes can be based on whichever flow is greater. The chance of peak foul and peak surface water flow coinciding is so remote that it is inappropriate as a design standard.

The foul drain flows given by this method assume that the flush from individual appliances is maintained as it passes down the drain. This is not the case and in fact the duration of the flow increases while the rate decreases as the flush passes further along the drain. The effect of this may be to increase the number of sanitary appliances that can be served by given sizes and falls of drain, but this is not firmly established and no practicable method exists for taking decay of the flush into account.

Drainage of basements

Provided a sewer at adequate depth is available, no special problem exists in the drainage of basements. It is important to remember, however, that many sewers become surcharged at times of peak flow and water levels may rise in manholes. This could cause flows of sewage into a basement from the drain. Special gullies and traps exist to overcome this, but they cannot be regarded as completely effective and if there is any rise of backflow it would be better not to have sanitary appliances in the basement. If their presence is unavoidable the flow can be collected in a sewage ejector and raised to ground level before being put into the drain. This arrangement will be inevitable if the basement is below sewer level.

Even where no sanitary appliances are present in a basement it is necessary to deal with seepages of water and water from leaks or draining down of pipework. These flows are catered for by draining ducts and boiler rooms to a sump in which, usually duplicate, sump pumps are sited. The sump pumps raise the water from the sump to a gulley at ground level. Figure 12.53 shows a typical installation.

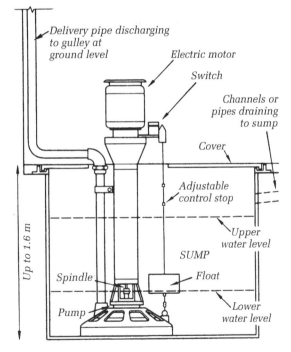

Figure 12.53 Sump pump installation.

13 Sewage disposal

13.1 Foul and its disposal

Where public sewers exist the simplest and best method for disposal of flows in foul drains is directly into the public sewer, and if there is a sewer within 30 m of the site boundary, local authorities can insist upon connection being made. If sewers exist but are further away than this, connection can still be required but the local authority may have to bear the additional cost involved.

Where no public sewers are available arrangements must be made to dispose of foul sewage. There are three methods:

- **Dilution**: the sewage is discharged to a sufficiently large body of water.
- **Conservancy**: the sewage is retained on the site and periodically removed.
- **Treatment**: a plant on the site renders the effluent sufficiently innocuous to be discharged to a stream or to be allowed to percolate into the soil by surface and subsurface irrigation.

Dilution

If the volume of water receiving a flow of sewage is sufficiently large, oxidation of the organic matter can be achieved by the oxygen dissolved in the water. In England it is not now considered appropriate to use this method with rivers since the population of the catchment areas would demand a much greater flow than is available. Many local authorities do discharge sewage directly into tidal waters, often after mechanical disintegration of the solid matter. In recent years this method has been increasingly questioned. In any case it is rarely practicable for individual buildings, since even if a seashore forms part of the site boundary the outfall arrangements required would involve considerable planning and expenditure.

Conservancy

Chemical closets may be used in temporary buildings. Their construction and siting is governed by the provisions of the Building Regulations 1985, the main stipulations of which may be summarized as follows:

- The closet may not open directly into a habitable room, kitchen or room used for manufacture, trade or business.
- The closet must be either entered and ventilated from the external air or ventilated by window or skylight with an opening area not less than one-twentieth of the floor area.
- The closet must be sited and constructed so as not to cause pollution or nuisance.
- There must be no outlet to a drain.

In permanent buildings not served by sewers there may be occasions when site treatment is not possible either because the size and configuration of the site do not permit the construction of a treatment plant and the disposal of effluent, or where the amount of flow is inadequate in volume and too intermittent to be treated successfully. In these cases a cesspool is appropriate. This consists of a watertight but ventilated underground container which can receive the flows from the drains and be pumped out at intervals. Local authorities normally provide an emptying service for which there is usually a charge. In order to reduce the cost involved cesspools are often allowed to overflow, or in suitable soils, even have their bases deliberately broken to allow seepage to take place. This practice is both contrary to the regulations and extremely undesirable.

Cesspools must be impervious not only to liquids inside but also to surface and subsoil water from outside. They must not be sited so as to create risks of polluting water supplies or nuisance or danger to health. The building bye-laws that preceded the present building regulations specified arbitrary figures for appropriate separation distances of cesspools from houses, wells, etc. Under the present regulations, however, cases can be considered on their own merits. It is unlikely that distances smaller than those previously required will be acceptable to local authorities. Cesspools will normally be sited at a lower level than the building so that drains can be laid to fall and pumping will not be required. There must be ready access for cleaning and emptying without passing through any residential or trading or business premises; this normally implies a road suitable for cesspool-emptying vehicles to a point close to the cesspool. Some vehicles carry sufficient suction hose to run for some distance but if this is to be depended upon in the siting of the cesspool the authority should be consulted to see what maximum run and lift is possible with their vehicles. Figure 13.1 is of a typical cesspool, showing statutory and other desirable features.

The capacity of a cesspool is often based on the estimate of a 45-day flow. This is considered to give a reasonable frequency of emptying. For dwellings 0.11–0.14 m³ per person per day is used to determine the input. The minimum capacity should not be less than 18 m³ and it is unusual to have cesspools larger than 45 m³. In fixing an actual size the capacity of the cesspool-emptying vehicle is often taken into account so that an exact number of loads will give the most economical emptying costs. Capacities of vehicles vary, but 3.5 m³ is usual. The local authority should be consulted.

Treatment

Various chemical and electrical means of sewage treatment have been developed, but virtually all sewage treatment plants in England, both large local authority plants and small private plants, are based on another system. The first step is to remove suspended solids using sedimentation, settling or septic tanks. Next comes the oxidation of any organic matter still contained in the liquid using biological agencies. Finally the treated effluent is discharged to a watercourse, or if that is impossible the effluent is discharged by surface or subsurface irrigation. The sludge accumulating during the first stage is removed periodically (6 months is the usual interval) and either dumped or used directly or in dried form as fertilizer. Large installations with constant supervision can have many refinements such as screening to arrest foreign objects, maceration of solids, chemical treatment to encourage flocculation and precipitation of solids, sludge-drying beds or digestion plant, facilities for recirculating effluent if desired standards have not been achieved, and more sophisticated procedures and design, but the general principles of operation are essentially similar. The following notes are mainly concerned with small plants such as might be used for a single building or small group of buildings.

Septic tank

The first stage of treatment in small plants is traditionally known as the septic tank. Sewage is allowed to stand in the septic tank, which will usually have a capacity of 16–48 h flow. Sludge will settle to the bottom, scum will form on the top and a clear liquid called liquor will overflow as new flows come in. Digestion may take place to some degree. This process is a breaking down of the organic content by the anaerobic bacteria, which can thrive under the conditions of a septic tank. This process reduces the quantity of sludge and renders the odours less offensive. At ambient temperatures this process can

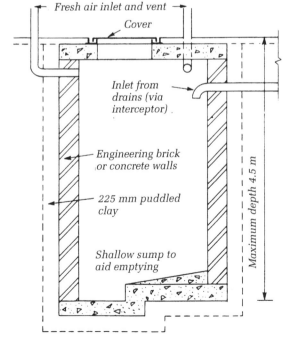

Fresh air inlet and vent
Cover
Inlet from drains (via interceptor)
Engineering brick or concrete walls
225 mm puddled clay
Shallow sump to aid emptying
Maximum depth 4.5 m

Figure 13.1   Typical cesspool.

occupy a period of two months or more, so it can only be partially effective in a septic tank. Special digestion facilities can be installed in large sewage works to produce useful quantities of sludge gas, which can be used for power and heating.

The size of septic tanks is governed by several factors. Tanks must be sufficiently large to ensure their contents are not noticeably disturbed by any entering flows. This will fix the minimum size at about 3.5 m³. The tank must be large enough to allow the sewage to remain in it for an adequate time. This is usually taken to be 16–48 h. The tank must also be large enough to contain the accumulation of sludge that will take place between emptyings, without restricting the necessary capacity; 0.8 l per person per day is the volume of sludge used for design. BS 6297 *Small domestic sewage treatment works and cesspools* gives a formula for sizing septic tanks which are to be emptied at 6-month intervals. In SI units the formula is:

$$\text{Capacity (m}^3) = \text{number of persons in full-time residence} \times 0.14 + 1.8$$

Septic tanks must be designed to allow flows to enter and leave without being affected by scum and to allow a gentle passage of liquid without short-circuiting. Figure 13.2 shows a septic tank for 10 people. The features of this installation are appropriate up to populations of about 30.

It is important to bear in mind that in the Building Regulations septic tanks are treated as cesspools so far as siting, emptying and construction are concerned.

Prefabricated units

It is possible to obtain prefabricated septic tanks formed in glass reinforced plastic (GRP). Agrément certificates have been issued confirming the performance and durability of some of these units. They offer considerable advantages for speed and ease of construction, particularly in poor and waterlogged subsoil conditions. A further stage of purification by aerobic bacteria is required before the effluent is acceptable. This may be achieved by a biological factor or, more usually with this type of unit, subsurface irrigation.

Figure 13.3 is based on the Entec unit.

Biological filters (often called percolating filters)

Traditionally known as biological filters, these devices treat the sewage not by filtration but by oxidation. The purpose of the material contained in the bed is to present a large surface area over which the liquid will spread and be exposed both to the air and to the action

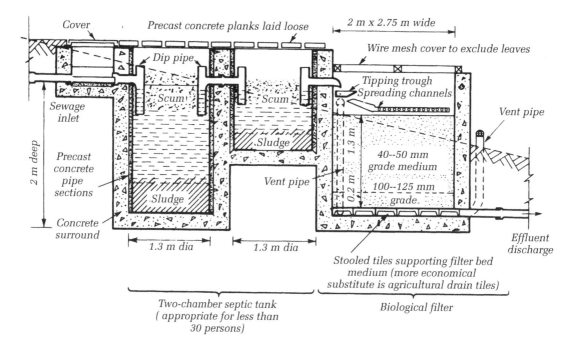

Figure 13.2 Typical small private sewage treatment installation. Dimensions are based on a population of 10; see text for details of other population sizes.

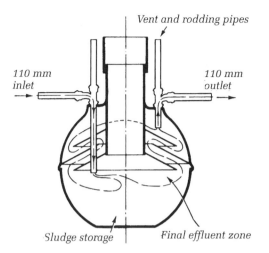

Vent and rodding pipes

110 mm inlet

110 mm outlet

Sludge storage   *Final effluent zone*

Figure 13.3 Prefabricated GRP septic tank.

of bacteria which form the main agency for oxidation. To function properly a biological filter must therefore have an adequate volume in relation to the flow, so the liquid will be adequately distributed; the bed must be well ventilated; and to ensure the liquid remains in the bed for sufficiently long, the depth must not go below 1.2 m, preferably not below 1.8 m. A spreading device is required to distribute the liquid over the surface of the bed. This usually takes the form of a series of channels with notched edges fed by a trough that tips when the liquid in it reaches the critical level, or by a rotating arm distributor operated by the water flow. The rotating arm requires a higher head.

For convenience biological filters are usually sited close to the septic tanks which serve them, although this is not essential. Unlike the septic tank, the outgo from which is at almost the same level as the input, a considerable loss of height occurs in the passage of liquid through a filter bed (nearly 2 m even in the shallowest cases). From both the points of view of ventilation and disposal of effluent it is desirable that the base of the filter should be as near ground level as possible. On sloping sites this is easily achieved. On flat sites it may be necessary to raise the flow to the filter by pump.

The materials forming the filter bed must be durable, strong enough to resist crushing and frost-resistant. Hard-burnt clinker, blast furnace slag, gravel or crushed rock are commonly used; clinker and slag give the best results because of the greater surface areas presented. Two sizes of medium are usually employed. At the bottom of the bed over and around the drain tiles, medium of 100–125 mm grade is

desirable for about 0.2 m; above this the rest of the bed can be formed of 40–50 mm grade.

The volume of material depends on its surface area and the strength of the sewage, etc. For small domestic plants, from 0.75 m³ per person for up to 10 persons, reducing to 0.5 m³ for larger installations, is recommended by BS 6297, with a further reduction to 0.4 m³ per person for part-time occupation.

It is neither necessary nor desirable to cover filter beds except by a wire mesh to exclude leaves.

Figure 13.3 shows a biological filter using a tipper and channels. In Fig. 13.2 advantage is taken of a sloping site. In some cases the sewage may have to be lifted by pump. Note how the filter beds are ventilated.

Many refinements to the oxidation process are found in large installations. The most notable is the activated sludge process. In this widely used method, instead of exposing a thin film of liquid to air over bacteria-laden filter medium, the bacteria are introduced to a volume of fresh liquid using sludge retained from the last treatment. Air is introduced either by blowing compressed air through the liquid or by paddles which agitate the surface and encourage absorption of oxygen. The method requires continuous attention and has a considerable energy consumption. On the other hand, the process can be continued until a satisfactory effluent is achieved. A degree of control is not possible with the normal filter bed where, if the first passage does not give satisfactory results, the effluent must be reprocessed. Considerable space is saved, the plant is not liable to harbour flies and the loss of head through an activated sludge works will be significantly less than a filter bed type which may save pumping in some cases.

Small packaged plant is beginning to come into private use.

Humus chamber

The effluent from a biological filter can be greatly improved if it is given a final treatment to remove any humus, a by-product of the bacterial action in the filter bed. For small installations without regular supervision the best method is to distribute the flow over an area of rough grass or scrub. An area of 1.0–3.5 m² per person may be needed, depending on the nature of the soil.

Provided regular weekly maintenance is possible a better method is to provide a chamber similar to the septic tank but of about one-quarter the capacity. Special care is desirable to achieve smoothly flowing, low velocity input and outgo. Humus chambers should be cleared once a week. The humus can be used as a fertilizer, being very much better suited to this purpose than the sludge from the septic tank.

Final disposal

If a watercourse is available and the approval of the appropriate authority can be obtained, this is a simple and effective method. Figure 13.4 shows a method of discharging an outfall drain to a watercourse.

Where no watercourse is available effluent may be disposed of below ground provided the soil is reasonably permeable and the water table does not approach too close to the surface (1.5 m is probably a minimum). A soakaway may be used but a system of agricultural drain tiles similar to that described under land and drainage is considered superior. The total length of drain and area of land is difficult to estimate, and local experience should be taken into account. Areas required can vary between 1 and 4 m² per head of population.

In siting subsurface irrigation disposal systems care must be taken to avoid pollution of water supplies which, in some circumstances, can take place over considerable distances. Water supply undertakings should be consulted when siting this type of installation.

Local authority standard details

In many areas local authorities have prepared standard details of sewage disposal plants which they consider particularly appropriate for use in their areas. It is clearly desirable to make enquiries of the local authority before embarking on the design of sewage disposal plant.

13.2 Surface water

Public sewers represent the easiest method of surface water disposal but in cases where they do not exist or where connection is impracticable, soakaways may be employed. Combined sewers may sometimes be overloaded due to extensive building development. Local authorities may then call for site disposal of surface water to minimize further overloading.

Permeable soil and a water table some way below the surface are needed for soakaways to be successful. In many cases a pit is dug, filled with rubble and the drain arranged to deliver water into the rubble, after which the hole is backfilled. Although it often enables site disposal of debris, this system cannot be recommended for soakaways. Silting up is possible but cleaning is not and the reservoir capacity is reduced. A more satisfactory method is to construct a chamber lined at the sides with perforated or open-jointed walling. In some types of soil no base is needed, but a cover slab with access for inspection and cleaning should be provided over the soakaway. Where the soil is not satisfactory, or where cleaning may have to take place, a concrete base is desirable. If no base is provided, some cushion to receive the inflows of water should be incorporated to prevent erosion.

The proper size of soakaways depends on the permeability of the soil, the area to be drained and the intensity of rainfall anticipated. Permeability can be tested but the capability of the ground to absorb moisture can vary from place to place and the varying level of the water table with the season of the year can affect the rate of percolation. Many successful soakaway installations have been based on a capacity equal to 12 mm depth of water over the whole area being drained. It is clearly sensible to site soakaways so that any overflow will not affect the building. Many authorities consider that several small soakaways are better than fewer larger ones.

BRE Digest 151 *Soakaways* deals with the construction and sizing of soakaways, describes a

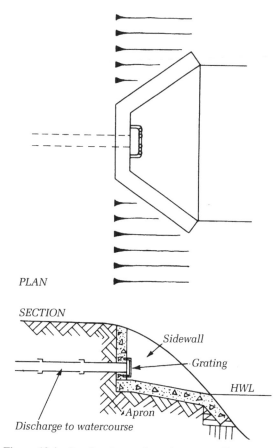

PLAN

SECTION

Sidewall

Grating

HWL

Apron

Discharge to watercourse

Figure 13.4 Details of outgo from the sewage disposal system to a watercourse.

method of testing soil for rate of percolation and gives a chart for sizing soakaways.

Land drainage

On some building sites it may be desirable to reduce the level (called the water table) of the subsoil water or to improve the drainage of surface water. The need may arise because of the planting or horticultural use of the site, because it is important to reduce surface flooding or because for health (damp penetration of cellars) or structural reasons it is necessary to keep down the level of groundwater near the building.

Before embarking on land drainage works the actual groundwater levels should be investigated by using trial holes. The water table is usually at its highest in the spring.

Subsoil drains are usually laid 0.6–1.0 m deep and the pipe layout will be arranged to follow the falls of the land. The spacing between drains will vary from 10 m for clay soils to 50 m for sand.

Figure 13.5 shows how subsoil drains lower the water table and demonstrates how drains may be more widely spaced in soils which allow water to permeate freely.

Sometimes pipes are laid round buildings in what is described as a moat system, intended to intercept the flow of water towards the foundations for more extension areas. However, it is usual to employ a herringbone pattern of pipes leading to main runs. The length of the subsidiary runs is normally limited to 30 m. It is not possible in most cases to make an estimate of likely flow. Pipe sizes are determined by connection. For building application the usual sizes are 75 mm diameter for the subsidiary drains and 100 mm for mains.

Subsoil drains have traditionally been formed from butt-jointed and usually porous pipes, laid in narrow trenches. The tops of the pipes are protected by turf or

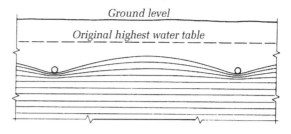

Figure 13.5 Reduction of the water table using subsoil drains.

building paper to prevent soil penetration and rubble, or broken stone laid around the pipe to aid in water collection. In cases where it is desired to catch water running on the surface, the rubble fill should be carried up nearly to the top of the trench, the top spit only being filled with topsoil. Polythene and PVC pipes are now being used extensively for subsoil drains. They are provided with holes on the underside, which allow water to enter while minimizing the penetration of soil which is a problem with normal drain tiles.

The outflow for laid drainage systems will normally be discharged to ditches or watercourses.

Sometimes the pipes are dispersed within land drainage work. A trench filled with rubble or broken stone (called a French drain) will give passage for water and is particularly effective at dealing with flows on the surface. Mole drains can be used in clay subsoils. They are formed by a plough drawing a cylindrical cartridge through the soil. A tube is formed in the soil which can remain operative for many years. This form of drainage does not have a life comparable to buildings and its use is more appropriate to agriculture.

BS 8301 *Building drainage* contains a table of subsoil drain depths and spacings for various types of soil.

14 Refuse collection and storage

Bins

The traditional, and still the cheapest, method of refuse storage in buildings is the dustbin. If adequate capacity is allowed, the placing of the bin, arranged satisfactorily both for user and the refuse collector, protection from the rays of the sun provided, adequate ventilation ensured and an impermeable and easily swept standing place provided, dustbins form an acceptable means of refuse disposal for many types of building. The noise inevitably associated with galvanized steel bins is largely overcome by the use of rubber and plastic bins. Small blocks of flats have them placed conveniently at ground level. Larger communal containers are sometimes used to simplify collection and make mechanical handling possible. Special collecting vehicles are required to handle and empty the containers, which are very much larger than dustbins. About $0.08 \, \text{m}^3$ of storage per dwelling is required for a weekly collection. British Standard dustbins occupy a space about 460 mm in diameter and 610 mm high and hold $0.09 \, \text{m}^3$. It is worthy of note that open fires and solid-fuel domestic boilers can be used for some refuse disposal and their absence in many new dwellings adds to the problem.

Paper sacks

The labour involved in collecting refuse can be substantially reduced by using paper sacks rather than bins, and since the sacks are not emptied, collection is considerably more hygienic and simpler vehicles can be used. The cost is at present high and hot ashes must not be put in. Increasing use is, however, being made of this method. Figure 14.1 shows a typical paper sack and mounting. The planning and hygienic considerations for bins apply equally to paper sacks.

Paladins

The paladin, shown in Fig. 14.2, is a very large cylindrical bin on wheels. It has a capacity of almost $1 \, \text{m}^3$, the equivalent of 10 dustbins. A special collection vehicle is required but most local authorities are equipped to collect paladins. Their use is in commercial and industrial premises but they are also used in blocks of flats.

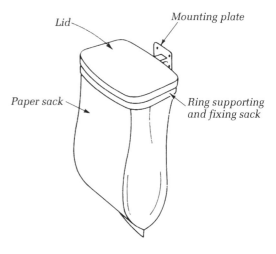

Figure 14.1 Typical paper refuse sack and its mounting.

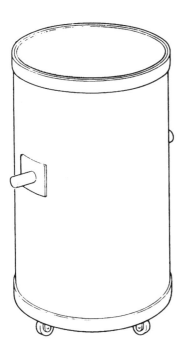

Figure 14.2   Paladin type of refuse container.

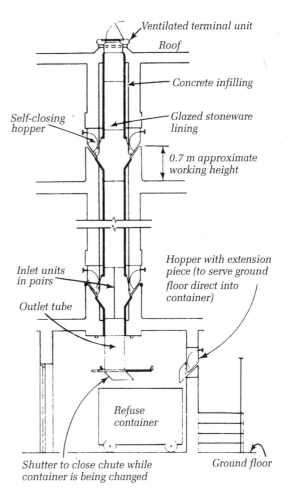

Figure 14.3   Typical refuse chute serving a block of flats.

Refuse chutes

In high residential buildings it is not practicable to require that all rubbish is carried by the occupants to containers at ground level. One way of overcoming this problem is to provide a chute running down the building which will take refuse from the upper floors and deliver it to a container at ground level. Figure 14.3 shows a typical refuse chute. Refuse, preferably wrapped in paper, is introduced by hopper flaps at each floor level and falls into the container, housed in an enclosure at ground level. The chute is carried up to roof level and vented. Refuse chutes have an influence on planning. They must rise vertically through the building, and dwellings will often be arranged in handed pairs to make maximum use of chutes. Delivery points are often arranged on communal landings. This sometimes creates nuisance from careless handling of refuse. Delivery points within the area of the dwelling encourage careful use and maintenance. The kitchen, although convenient, is not acceptable from the viewpoint of hygiene. Delivery points on a private balcony have much to recommend them. The ground level containers must have access by a special refuse collection vehicle; they are larger than the chute itself, so they may affect the groundfloor planning. Complaints of smell often arise from dwellings near the containers.

Chutes are constructed of large-diameter pipes; 380 mm has been usual but 450 mm is the present trend, which reduces the risk of blockage. Vitrified clay, concrete or asbestos cement pipes are used, set into a builders' work duct with specially designed hopper units for delivery. Provision should be made for hosing down the container enclosure but not for washing the chute itself, since this appears to make things worse rather than better. The possibility of a fire being started in the refuse must be borne in mind and also the noise which is inevitable when solid items fall down the chute and perhaps break in the container. A container capacity of 0.14 m³ per dwelling for weekly collection is thought appropriate for this method. Any type of refuse can be dealt with provided it will go into the hopper. Larger items will have to be carried down by the occupants to a special communal container.

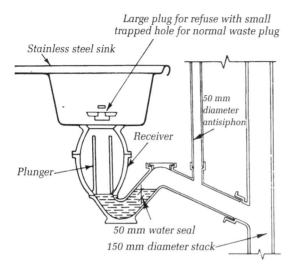

Figure 14.4 Garchey disposal system: section through the sink unit showing the receiver for refuse, the plunger for disposal and the trapped outgo leading to the stack. The stack takes waste flows only.

Chutes are used mainly for domestic buildings. Offices present special problems in relation to their use, having large volumes of paper waste with a high fire risk but little putrefiable matter.

Garchey system

The Garchey system was developed in France but British rights are held by a British firm. In this system refuse is introduced into a receiver under the kitchen sink through a large plug in the sink. A smaller plug inset in the larger one deals with ordinary sink waste discharge. When sufficient refuse and water have accumulated in the receiver, a plunger is raised and the contents of the receiver are discharged through 150 mm or 180 mm diameter cast iron drainpipes running down the building. Waste appliances, but not soil appliances, may also be discharged into this stack. Figure 14.4 shows a sink unit. In the original system the refuse was collected in pits near the foot of each stack then drawn by suction to a central plant for dehydration and burning in a furnace intended to produce heat for laundry use. The central plant is expensive, limiting the use of the system to large installations, and the utilization of the heat from burning is not always very effective. The British patentees have therefore developed a collecting vehicle which replaces the central plant and itself draws the refuse from the collecting pits. The vehicle is very much cheaper than the plant and can serve a

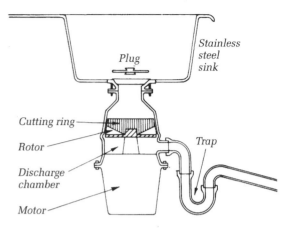

Figure 14.5 Sink grinding unit for refuse disposal.

number of buildings. Provided the vehicle is available quite small buildings can be economically served by this method.

Any type of refuse can be handled, including bottles and tins, but the size is limited. Large items must be carried down and deposited in a communal container. The system is very much more expensive than chutes or bins but is much more hygienic and gives a high proportion of satisfied users.

Grinders

Figure 14.5 shows a grinding unit fitted under a sink. It can reduce organic waste to fine particles suspended in water; the suspension is carried away in the normal drainage pipework. Tins and bottles cannot be dealt with and normal bins must be provided, but all refuse which could putrefy can be disposed of in this way. In individual dwellings or small buildings, where chutes or Garchey systems would be impracticable, grinders are the only way of improving on the service given by bins. Care must be taken when siting the grinder then mounting and connecting it to minimize the risk of trouble from noise.

Incinerators

Small incinerators for particular purposes such as hospitals and women's lavatories are widely used and are sometimes installed in blocks of flats without solid-fuel heating appliances. Larger incinerators capable of dealing with all the refuse are often employed in America in domestic buildings, frequently in conjunction with refuse chutes which automatically feed the incinerator. In this country at present on-site incineration of refuse is largely confined to hospitals and industrial premises.

Compactors

One of the problems of modern refuse is the low density and large volume. Small compactors, similar in size to a domestic refrigerator, are now available. Refuse is powerfully compacted into solid blocks contained in plastic bags. Small compactors are electrically driven.

Non-domestic premises

Trade refuse presents a special problem to be considered in each case on its merits. Apart from this – unless incinerators are used – refuse disposal from buildings other than dwellings is usually by bin or container. The general considerations involved are identical with those described for domestic premises. The main problem with non-domestic use is in establishing a proper volume of storage to suit the accumulation of refuse and the frequency of removal. It is not possible to lay down general rules; careful thought should be given to the number of bins that will be required and the possibility of increasing them if necessary. Many buildings are rendered unsightly and unhygienic by inappropriate and inadequate spatial provision for bins or containers.

BRE Digest 40 (second series) *Refuse disposal in blocks of flats* describes a study of user experience of several systems of refuse disposal and gives useful tables of capital and running costs.

15 Electricity and Telecommunications

Installations using electricity for power supply, lighting, environmental control, communications, security and computing are now essential provisions in all but the simplest buildings. Although the sizes of cables and control equipment are modest compared with most other services the need for access to all parts of the building and the requirements for flexibility mean that these installations exercise a very important influence on planning and space requirements for cable distribution. In some large commercial buildings the pattern of layout and use made of electronic equipment has called for very large open spaces with cable access to every point and substantial air-conditioning because of the size of the space and the heat generated by equipment and lighting. At the other extreme, the ability to transmit data by telephone line opens the possibility of replacement of the office by homeworking.

Although the design and installation of electrical and electronic systems is a specialist activity, it is important for everyone concerned with the design, construction and use of buildings to be conscious of the spatial and environmental requirements, the principles of operation and, in particular, the safety requirements for electrical and electronic installations.

15.1 Basic electrical phenomena

Electrons are basic particles of matter; together with protons and neutrons they are the constituent parts of atoms. In some substances electrons are readily detached from the atoms of which they form part so that free electrons entering the substance can cause progressive displacement of electrons through the substance and the appearance of free electrons at the far end. Materials of this type are known as *conductors*, whereas materials which do not allow electrons to move so freely are non-conductors or *insulators*. Semiconductors also limit the flow of electrons but do not have direct application to electrical installations in buildings.

In the early days of electrical discovery the electrical charge of the electron was described as *negative*, and materials containing additional electrons are known as negatively charged. A deficiency of electrons results in the phenomenon of positive charge. A conductor linking a negatively charged zone to a positively charged one will allow electrons to pass from the negative to the positive zone until equilibrium is reached. The flow of electrons is described as an electric *current* (I). For historical reasons the current is normally regarded as flowing from positive to negative. The unit of current is the *ampere* (A). The difference in level of electrical concentration which gives rise to the current is described as *electromotive force* (V) and is measured in *volts* (V). Electromotive force (EMF) is usually applied to the voltage of a source of electricity. Voltage differences between parts of a conducting system are described as *potential difference* (PD), also measured in volts. Horizontal cables rest on the building fabric; vertical cables are supported every 5 m. The current through a conductor varies directly with the potential difference. Conductors do not, however, have a perfect performance in transmitting currents. They present some obstacle to flow, even if this is small. The phenomenon is known as *resistance* (R). Some of the electrical energy passing along the conductor is transformed into heat as a result of resistance. Considerable heat is produced by resistances and advantage is taken of this phenomenon in many types of electrical appliance. Current will vary inversely with resistance. The relationship between potential difference, resistance and current is defined by Ohm's law:

$$Current = \frac{voltage}{resistance}$$

$$I = V/R \quad or \quad V = IR$$

A relationship of more immediate use to non-specialists connects power, voltage and current. Power, measured in watts (W), is the product of voltage and current:

$$Power (W) = voltage (V) \times current (I)$$

$$P = VI$$

For a known voltage, which is the case in most building situations, it is simple to relate the current-carrying capacity of a conductor to power requirements.

Since Ohm's law tells us that voltage equals current multiplied by resistance, it follows from the equation above that:

$$W = VI$$
$$but \quad V = IR$$
$$\therefore W = (IR)I$$
$$= I^2 R$$

This expression enables the heat generated in electrical circuits to be calculated, which is important not only for consideration of heat-generating apparatus but also for estimating the heat generated and consequently the temperature rise in normal cables. The temperature rise must be kept within acceptable limits and this can limit the current-carrying ability of cables, particularly in conduits and enclosed ducts, where heat cannot readily escape. As well as the heating effect, voltage potential will decline as currents pass along a cable. If the wiring installation caused a substantial *voltage drop* the performance of electrical equipment would be affected. The distribution system of electricity companies is subject to the same problem and companies have a statutory responsibility to maintain their supply within 6% of the declared voltage. Within buildings the voltage drop under full load conditions should not exceed 2.5% of the declared voltage.

For installations using current flowing steadily in one direction, called *direct current* (DC), the concepts described might suffice. However, while battery-operated energy supply systems are used in many buildings, especially for emergency lighting, the normal mains supply is in the form of *alternating current* (AC). In this type of supply the direction of flow of electrons changes rapidly. For circuits and apparatus which present only resistance to the flow of electricity the relationships described above are not affected and the performance of the installation will be the same whether direct current is used or an equivalent alternating current. Alternating current varies in a repetitive fashion in which the voltage change shows a sinusoidal variation in time as shown in Fig. 15.1. The equivalent DC voltage that would have the same heating effect in a given resistance is described as the *root mean square* (RMS) voltage and is calculated, as its name suggests, by taking for one half-cycle, the squares of the voltage values at each instant of time, finding the mean of the squares and then the square root of this mean. Figure 15.1 shows the variation of voltage in a 240 V, 50 Hz AC supply and the equivalent DC voltage pattern. In resistance circuits the power will follow the same pattern as the voltage.

The use of alternating current brings into effect several other important considerations.

Transformers

A current flowing through a conductor generates a magnetic field. In the present context this is not particularly significant for direct current applications. With alternating current, however, the situation is very different. A varying magnetic field can produce a flow of current in an adjacent conductor and a varying current will produce a varying magnetic field. If a conductor is formed into a coil the magnetic effects will be concentrated and this is further intensified by providing a core which has high magnetic permeability. A second coil of wiring located within this magnetic field will have an alternating current induced within it when an alternating current is passed through the first coil. This arrangement, which only works with alternating current, is known as a transformer. Figure 15.2 shows a simple transformer.

A perfect transformer would transmit the whole of the power in the primary coil to the secondary coil. In practice transformers are not 100% efficient but the power losses are remarkably small at the design loads. If the number of turns in the two coils is the same, allowing for losses, the voltage, current and power in the secondary winding will be the same as in the primary winding. If the number of turns in the secondary winding is reduced the power in the secondary coil will still be the same but the voltage will be reduced and the current increased. Conversely if the number of turns in

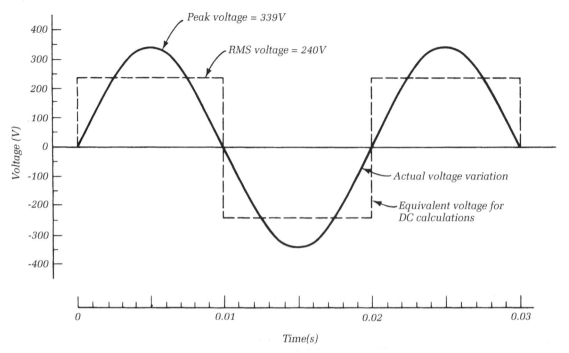

Figure 15.1   Voltage variation in single-phase AC.

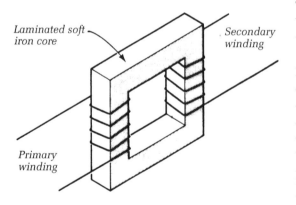

Figure 15.2   Schematic of a transformer.

the secondary winding is increased the voltage will be increased and the current reduced. Transformers of these types are known as step-down and step-up transformers respectively. The relationships between numbers of turns in primary and secondary windings in transformers and voltage variations can be expected.

$$\frac{\text{number of secondary turns}}{\text{number of primary turns}} = \frac{\text{secondary voltage}}{\text{primary voltage}}$$

Transformers are important in economical power transmission and in reducing voltages for special applications.

Inductance

To a DC current the resistance of a coil is the same as that of the same length of straight cable. For an AC current, however, the magnetic flux from the basic current causes an EMF in the coils, which tends to oppose the basic current. This resistance becomes greater with the number of coils or with increase in frequency. The effect is known as *inductance*. The unit of inductance is the henry (H).

Capacitance

Two sheets of conductive material in close proximity can store an electric charge because of the attraction between positive and negative charges. Figure 15.3 shows the principles involved. If the two plates are connected by a conductor the electrical equilibrium will be restored and the charge will be lost. If the conductor has a resistance included this will limit the flow of electricity and the process of discharge will be spread over a period of time. The phenomenon is known as *electrical capacitance* and special devices

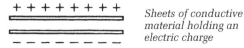

Sheets of conductive material holding an electric charge

Figure 15.3 Schematic of a condenser.

that exploit it are called capacitors. Electrical equipment not specifically intended to provide capacity, which is the case in most building installations, may nevertheless have this effect to some degree because of its configuration. Direct current will not flow through capacitors but, in effect, alternating current will do so. The unit of capacitance is the farad (F), which represents:

$$\frac{\text{charge}}{\text{voltage}} = \frac{\text{amperes} \times \text{seconds}}{\text{volts}}$$

One farad is too large for most practical purposes and devices are frequently used with capacities measured in microfarads or picofarads.

Power factor

Inductive circuits and capacitative circuits affect alternating current supplies. In a purely resistive circuit the voltage, current and power vary together (in phase) with either AC or DC current. Inductive circuits provide a resistance (reactance) to the flow of alternating current, which causes the flow of current

itself to lag behind the voltage. In capacitative circuits the rapid initial flow of current causes the voltage to lag behind the current. The current and voltage are described as being out of phase. The degree of displacement is defined in terms of angular measure with the full sinusoidal cycle representing 360°. Since the actual power at any instant is the product of voltage and current, the effect of the out-of-phase variation is a reduction in the power that can be provided by the circuit. The ratio

$$\frac{\text{available power in out-of-phase circuit}}{\begin{array}{c}\text{power which would be available in resistive}\\\text{in-phase circuit}\end{array}}$$

is known as the *power factor*. Available power is measured in *kilowatts* (kW) whereas the power which could be available, the apparent power, is defined as volt-amps, or more usually *kilovoltamps* (kVA). These terms distinguish the measurable, actual power existing in a circuit and the theoretical maximum. Figure 15.4 shows typical current, voltage and power variations in resistive, capacitative and inductive circuits.

Installations with low power factors take current from the mains but only utilize a reduced proportion. This is normally a problem only for some commercial and industrial buildings. Electricity companies may, in particular cases, call for correction of low power factors and make an adjustment to the charge for electricity.

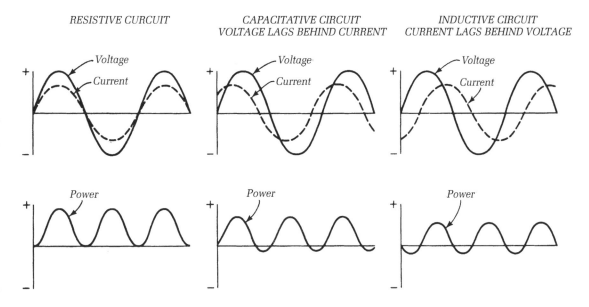

Figure 15.4 Voltage and current waveforms for three perfect impedances: a resistor, a capacitor and an inductor.

15.2 Generation and distribution

The principle of induction is used in electrical generators where currents in coils of wire are induced by moving negative fields. Alternating current has considerable advantages for power transmission over distances. The size of conductors is governed by the current carried not the voltage. Power, the product of current and voltage, can therefore be transmitted over distances by high voltage, low current systems and transformed locally to give safer, low voltage supplies for use in buildings. Transmission cables are used above ground, supported by the familiar pylons. Distribution below ground is less economical; apart from the obvious costs of underground ducting, cable loadings may have to be reduced to prevent overheating, and inductive effects may reduce the efficiency of the system. Inductive effects can be avoided by using direct current and the cross-channel link to France uses direct current. Local AC distribution in towns at low voltage presents less of a problem and the convenience and neatness of underground supplies is the deciding factor.

In Great Britain a national grid enables current to be transmitted throughout the country with considerable economic advantages since large power stations can be sited at convenient positions remote from towns and to balance economical generation arrangements with varying patterns of load. The major lines, sometimes called the supergrid, operate at 400 kV (400,000 V); via transformer substations, they feed a 275 kV system and some power stations generate this voltage. In its turn the 275 kV network feeds a 132 kV network and this feeds a 33 kV system, which in urban areas may be underground. Transformer stations make a further reduction to an 11 kV network. Some industrial consumers purchase electricity at 11 kV and either use it at this voltage or make their own arrangements for further modifications. The 11 kV supply is further reduced to 415 V, and it is this network which supplies power to the great majority of buildings.

415/240 V 50 Hz three-phase four-wire supply

The final distribution network usually consists of four conductors. The neutral conductor is earthed at the substation. Each of the other conductors carries an alternating current which completes 50 complete cycles in each second. The supplies in the three conductors are out of phase. The RMS voltage between each separate conductor and neutral is 240 V, between conductors it is 415 V. Figure 15.5 shows these relatively. Figure 15.6 shows the pattern of voltage variation between the three phases and neutral and Fig. 15.7 shows the variation of voltage difference between two phases.

The electricity supply to small buildings is usually provided by connection of one phase wire and the neutral. This is described as a *single-phase* supply. To balance the load between the phases buildings are connected sequentially to the three different phases. The supplies to large buildings comprise all three of the phase wires and the neutral. This is described as a *three-phase* supply since it involves all three phases. The building is then divided into zones each of which is provided with a two-wire, single-phase supply. Most small electrical apparatus requires only a two-wire, single-phase supply. Some large apparatus, particularly electric motors, requires a complete three-phase supply connected directly to the apparatus.

15.3 Basic circuits and safety

Early electrical installations often consisted of bare conductors laid in grooves in wooden mouldings and secured by wooden cappings. Conductors could become displaced and come into contact. Very heavy currents could then pass through the conductors, possibly raising their temperature to a point which set fire to the adjacent timber. Many fires resulted. It is also apparent that such an installation, even if no short circuits developed, could present a risk of electric shock to the occupants of the building.

Insurance companies moved quickly and effectively to establish standards for layout, materials and work and these ultimately led to the Institution of Electrical Engineers Regulations for Electrical Installations in Buildings. This document established standards for the design and installation of electrical systems and was kept up to date by 16 editions. These regulations have now been incorporated into BS 7671.

Present-day installations have many features designed to give personal and fire safety. The installations covered by BS 7671 are described as 'low voltage'. In buildings served by mains electricity this means that the maximum voltage between conductor and earth is 600 V AC, which covers both the single-phase and the three-phase supplies provided in the United Kingdom. A special case described as SELV (extra low voltage) is identified. This system provides a low voltage, electrically separated from earth and other electric systems in such a way that the risk of electric shock is eliminated. In bath and shower rooms the supply is limited to 12 V and enables the use of shavers and hairdryers without risk of electrocution.

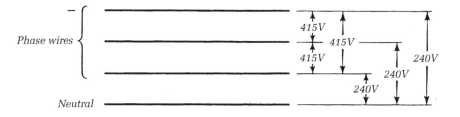

Figure 15.5 RMS voltage in three-phase four-wire supply.

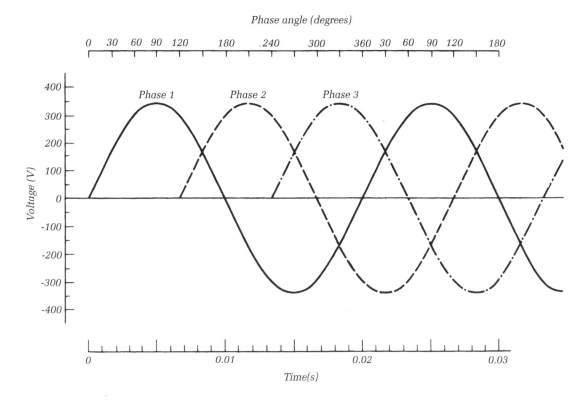

Figure 15.6 Voltage variation in a 50 Hz three-phase supply.

As well as the metalwork enclosing electrical appliances, all pipework and metalwork which is connected to earth should be bonded to the electrical system earth to avoid any risk of potential difference. Each conductor is surrounded by insulation and related to the voltage carried further enclosed by an additional protective sheath or casing (Fig. 15.8). Switches are placed in the live (called line) conductor so that when the switch is turned off the apparatus controlled is not live. Fuses, which are short lengths of fine fusible wire, are located in circuits to ensure that the maximum possible current through the circuit does not exceed the capacity of the wiring. Large currents will cause the fuse to melt and cut off the current. Fuses are also always located in the line side of the installation so that when they melt the wiring will cease to be live. Metal cases or parts of electrical equipment which might conceivably come into contact with a displaced or frayed wire, thus becoming live and constituting a danger, are connected to a local earth through a low resistance circuit. A heavy current will then pass to earth if the casing becomes charged, and the fuse will isolate the fault (Fig. 15.9).

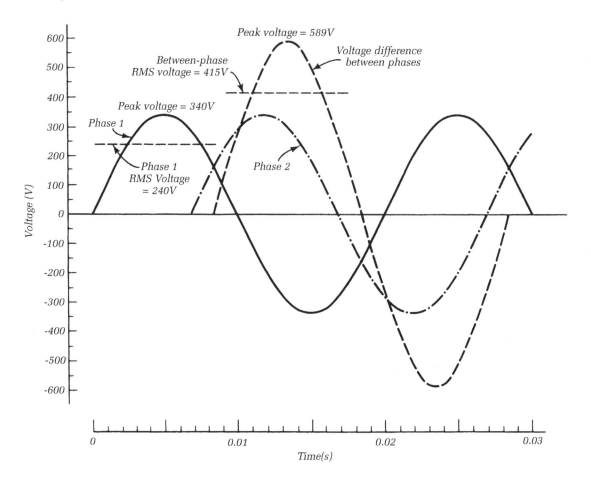

Figure 15.7   Voltage difference between phases in 240/415 V, 50 Hz three-phase supply.

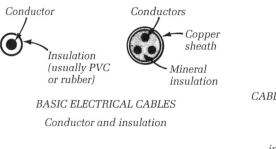

Conductor

Insulation
(usually PVC
or rubber)

Conductors

Copper
sheath

Mineral
insulation

BASIC ELECTRICAL CABLES

Conductor and insulation

CABLES PROTECTED IN CONDUIT

*Seven 1.38 mm² cables
with PVC insulation,
overall diameter 3.1 mm,
in 16 mm light steel conduit;
this is the maximum capacity
which complies
with the
IEE Wiring Regulations*

Figure 15.8   Insulation and conduit protection for electrical safety.

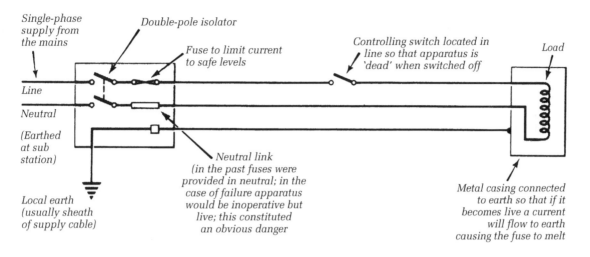

Single-phase supply from the mains

Double-pole isolator

Fuse to limit current to safe levels

Controlling switch located in line so that apparatus is 'dead' when switched off

Load

Line

Neutral

(Earthed at sub station)

Local earth (usually sheath of supply cable)

Neutral link (in the past fuses were provided in neutral; in the case of failure apparatus would be inoperative but live; this constituted an obvious danger

Metal casing connected to earth so that if it becomes live a current will flow to earth causing the fuse to melt

Figure 15.9 Overload protection, polarity and earthing for electrical safety.

Further protection measures

Although many installations use exactly the measures described, more sophisticated equipment has been developed to improve performance and to meet special needs. Testing is recommended every five years for commercial systems and every ten years for domestic installations.

Excess current protection

Excess currents can arise as a result of the connection of too many pieces of electrical apparatus to a particular circuit or as a result of faults in the wiring. The ideal protection will not be affected by the transitory surges of current associated with the switching on of apparatus but will operate very quickly when a heavy current flows.

Three types of equipment are used for excess current protection: rewireable fuses, cartridge fuses and miniature circuit-breakers (MCBs). A fuse is a short length of fine wire which will melt and disconnect the circuit served if an excessive current passes. MCBs are electromechanical or electrothermal devices which break the circuit in the case of excess current. The Institution of Electrical Engineers recognizes two categories of excess current protection, coarse and close. Rewireable fuses and some cartridge fuses fall into the coarse category whereas high breaking capacity (HBC) cartridge fuses and MCBs give close protection. Both types are acceptable from the point of view of safety but the close types allow higher current-carrying ratings to be given to cables with consequent economy of wiring.

Apart from economy of wiring there is much to be said in favour of the general use of close protection devices. Rewireable fuses can deteriorate with time and fail for no apparent reason, their performance is affected by the way in which new fuse wire is fixed, rewiring is troublesome and incorrect wire can be used, thereby destroying the protection. Cartridge fuses have some advantage and the HBC types also provide 'close' protection. The improved performance is achieved in some cases by packing the cartridge with inert material and the use of specialist alloys for the fuse. MCBs which provide close protection also give great convenience since resetting requires only a simple switching operation and no separate wire or cartridge is needed. In industrial environments, where variations in the use of electrical equipment may, in normal operation, cause transitory overloads, the ease of resetting is an important advantage. Large lighting installations, not subject to frequent switching, can be protected and directly controlled by MCBs.

Residual current circuit-breakers

To ensure freedom from risk of shock it is important that the earth connection linking the metalwork of electrical equipment has a very low resistance. It is not easy in all types of soil to achieve an adequate earth and this may be particularly difficult where overhead electrical distribution is used since there will be no earthed cable sheathing.

Residual current circuit-breakers provide a possible solution. The current-operated type compares the current flowing in the line and neutral of a circuit. In

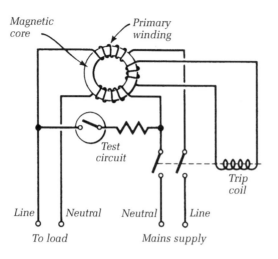

Figure 15.10   Schematic of residual current circuit-breaker.

the case of a fault the current flow will no longer be equal and the device will disconnect the supply. Figure 15.10 shows the principle of operation.

Protective multiple earthing

Protective multiple earthing (PME) can only be used with the approval of the local electricity company and only in an area where all the buildings conform to the system. In this method each building has its own earth which is connected to the neutral conductor. All metalwork which would normally be earthed is connected to this terminal and consequently to the neutral conductor. Although the earth to each separate building may be inadequate to operate normal protective devices the combined resistance of many

earths linked by the mains neutral is likely to be low and to allow sufficient current to flow to rupture the fuse or operate the excess current circuit-breaker. In this system there must be no switches, links or breaks in the neutral conductor.

15.4 Wiring systems

The current-carrying capacity of cables is described as the rating. The main factor governing the rating is the need to limit the normal continuous working current to a level that will prevent the temperature of the cable rising to a level which will present damage to the insulation. The rating will therefore vary according to the ambient temperature round the cable and also according to whether the cable is exposed and can lose heat readily or is enclosed. The nature of the excess current protection also influences the rating cables. Close protection ensures that an excess current of 1.5 times the designed load current cannot be sustained for as long as 4 h whereas coarse protection will allow this excess to continue for longer periods. Wiring with close protection is allowed a higher current rating. Table 15.1, taken from information in the 14th edition of the IEE Wiring Regulations shows the typical sizes of small copper cables and their ratings, unenclosed and enclosed for coarse and close excess current protection. The assumed ambient temperature is 30 °C. Rating factors can be applied to give values for cables subject to higher temperatures. For each electrical installation it is necessary to decide upon the type of cables to be used and the system which is to be adopted for their protection and distribution through the building.

Two general considerations affect this decision. It is cheaper to expose wiring on wall surfaces than to

Table 15.1   IEE ratings for twin cables with copper conductors and PVC insulation and sheathing, with or without earth continuity conductor (BS 6004): ambient temperature 30 °C

Conductor			Current rating (A) using excess current protection: unenclosed (enclosed)		Voltage drop (mV) per amp per metre for both conductors
Nominal cross-sectional area (mm²)	Number of wires	Diameter of each wire (mm)	Coarse	Close	
1.0[a]	1	1.13	12 (11)	16 (14.5)	40
1.5	1	1.38	15 (13)	20 (17.3)	27
2.5[b]	1	1.78	21 (18)	30 (23.9)	16
4	7	0.85	27 (24)	35.9 (31.9)	10
6	7	1.64	35 (30)	46.6 (39.9)	6.8
10	7	1.35	48 (40)	64.8 (53.2)	4

[a] Typically used for lighting circuits.
[b] Typically used for ring-main circuits.

conceal it within the construction. Sometimes exposed wiring will have to be protected against mechanical damage but even in these cases the cost of a surface system is likely to be lower than a concealed system. The appearance of a surface system is, however, normally unacceptable except in industrial buildings. Surface systems are readily rewireable. In concealed systems a decision must be made whether it is acceptable to install a system that will present difficulties when rewiring becomes necessary or modifications are needed. If this limitation cannot be accepted it will be necessary either to select cables, such as mineral-insulated cables, which do not deteriorate significantly with time, or to run the cables in conduit or trunking, which will allow new ones to be introduced or old ones to be withdrawn and replaced.

Types of cable and their applications

Colour coding

Each single-core cable and each separate core of multicore cables must be identified by either numerical or colour coding. Colour is the most common method and both rubber and PVC insulation can be produced in different colours. It is preferable to have the identification running the whole length of the cable, but in cases where this is not possible, such as mineral-insulated cables, bare conductors in bus-bar systems and earth continuity conductors in sheathed cables, sleeves of the appropriate colours should be slipped on to the terminations of the conductors. The identifications are given in Table 15.2.

Sometimes, as in wiring two-way switches, the function of a particular core will not always correspond with the colour coding. Green and yellow are, however, reserved exclusively for earth conductors.

Flexible cables linking movable apparatus to socket outlets have a different colour coding. This has been

Table 15.2 Contemporary cable colours

Function of core or conductor	Non-flexible cable		Flexible cable
	Colour code	Numerical code	
Earth	Green and yellow	–	Green and yellow
Live			
(single phase)	Red	1	Brown
Neutral	Black	0	Blue
Other phases of	Yellow or blue	2	–
three-phase supply		3	–
Special purpose	Orange	4	–

Table 15.3 Cable colours: past versus present

Function of conductor	Present colour coding	Old colour coding
Live	Brown	Red
Neutral	Blue	Black
Earth	Green and yellow	Green

introduced recently and many pieces of existing apparatus still have the old coding (Table 15.3).

PVC-sheathed cables

This type of cable consists of two or three insulated cores, with or without an earth continuity conductor, or single- or four-core cables without earth conductors. In each case the cores and earth conductors are contained within an outer protective sheath of PVC. The cables are easily manipulated and installed and are resistant to moisture and most environments normally encountered. This type of cable is now used very much more than its rubber-sheathed predecessor.

TRS (trough rubber-sheathed)

This form of cable is no longer used but may be found in many buildings. It is essentially similar to PVC-sheathed cable, which has superseded it to a great extent. Rubber is not as resistant as PVC to sunlight and some forms of chemical.

The life of both PVC-sheathed cable and TRS cable is limited and installations will need to be replaced during the life of most buildings. If properly installed and not subject to high temperatures, either as a result of overloading the circuit or running the cable too close to other sources of heat, sheathed cables can give excellent service over a long period of time.

Installation of sheathed cables

Sheathed cables are widely used for surface wiring and almost universally used in houses. They are not normally directly buried in concrete or embedded in mortar, but are fixed to surfaces by clips (Fig. 15.11). For most cables except the largest the spacing of supports is 400 mm for vertical runs and 200 mm for horizontal runs.

Timber floors and roof spaces within houses provide the opportunity to run sheathed cables quite freely while keeping them concealed. Cables running down walls beneath the plaster finish are protected by metal cover strips (Fig. 15.11). If any short lengths of cable must be buried in concrete or covered by screed, lengths of conduit may be built in to the construction

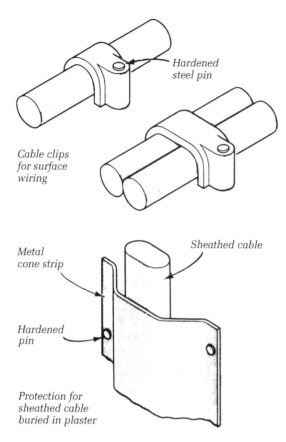

Cable clips
for surface
wiring

Hardened
steel pin

Metal
cone strip

Sheathed cable

Hardened
pin

Protection for
sheathed cable
buried in plaster

Figure 15.11 Cable clips and protection for sheathed cable buried in plaster.

to provide a route for the cables. The conduit must be large enough and have easy bends so that the sheathed cable can be easily passed through. Short lengths of conduit should also be used where cables pass through walls. In all cases the ends of conduit should be bushed to prevent damage to the cable from sharp edges. This precaution is often neglected in practice.

Where cables have to go through a timber, any holes should be made as small as possible at the centre of the joist's depth and preferably not near the centre of the span. In no circumstances should such a hole be nearer than 50 mm to the upper surface of the joists.

Cables should be kept away from other services and direct contact always avoided. Leakage from and condensation on water pipes should be considered when cables approach pipes carrying water. Where cables must cross the lines of hot-water pipes cables should be kept below the pipe to minimize the heating effect. Cables should not be covered by insulation.

15.5 Conduit systems

An electrical conduit is simply a tube into which cables are drawn, normally single-core cables. They can be installed on the surface where they provide protection for wiring or concealed when they make rewiring practicable. Traditionally conduit was of metal and provided both mechanical protection and an earth continuity conductor. Plastic conduit is now becoming popular and, in this case, an earth cable has to be added to those which would be contained in a metal system. Flexible conduit is also available. It cannot be used as an earth conductor. The expense of this type, however, precludes its use to simplify the installation process and flexible conduit is mainly used as a final link to pieces of free-standing apparatus. It is clearly not practicable to fill conduit with cables and normally the number of cables is limited to 40% of the theoretical maximum. In laying out conduit it is necessary to ensure there are sufficient access points through normal switch and junction boxes to allow the cables to be drawn in. If access is not adequate, special junction boxes and access elbows must be incorporated in the installation.

Steel is the standard material for conduits. It is available in light or heavy gauge. Both types are normally formed from bent strips with the joint welded and brazed. Light gauge can be obtained with a simple butt joint. Heavy-gauge conduit is connected by screwed joints, whereas light-gauge conduit is jointed by mechanical grip devices. Both types of conduit are protected against corrosion by black enamel. Heavy gauge can be galvanized; it can also be bent to simplify installation. Bending is more difficult with lightweight conduit and manufactured bends are used. Heavy-gauge conduit is indicated where great resistance to damage is needed, where a screwed system is necessary, as in flame-proof installations, and where the environmental conditions dictate that a galvanized finish is needed. Apart from these considerations, light-gauge conduit is used for most normal single-phase installations in buildings and heavy-gauge conduit for three-phase installations.

Conduits are sometimes made of copper, which is particularly resistant to corrosion, has excellent conductive properties and is smooth for drawing in cables. Joints in this case can be compression, capillary soldered or welded. Aluminium is also used in conjunction with screwed steel fittings. Care must be taken that aluminium conduit does not come into contact with concrete, mortar or plaster when any moisture is present. Both copper and aluminium conduits can be bent readily.

Plaster conduits are now widely used. They are available in rigid and flexible form. Rigid conduit comes in heavy gauge or light gauge; heavy gauge is screw jointed and light gauge is solvent weld jointed. They can be bent if heated. They are resistant to most normal environmental conditions in buildings and can be buried in concrete. Oval-shaped conduits are available for use in screeds and plaster finishes where space for conduit is limited. Plastics have very high thermal expansion and straight lengths of plaster conduit screeding of 5 m may require provision for expansion and contraction. Special couplings that allow relative movement enable expansion to be taken up neatly and economically.

Installation of conduit systems

Both steel and plaster conduits can be obtained in 16, 20, 25 and 32 mm diameter and some manufacturers make larger diameters. Timber-joisted floors are usually a feature of houses, and sheathed systems will be used in most cases where they occur. It is not easy to run rigid conduit across joisted floors but if necessary steel conduit can be used in notches. The notches must be near the supports of the joists and should be as shallow as possible. Figure 15.12 shows the method of installation. Conduits are more usually

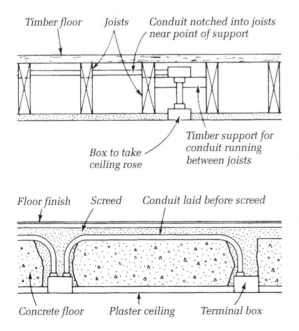

Figure 15.12 Conduit installed in timber and concrete floors.

laid on structural concrete floors and covered by the normal floor screed. Vertical runs may be concealed in ducts and cupboards and may be fixed in a shallow chase in the wall and covered by the normal plaster finish. It is possible to embed conduit in structural members themselves but this often causes disruption in the progress of each trade and is best avoided if possible.

The sharpest bend allowable in conduit has an inner radius of not less than 2.5 times the outside diameter of the concrete. Manufactured elbows allow sharper curves than this when needed and also provide access, probably essential at a very sharp bend. Fixing of conduit is sometimes done with pipe nails but a more satisfactory method is the use of saddleband clips as shown in Fig. 11.9.

Work standards are very important in conduit installations. In steel systems the electrical continuity of earthing depends on the soundness of jointing, and in light-gauge systems it is essential that the black enamel is removed from the pipe at the junction and the securing lug firmly tightened. It is not desirable, however, to strip the enamel too far or to extend threads beyond the essential minimum since this opens up the unprotected conduit to rusting and corrosion. It is important that burrs are removed from cut ends of conduit and that bushes are provided to give a smooth surface at the end of the conduit. The effect of sharp edges is to damage the insulation as cables are drawn in to conduits. Where conduit or cables pass through walls and floors it is important that the holes are well filled to minimize penetration by fire.

Trunkings

As the number of cables increases, with the size and complexity of buildings and their electrical installations, it becomes less convenient to use conduit. The amount of work involved in drawing in the large number of cables would be considerable and the largest single conduits will often prove unable to accommodate the number of cables. Cable trunkings overcome this problem. They can be obtained in much larger sizes than conduit and have continuous access via a cover strip, which simplifies wiring. A space factor of 45% is allowable, slightly more than the 40% appropriate for conduit. Metal trunking can provide an earth conductor. Plastic types have to have separate earth conductors. Each manufacturer of trunking provides a range of junctions and fittings.

In addition to rectangular trunking, which presents some problems of accommodation and concealment

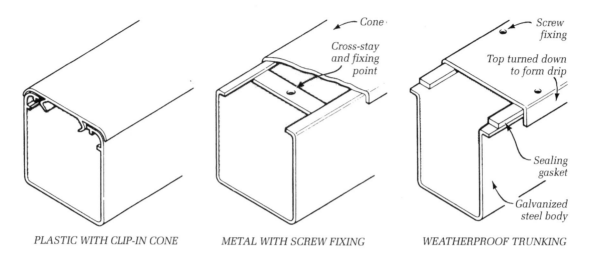

PLASTIC WITH CLIP-IN CONE   METAL WITH SCREW FIXING   WEATHERPROOF TRUNKING

Figure 15.13   Types of trunking for cables.

except where service ducts and suspended ceilings are present, special trunkings have been specially adapted to allow incorporation into structural elements. Floor and skirting trunkings are the most common. Floor trunking systems are particularly important in large open office areas where cables must be brought close to work positions. In industrial environments overhead distribution can take place in similar circumstances with terminals for use being provided on pendants hanging from an overhead trunking system. Figure 15.13 shows several forms of trunking.

Mineral-insulated cables

This type of cable, usually made in copper although aluminium is now often used, consists of a metal tube containing densely packed magnesium oxide insulation in which there run single-strand conductors. Cables with one, two, three, four and seven conductors may be obtained. As in the case of conduit the outer layer of the cable provides an earth conductor. The cable is of small dimensions in relation to its rating and has great resistance to heat and deterioration. The insulation properties of the magnesium oxide are rapidly reduced with increasing moisture content. The ends of mineral-insulated cables must therefore be sealed to prevent moisture penetrating the insulation. Figure 15.14 shows the arrangement for terminating a mineral-insulated cable and for coupling the terminal to switch or socket boxes.

Mineral-insulated cables are often used for three-phase wiring and for submains in larger buildings. They are also often employed in industrial situations where neatness and heat resistance are important. In other installations, including houses, they are used for exposed or buried wiring or other circumstances where the superior durability of mineral-insulated cables is advantageous. This type of cable cannot suffer from condensation, which may be a problem in conduit systems running externally.

Mineral-insulated cables require substantially fewer supports than sheathed cables. Even the smaller diameters can have 600 mm horizontally and 800 mm vertically between supports. The cable can be readily bent. The minimum internal radius must not be less than 6 times the cable diameter.

Power and armoured cables

Large cables supplying the current to complete buildings or groups of buildings are constructed somewhat differently to cables for internal use. The conductors are multistranded, very often of aluminium, and insulation materials are employed which are more tolerant of overheating. Paper impregnated with mineral oil is often used. The insulated conductors are bedded in further insulating material, itself surrounded by an armoured sheath perhaps formed of galvanized wire, and the whole is then covered with an outer sheath. Cables of this type are clearly strong and resistant to moisture. They are difficult to handle and bend. The minimum internal radius is likely to be 15–18 times the diameter. The entry of this type of cable requires special consideration. It will often be leading to a transformer substation either outside or inside the building. Ducts of 1 m × 1 m may well be required to

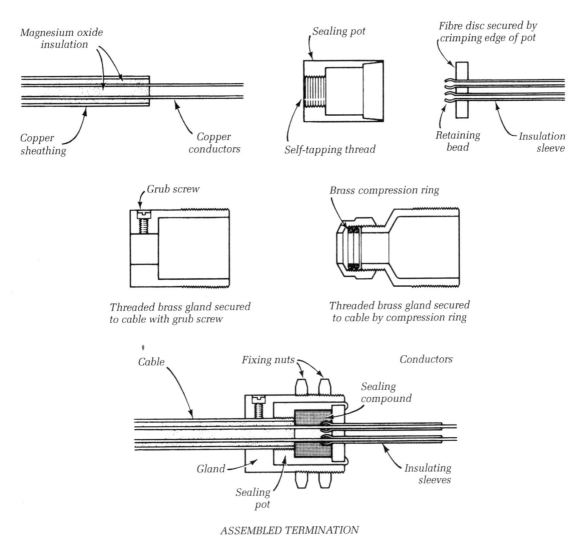

Magnesium oxide insulation

Copper sheathing

Copper conductors

Sealing pot

Self-tapping thread

Fibre disc secured by crimping edge of pot

Retaining bead

Insulation sleeve

Grub screw

Threaded brass gland secured to cable with grub screw

Brass compression ring

Threaded brass gland secured to cable by compression ring

Cable

Fixing nuts

Conductors

Sealing compound

Gland

Insulating sleeves

Sealing pot

ASSEMBLED TERMINATION

Figure 15.14   Terminal for mineral-insulated cables.

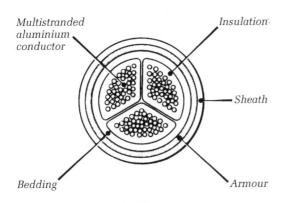

Multistranded aluminium conductor

Insulation

Sheath

Bedding

Armour

Figure 15.15   Armoured cables.

accommodate the cable and allow it to be turned and connected to adjacent apparatus. Figure 15.15 shows a section through an armoured cable, and Fig. 15.16 a cable entry to a dwelling.

15.6 Distribution circuits

Special arrangements of wiring fuses, switches and socket outlets have been developed to enable the electrical needs of buildings to be met neatly and economically. In houses and small buildings the main control element is the 'consumer's unit', which consists of a service intake, a fuse under the control of the electricity company, a meter, a main socket and

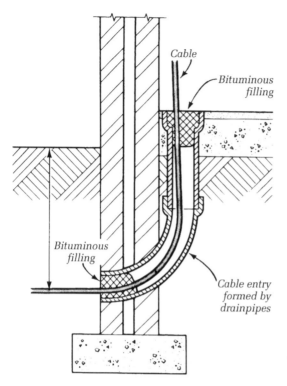

Cable

Bituminous filling

Bituminous filling

Cable entry formed by drainpipes

Figure 15.16 Cable entry to dwelling.

a battery of fuses or miniature circuit-breakers leading to individual circuits serving lighting, socket outlets and the individual appliances requiring their own circuits, such as cookers and immersion heaters. Figure 15.17 shows a typical control unit for a dwelling and Fig. 15.18 is a schematic diagram (with earth omitted) of the parts of a complete installation.

Figure 15.19 shows typical lighting circuit arrangements which permit simple wiring but allow several light fittings to be individually controlled from a single basic circuit.

Figure 15.20 shows a ring-main circuit which supplies socket outlets for portable apparatus. Many buildings still have installations where each socket is served by an individual circuit and fuse. In order to allow proper fuse protection for different types of apparatus three sizes of socket, 2, 5 and 15 A had to be provided and apparatus could only be used in its own size of socket. In addition, the number of cable runs and the total amount of cable is considerable in this system. In 1947 a new system was introduced, the 13 A ring main. With this system there is only one type of socket. The plugs are distinguished by rectangular pins and provided with fuses (3, 5 and 13 A) according to the type of apparatus connected. Thus a single circuit

and standard sockets can serve different pieces of apparatus with different wiring while retaining fuse protection appropriate to the wiring. In the case of a fault in a piece of apparatus its own fuse will blow, putting it out of action but leaving the socket to which it was connected still operative. Thus the ring-main system provides economy of wiring, flexibility in the location of apparatus, since all items can be plugged into all sockets and robustness in that a fault in a piece of apparatus does not put the whole circuit out of use until the main fuse is repaired. In domestic buildings it is considered that a single ring main can serve an area of $100 \, \text{m}^2$ without risk of overloading. Any number of socket outlets can be provided since more sockets give greater flexibility in placing apparatus but, in dwellings, do not contribute to greater loads. To economize in wiring it is not essential to run the ring wiring to every socket; spurs serving not more than two sockets can branch off the main ring, provided that not more than 50% of the sockets are served by the spurs.

The 13 A ring-main system has served well for some 30 years, although many of the previous types of installation are still in use. At present a new 16 A international plug and socket is being discussed which may eventually supersede the 13 A sockets.

Pieces of fixed apparatus such as water heaters or cookers usually have their own individual circuits and fuses. There should be an isolating switch adjacent to each piece of apparatus so that work on it may be carried out in safety.

Bathrooms represent a special safety risk. The presence of water makes the chance of the consequences of electric shock much more serious than would be the case in other rooms. Socket outlets are not permitted. No switches may be located where they could be touched by anyone using a bath or shower. Switches are sometimes placed externally to overcome this problem, but a better solution appears to be the use of a cord-pull ceiling switch. No electrical apparatus with a metal casing may be located where it can be reached by anyone in the bath or shower. Lampholders must be constructed of insulating material and the terminal protected against accidental contact by a protective shield. The bath and all other metal apparatus in bathrooms must be electrically bonded by special bonding leads and connected to earth. Where showers occur in rooms other than bathrooms, these considerations apply within reach of anyone using the shower; socket outlets may not be closer than 2.5 m to the shower cubicle. It is permissible to have a point for an electric shaver, provided it is a shaver supply unit complying with BS 3052.

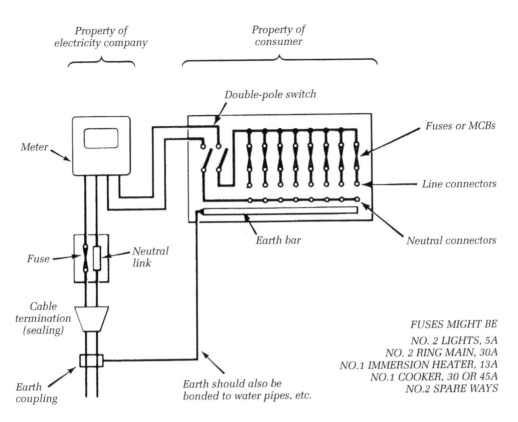

Property of
electricity company

Property of
consumer

Double-pole switch

Fuses or MCBs

Meter

Line connectors

Neutral connectors

Earth bar

Fuse

Neutral
link

Cable
termination
(sealing)

FUSES MIGHT BE

NO. 2 LIGHTS, 5A
NO. 2 RING MAIN, 30A
NO.1 IMMERSION HEATER, 13A
NO.1 COOKER, 30 OR 45A
NO.2 SPARE WAYS

Earth
coupling

Earth should also be
bonded to water pipes, etc.

Figure 15.17   Typical controls at the entry to a dwelling.

Distribution in larger buildings

The final circuits of larger buildings are normally very similar to those described above and local control and distribution boards are provided which serve lighting, ring mains and individual appliances. One of the major differences in large buildings is that a system of mains and submains is required to run from the point of entry of the supply and the main switchgear, to serve the various local distribution boards. Figure 15.21 shows schematically a typical system involving main, submain and final distribution circuits; where heavy loads are involved special attention must be given to the mains and submains not only to ensure their current-carrying capacity but also to facilitate additions and modifications to the subcircuit. One method specially designed for this purpose is the cable tap system. In the arrangement the four heavy cables for a three-phase supply are supported in a metal trunking and connections can be made by stripping the insulation and taking leads from special connectors to fused tap-off units. Figure 15.22 shows a typical example. For heavier loads, or where it is important to

be able to make connections without interrupting the supply, a bus-bar system may be used. Solid conductor rods, without insulation are supported in a metal trunking and special connecting units enable leads to be run from the conductors to tap-off boxes without switching off the current (Fig. 15.23).

Several different fuses are required to give progressive protection to the wiring as the current rating reduces towards the final circuit. It is important that the fuses and wiring are selected to ensure there will be adequate discrimination in the operation of the fuses and any fault will only affect the fuse controlling the circuit involved without putting larger sections of the installation out of action.

Larger buildings often require substantial switchrooms and may also have to accommodate transformer substations if the electricity company requires one to supply the building. Substations are best located on the site near the entry of the main supply but, if necessary, they can be incorporated into the building. Ready access must be provided for delivery of the heavy transformer and for electricity company maintenance. Considerable heat is generated

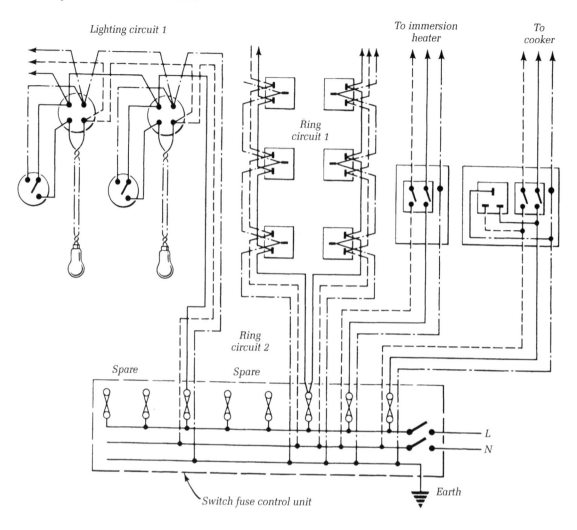

Figure 15.18 Circuits served by control unit.

and very adequate ventilation must be allowed. Transformers are normally weatherproof and do not require special protection from the weather. There is a risk of fire, so adequate protection must be incorporated into the construction of any transformer chamber.

Some apparatus in larger buildings may require three-phase supply. Lifts are a common example. Warning notices must be provided to apparatus with voltages exceeding 250 V so they will be observed before access is gained to any live parts. Large buildings will also have different phases serving different areas of the building. If the pieces of apparatus on different phases are within reach of each other, a warning notice must be displayed. All socket

outlets in a single room should be on the same phase. If this is not possible, sockets on different phases should be separated by at least 2 m.

Emergency supplies

Emergency lighting should be provided in all buildings for public entertainment and many others where failure of the main lighting might give rise to danger. The lighting should illuminate escape routes. The levels can be very low. A normal standard is 0.2 lx and this can be reduced in auditoria where a higher level might interfere with performances. In some cases emergency lighting operates continuously but it is more usual to provide a system that is automatically switched on in

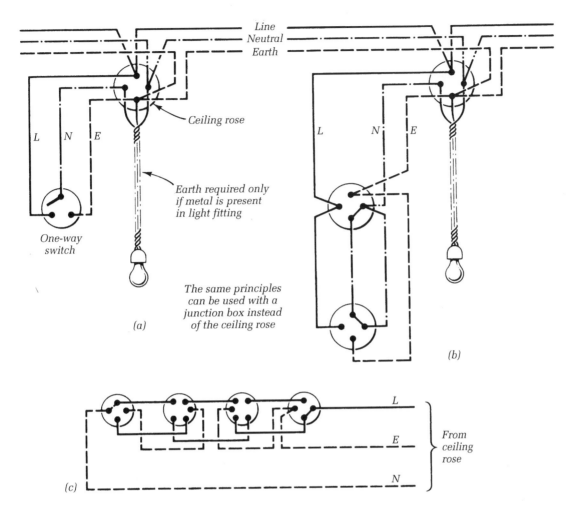

L
Neutral
Earth

Ceiling rose

L N E

*Earth required only
if metal is present
in light fitting*

*One-way
switch*

*The same principles
can be used with a
junction box instead
of the ceiling rose*

(a)

L N E

(b)

L

E

*From
ceiling
rose*

N

(c)

Figure 15.19 Light controls: (a) one-way switching; (b) two-way switching; (c) intermediate switching.

the event of mains failure. The source of electricity is by batteries which are automatically kept charged by the mains. Battery rooms require very good ventilation, access for carboys and acid-resisting surfaces. In smaller installations where skilled maintenance might be a problem nickel/alkali batteries have a longer life and do not present so many corrosion problems. Small installations can be contained in a metal cabinet little bigger than a filing cabinet and even individual lamp units can be obtained. Very large buildings may need big rooms devoted wholly to batteries.

Some buildings will use generators to supply power in case of mains failure. Hospitals often have emergency generators and very tall buildings may also justify emergency generators.

The wiring for emergency circuits should be

separate from the normal wiring and should ideally be in screwed conduit or mineral-insulated cable to give fire resistance.

Raised-floor systems are employed in computer centres or other rooms with concentrated equipment requiring complete flexibility of location and ease of rewiring. The whole floor is covered by trays containing the floor finish and supported to give a clear space beneath. The trays can then be raised to give total access to the wiring space below.

15.7 Telecommunications

In addition to systems for the supply of electric power, electric wiring is used for the transmission of information. Such systems include British Telecom and internal telephones, teleprinters, public address,

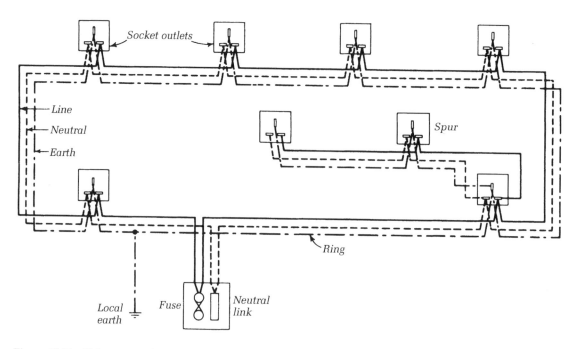

Line
Neutral
Earth

Socket outlets

Spur

Ring

Local earth

Fuse

Neutral link

Figure 15.20 Thirteen-amp ring main circuit.

staff location, bells, fire alarms, security, impulse clocks, radio and television cable distribution monitoring and control. Normal telephone circuits are also provided for many data transmission applications.

The voltages and currents in these installations are very low and they are not directly connected to the mains. Normally the wiring must be separated from mains cables and not drawn into any conduit or trunking containing them. If there is occasion to include both types of wiring in one trunking it may be possible, provided the insulation of all cables meets the requirements of the highest voltage present. Fire alarm circuits must in no circumstances be included with mains wiring. British Telecom make it clear that their telephone cables are not insulated for mains voltages and that they must be segregated from cables carrying mains voltages. Mineral-insulated metal-covered cables can be regarded as providing adequate segregation in themselves and may be located in the same trunkings as telecommunications cables.

A clearance of 50 mm should be allowed where British Telecom telephone cables cross cables carrying mains voltages and neither is contained in conduit or trunking, or if this is not possible 5 mm of insulation should be provided between the cables.

British Telecom carry out the complete wiring and installation of British Telecom telephones. For the other types of installation and internal telephone systems the normal electrical subcontractor can be employed, although in some cases particular specialists may be brought in. British Telecom does not carry out any work other than the installation of its own cables and instruments and all trunkings, conduit and other facilities for concealing cables must be provided ready for British Telecom engineers to carry out their work.

Telephone installations vary considerably and the essential problem in all types of building is to provide the greatest possible freedom for location and movement of instruments. This involves consideration in the early design stages, consultation with British Telecom and the provision of routes for telephone cables during planning construction (Fig. 15.24).

Office buildings have large numbers of telephones. The cable entry will normally be below ground via a duct supplied by British Telecom, which they will subsequently waterproof. The incoming cable will terminate in a distribution case, or for large buildings, a distribution frame for which special accommodation is required. In small offices a receptionist or typist may operate a desktop unit, and a few conduits provide routes for calls. In large offices having private exchanges an apparatus and battery room and a switchroom will be needed. The equipment is heavy and may require headroom of some 3 m. British

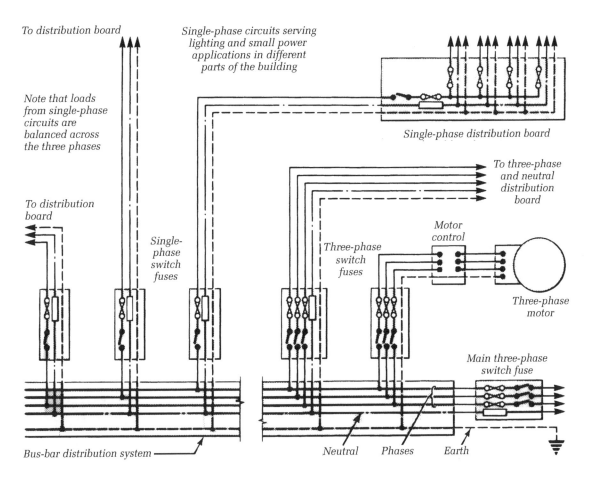

To distribution board

Single-phase circuits serving lighting and small power applications in different parts of the building

Note that loads from single-phase circuits are balanced across the three phases

Single-phase distribution board

To distribution board

Single-phase switch fuses

To three-phase and neutral distribution board

Motor control

Three-phase switch fuses

Three-phase motor

Main three-phase switch fuse

Bus-bar distribution system

Neutral   Phases   Earth

Figure 15.21 A typical distribution system: main, submain and final circuits.

Telecom suggest that a manual installation with 25 exchange lines and 300 extensions will need four operators and that ideal areas for switchroom and apparatus room would be 12 m² and 20 m² respectively. Natural lighting is particularly desirable in the switchroom and also if possible in the apparatus room where, in large installations, full-time maintenance staff may be employed. In offices with multiple occupation small switchboards may be distributed throughout the building and the important provision will be an adequate system of trunkings.

The British Telecom recommendation for distributing trunkings in offices envisage that no telephone outlet will involve more than 30 m of cable to link it to a main vertical riser. Taking into account normal routing this means that risers are likely to be required approximately every 40 m along the length of the building. The recommended dimensions of risers are shown in Fig. 15.25. It is recommended that the

horizontal trunking leading from vertical risers should be at least 1875 mm² and the branches leading to outlet not less than 1250 mm².

In small offices it is very convenient and often quite satisfactory to carry the horizontal cable runs in a skirting trunking. If this follows the external walls there will be no interruption from doors, and no problems will arise when desks are reasonably near the wall. In larger offices, where desks will be remote from the wall, a grid of underfloor trunking will be needed with special floor outlets at desk locations (Fig. 15.26). The floor outlets will supply power as well as telephones.

Subject to agreement most of the other extra low voltage cables required for services such as internal telephones, alarms, clocks and public address can be carried in the same trunkings as the British Telecom cables, provided they are big enough and suitably located.

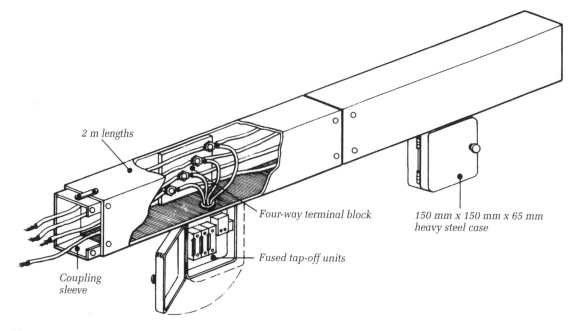

Figure 15.22   Cable tap system.

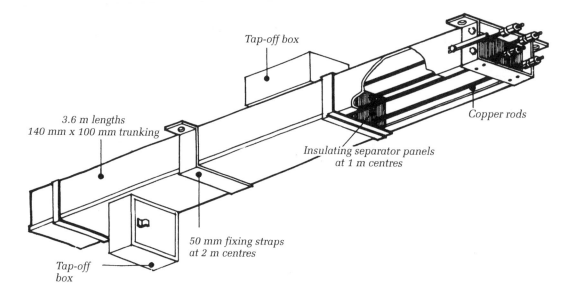

Figure 15.23   Bus-bar system.

Staff location is often carried out via a public address system. Illuminated signs at selected locations are also used. These systems can be served through the normal trunkings and conduits. Some staff location systems are, however, based on individual receiving units carried by the personnel concerned. These units may be called by a radio transmitter but more usually by an inductive loop of cable running round the areas covered by the system. This cable cannot be contained in trunking or conduit and must be specially routed.

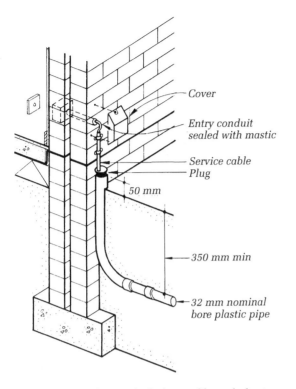

Figure 15.24 Underground telephone cable: typical entry to dwelling.

Telephone cables for dwellings are often distributed at high level. A short length of conduit adequately protected against damp penetration makes a suitable entry. British Telecom recommended that conduits as skirting trunkings should be provided as possible telephone locations.

British Telecom should be consulted at the earliest possible opportunity in the design of a new building, not only to discuss installation problems but also to ensure that the exchange can deal with the new telephones required.

British Telecom issue two useful booklets giving information about telephone installations. They are, *Facilities for telephones in new buildings* and *Provision for telephone facilities on new housing estates*.

Interference

As electrical and electronic equipment becomes more sophisticated and sensitive, it becomes more susceptible to interference. This may take place through the wiring system where fluctuations in supply (transients) occur, by electrostatic discharge or electromagnetic coupling with radio transmitters or

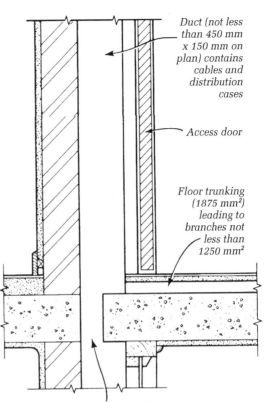

Cable hole through structural floor: located at one side of the duct, of dimensions not less than 150 mm x 75 mm

Figure 15.25 Vertical duct for British Telecom telephones: precautions against the spread of fire may be necessary. Stopping in the floor is usually acceptable but the ducts may require fire floors if stopping is not practicable.

equipment which produces radio frequency emissions. Highly specialized technical advice is needed to deal with problems of this sort. Some general precautions should be taken where there is any possibility of equipment being affected:

- Equipment should be located as far as possible from any sources of interference.
- Proprietary devices are available to reduce transients.
- Power and signal cables should be separated.
- Signal cables should be screened and earthed.
- Earthing should be effective.
- Static should be reduced (increased humidity, natural fibres for carpet).

Equipment can be provided with a conductive enclosure in highly critical cases of radio frequency interference.

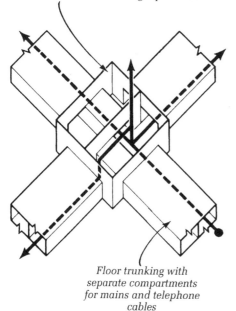

Junction box providing cable crossing and access while maintaining separation

Floor trunking with separate compartments for mains and telephone cables

Figure 15.26   Underfloor trunking junction box showing how different types of cable may cross and have access while maintaining separation.

Electrical layout drawings, installations and standards

It is unusual for detailed drawings of electrical circuits to be prepared for normal installations. In large buildings the services engineer will prepare plans showing the location and type of mains switch and distribution boards, the sizes and location of the main cable runs serving them and the sizes and location of cable trunkings. Among other aspects, a specification will determine the type of wiring to be used and whether the wiring is to be concealed or run on the surface. Drawings and schedules showing the location and nature of the electrical appliances will be required. The final cable and conduit runs in large buildings and the whole installation layout in smaller buildings are usually left to the discretion of the electricians carrying out the work.

Few problems normally arise but architects wishing to have wiring concealed (e.g. in screeds, suspended or timber floors and plaster) must ensure that practicable wiring routes exist to all the pieces of electrical apparatus. Figures 15.27 and 15.28 show a typical layout of electrical fittings for a dwelling and a typical schedule.

15.8 Electrical terms

- **Volt (V)**: measure of electrical pressure.
- **Ampere (A)**: measure of flow of charge (current).
- **Watt (W)**: measure of electrical power.
- **Kilowatt (kW)**: equal to 1000 W (volts × amps = watts).
- **Unit**: short for the commercial measure of electrical power, the Board of Trade Unit which is equivalent to 1000 W for one hour (or 1 kilowatt-hour, kWh).
- **kVA**: measure of apparent electrical power in alternating current systems (line voltage multiplied by line amperage).
- **Voltage drop**: if a long length of wire is used its resistance to the flow of electricity may give rise to a drop in voltage. The permissible amount of this drop in electrical installations is limited.
- **Power factor**: ratio of actual power (kW) to apparent power in alternating current system.

The balance of inductance and capacitance in an electrical installation may cause the voltage and current fluctuations to become out of phase with each other. This causes the actual power to be less than the apparent power and gives a power factor of less than unity. It is desirable that power factors should be as near unity as possible since otherwise the efficiency of the distribution system is reduced and electrical costs are increased. In large installations additional inductance or capacitance may be purposely added to balance the power factor.

15.9 Lightning protection

Strokes of lightning coming from thunder clouds to the ground may strike anywhere. The peak currents average 20,000 A and may reach 200,000 A. If lightning strikes a building substantial structural damage can be caused and fires may be started. The risk of lightning striking buildings is not so great that every building has to be protected. Protection should be provided, however, in cases where the likelihood of lightning striking is greater than normal or where the consequences would be particularly unfortunate.

BS 6651 *The protection of structures against lightning* gives details of the technical requirements for lightning protection and also gives information and tables to assist in deciding whether a lightning protective system is required for the building.

The method of deciding whether protection must be given takes into account the likelihood of thunderstorms in the various parts of the country, the use to which the building is put, the type of construction, the contents of the building, the degree

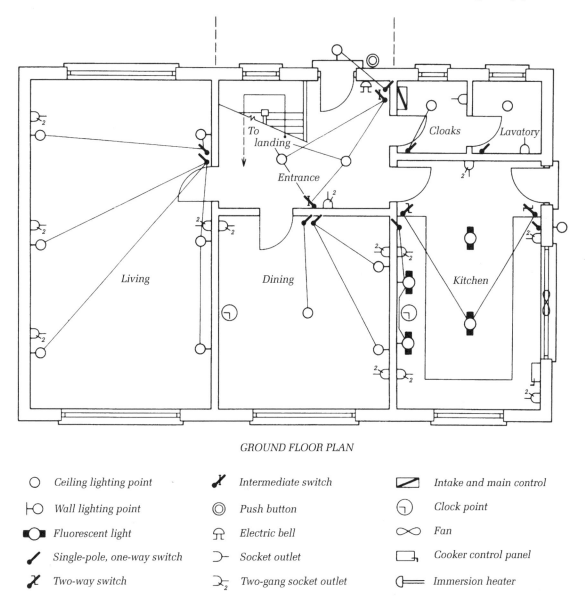

GROUND FLOOR PLAN

○	Ceiling lighting point	✗	Intermediate switch	▱	Intake and main control
⊢○	Wall lighting point	◎	Push button	⊙	Clock point
◀▮	Fluorescent light	⋒	Electric bell	∞	Fan
✒	Single-pole, one-way switch	⊃⊦	Socket outlet	⊐	Cooker control panel
✗	Two-way switch	⊃⊱₂	Two-gang socket outlet	◖▭	Immersion heater

Figure 15.27 Typical electrical layout drawing and key.

of isolation of the structure, the nature of the topography and the height of the structure. The following factors generally lead to increased need for protection. High buildings present a special risk. Buildings rising above 53 m require protection in all cases. Isolation from other structures which results in the building rising above its surroundings increases the risk, as does a location in mountainous or hilly country. Particularly valuable or dangerous contents give high priority for protection. Domestic buildings have the lowest priority, whereas high priority is given to power stations, gasworks and similar buildings. Ancient monuments are awarded an even higher level of priority and the maximum priority for usage is given to schools, hospitals and places of assembly. From the constructional point of view thatched roofs present the worst risk. Brick and timber construction offer lower risk, but reinforced concretes and steel-framed buildings offer even lower risk. To determine the need for a particular type of building in a particular location

LOCATION			LIGHT FITTINGS							SOCKET OUTLETS				SPECIAL FITTINGS
Block	Floor	Room	Unit no	Description	Cat ref	Lamps	Height	Switch	Sw Ht	Unit no	Type	Cat ref	Height	

Figure 15.28 Typical electrical schedule.

the tables and details in BS 6651 should be consulted.

Lightning-protective installations have three major parts: air terminations, down conductors and earth terminations.

Air terminations

Air terminations are located over the highest points of the building to intercept lightning strokes before they strike the fabric itself. A single-point air termination is considered to give a zone of protection defined by a cone; the apex of the cone is formed by the air termination and the base of the cone has a radius equal to the height of the air termination. Vertical features such as spires can be protected by a single vertical air termination above their highest point. Generally, however, air terminations will be horizontal conductors running along critical edges of the roof. For these horizontal terminations a zone of protection of similar proportions to those for single terminations extends along the full length of the horizontal member. Thus a pitched roof can be regarded as protected by an air termination running along the ridge whereas a flat-roofed building would require an air termination round its perimeter. For large roofs a grid of parallel terminations should be provided so that no part of the roof is more than 9 m from protection. Some buildings have roofs at several levels. The lower levels will only require protection when they project beyond the zones of protection provided by the higher air terminations. Additional terminations must then be located to cover the unprotected roofs.

Down conductors

Down conductors must be provided to carry the currents arising from lightning strikes from air terminals down to earth terminations. They should follow a simple path down the buildings. Bends are permissible but re-entrant loops should be avoided. A building of up to 100 m² in plan can be served by one down terminal and one additional down terminal should be provided for every 300 m² of area in excess of 100 m².

Earth terminations

Earth terminations are required to give a low resistance path to earth for the current. They should be near the bottoms of down conductors and the total resistance of the lightning-protective system to earth should not exceed 10 Ω. A system having a single down conductor and earth termination must have a resistance not exceeding this value. In multiple systems individual

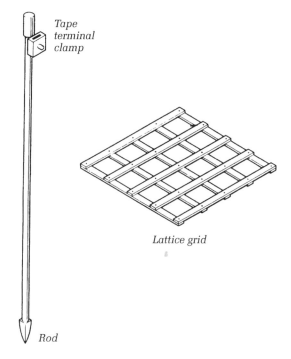

Tape terminal clamp

Lattice grid

Rod

Figure 15.29 Earthing rod and grid for lightning conductor installations.

resistances may be higher provided the combined resistance does not exceed 10 Ω. The preferred form of earth termination is to use metal rods driven into the ground (Fig. 15.29).

Air terminations and down conductors are made of corrosion-resistant metal in either strip, rod or stranded form. Aluminium, aluminium/copper alloy or phosphor bronze rods of 10 mm diameter are frequently used for air terminations and down conductors. Rods of 10 or 12 mm diameter are used for earth terminations. They may be made of hard drawn copper for driving into soft ground, phosphor bronze for harder ground and copper-clad steel for very hard ground.

Other metal on the face of a building not forming part of the lightning-protective system must either be sufficiently isolated from the lightning system to prevent side flash occurring or be electrically bonded to the lightning system.

A testing joint should be provided in each down conductor. It should be out of normal reach to prevent interference. Its purpose is to enable the system to be isolated and a connection made so that resistance to earth can be tested.

In addition to the problems of normal buildings, BS 6651 gives advice dealing with structures containing explosive or flammable contents, fences, tents and marquees and trees near structures.

16 Gas supplies

Gas has the virtues of giving almost pollution-free products of combustion, entirely acceptable in clean air zones, and of flue requirements which are smaller and often very much simpler than those for solid fuel or oil. Recently the use of gas as a fuel has been revitalized by the discovery of natural gas under the North Sea. This has happened at a time of increasing expense and shortage of other fuels, including manufactured town gas. As a result all gas supplies in the United Kingdom are being converted to the use of natural gas.

Town and natural

Town gas, originally made from coal, and more recently from oil has a high proportion of hydrogen and a significant proportion of carbon monoxide which made it poisonous. Natural gas is mainly methane and is virtually free of carbon monoxide. Although this means that natural gas is not toxic, it can cause asphyxiation when breathed in high concentrations. The burning velocity (or the rate at which gas ignites after leaving the jet) for natural gas is lower than for town gas but its calorific value is higher. Consequently, different burners are required for town and natural gas. Natural gas is distributed at higher pressure than town gas and this coupled with its higher calorific value per cubic metre means that smaller main and distributing pipes in buildings can be used.

Explosion

The calorific value of natural gas is $37\,MJ/m^3$ (for town gas it is $18.5\,MJ/m^3$) and the resulting pressure supplies to domestic and other non-industrial users are at 1–$2\,kN/m^2$. As was the case with town gas, mixtures of natural gas and air can explode if ignited in a confined space. This risk can be minimized by proper materials, work standards and inspection for gas installations.

Pipes

Unplasticized PVC is now becoming more usual for gas mains but fire risks generally preclude its use internally. Copper is widely used but the most common material for gas pipes in buildings is mild steel, to BS 1387.

Installation

A typical domestic gas installation is shown in Figs 16.1 and 16.2. Notice that the pipe arrangements from the main to the building are quite similar to the water supply, although the gas pipe must fall towards the main so that condensation may drain. Inside the building a main control cock precedes the gas meter, which must be sited conveniently for access. Sometimes a pressure governor is fitted before the meter and sometimes, particularly in larger installations, a bypass to the meter is included. For a small domestic installation a meter space 550 mm × 550 mm × 300 mm deep might be appropriate. Meters for large installations can be very big, requiring special accommodation and sometimes lifting tackle. The gas company should be consulted at an early stage in design.

Underground service pipes must be protected by bituminous or other wrapping.* Pipe runs after the

* During building an entry for the gas pipe should be formed in the foundation. Vitrified clay pipes may be used and the gaps after installation packed with bituminous filling.

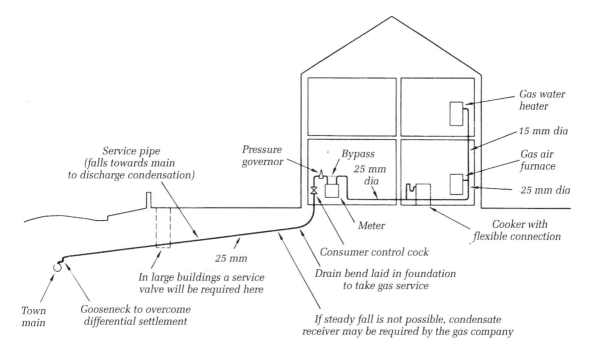

Figure 16.1 Typical domestic gas installation.

meter can be dictated by convenience and neat appearance. It is acceptable to bury gas pipes in floor screeds and chases in walls, but mild steel pipes should be wrapped to prevent corrosion if they are buried in plaster. Plastic pipe sleeves should be used wherever gas pipes pass through walls and floors, not only for neatness but also to ensure that thermal movement and settlement stresses are prevented.

Where gas pipes run in horizontal and vertical ducts a grille must be provided at the highest point with a free area of at least 3000 mm^2.

Most gas appliances are connected directly to the pipework. In the case of cookers and refrigerators, however, it is becoming increasingly usual to provide flexible connections to allow the appliance to be moved for cleaning.

Flues

Very small gas appliances and cookers are permitted to discharge their products of combustion directly into the air of the rooms in which they are sited. All other appliances must have flues, but flues for gas appliances are very much simpler than for solid fuel or oil. All gas appliances have to be able to function without a flue, so the flue itself is merely a discharge

for products of combustion. It does not have to provide a draught to aid combustion. No cleaning is required for gas flues, and flues formed of hollow blocks set into walls can be used for single appliances. Products of combustion from gas appliances contain substantial quantities of water vapour and care must be taken that condensation does not cause problems. Vitreous clay and asbestos cement pipes make excellent flues. There are, however, two major devices that simplify gas flue requirements. One is the balanced flue shown in Fig. 7.26.

Where gas appliances can be fixed on an external wall it is possible to use models with a sealed combustion chamber, which draws air for combustion from outside and discharges the products of combustion through the same hole in the wall. The vertical flue is totally eliminated. Steam is likely to be produced by any gas flue in cold weather and balanced flues should not be positioned under windows. In multistorey buildings it is possible to use a single duct running from top to bottom to deliver air for combustion to gas appliances and also take the products of combustion. This arrangement is known as an SE duct and is shown in Fig. 16.3. Sizes of SE ducts are governed by the height of the building and the number and type of the appliances. The gas

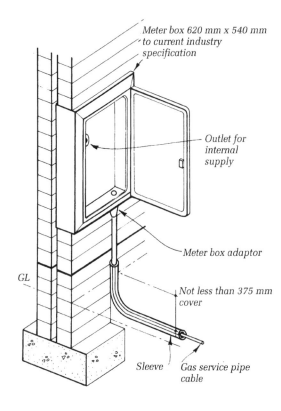

Meter box 620 mm x 540 mm
to current industry
specification

Outlet for
internal
supply

Meter box adaptor

GL

Not less than 375 mm
cover

Sleeve   Gas service pipe
cable

Figure 16.2   Typical gas service entry with externally accessible meter cupboard.

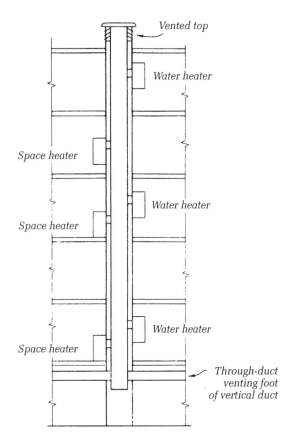

Vented top

Water heater

Space heater

Water heater

Space heater

Water heater

Space heater

Through-duct
venting foot
of vertical duct

Figure 16.3   SE duct.

company should be consulted. Although the SE duct offers substantial simplification of the requirements, it governs the placing of gas appliances and must be considered during the early planning stages of building design. BS 3528 : 1977 provides a detailed specification of requirements for gas installations.

17 Mechanical conveyors

Mechanical systems for the movement of people and goods are now essential equipment in many types of building. Their spatial and functional requirements and the need to make the best and most economical use of the installation often exercise considerable influence on construction and planning and they may, as in the case of lifts for multistorey buildings, make new building forms possible. In common with many other service installations, mechanical conveyor installations must therefore be taken into account from the very first stages of design. A bewildering range of devices exists, particularly for the conveyance of goods, but the underlying principles of operation are relatively few and can be found applied both to the movement of people and goods. Table 17.1 shows the most significant forms of conveyors and typical applications.

17.1 Mechanical movement of people

Four types of installation can be used in buildings. Lifts, escalators, paternosters and travolators. Lifts are widely employed to make vertical circulation quicker and easier and they make possible buildings rising above the four or five floors which would be a maximum for reasonable access on foot. Escalators occupy considerable floor space and, being inclined, they raise people through a limited height, but they will deal with large numbers (a 900 mm wide escalator is capable of handling 6000 people per hour). Paternosters are rare in this country but they are widely used in Europe. They are particularly appropriate where internal circulation is the critical feature rather than inward or outward flow. Travolators assist in the movement of large numbers of people horizontally or up very limited inclines.

Lifts

The essential features of a lift installation are a car to carry passengers supported on a cable running over a pulley (traction sheave) and balanced by a counterweight (Fig. 17.1). Both the car and counterweight rise and fall between steel guides in a shaft formed in the building, known as the lift well. The traction sheave is housed, together with the motor and control gear, in a motor room over the lift well. The well itself is extended some distance below and

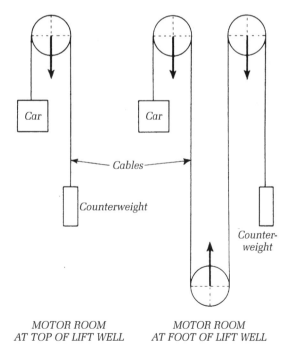

MOTOR ROOM
AT TOP OF LIFT WELL

MOTOR ROOM
AT FOOT OF LIFT WELL

Figure 17.1 Operation of a lift.

Table 17.1   Types of conveyor

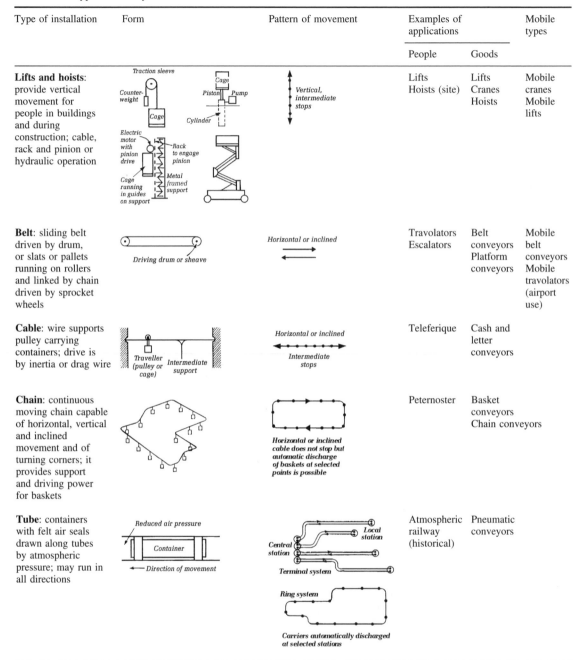

Type of installation	Form	Pattern of movement	Examples of applications		Mobile types
			People	Goods	
Lifts and hoists: provide vertical movement for people in buildings and during construction; cable, rack and pinion or hydraulic operation	*Traction sleeve* *Counter-weight* *Cage* *Electric motor with pinion drive* *Cage* *Piston* *Pump* *Cylinder* *Rack to engage pinion* *Metal framed support* *Cage running in guides on support*	*Vertical, intermediate stops*	Lifts Hoists (site)	Lifts Cranes Hoists	Mobile cranes Mobile lifts
Belt: sliding belt driven by drum, or slats or pallets running on rollers and linked by chain driven by sprocket wheels	*Driving drum or sheave*	*Horizontal or inclined*	Travolators Escalators	Belt conveyors Platform conveyors	Mobile belt conveyors Mobile travolators (airport use)
Cable: wire supports pulley carrying containers; drive is by inertia or drag wire	*Traveller (pulley or cage)* *Intermediate support*	*Horizontal or inclined* *Intermediate stops*	Teleferique	Cash and letter conveyors	
Chain: continuous moving chain capable of horizontal, vertical and inclined movement and of turning corners; it provides support and driving power for baskets		*Horizontal or inclined cable does not stop but automatic discharge of baskets at selected points is possible*	Peternoster	Basket conveyors Chain conveyors	
Tube: containers with felt air seals drawn along tubes by atmospheric pressure; may run in all directions	*Reduced air pressure* *Container* *← Direction of movement*	*Local station* *Central station* *Terminal system* *Ring system* *Carriers automatically discharged at selected stations*	Atmospheric railway (historical)	Pneumatic conveyors	

above the furthest extent of the car movement to allow for inaccuracies of control. These spaces are known as the pit and overrun respectively. The top of the lift shaft is vented to the fresh air. This ensures that in case of fire the lift well acts as a flue and does not encourage the spread of smoke and fire from one floor to another. Figure 17.2 shows these features.

Lift speeds vary from 0.5 m/s for small, slow installations to 6–7 m/s for very high buildings. This higher speed is governed by the capacity of the human ear to adapt to changing atmospheric pressures. It is conventional in this country to use speeds of 1.2–1.5 m/s for buildings of six to eight floors and 2.0–2.5 m/s over this height, except for very high buildings where greater speeds may be considered. For low speed lifts an AC motor driving the traction sheave via a worm gear provides the motive power. In faster installations better control, quieter operation, smoother and more rapid acceleration and other technical advantages can be achieved with a DC motor directly coupled to the sheave and operating on a variable-voltage system.

The DC current is usually generated from the mains supply by individual generators for each lift motor. The size for the motor room is therefore increased. Since lift speeds are higher, pit and overrun requirements are also greater.

Control of lifts

Control of lift operation is important. Many arrangements are in use but there are five main control possibilities.

Operator This system is rarely employed except in special circumstances, such as prestige or department stores, since other systems can give equal or better service at lower cost. It operates with call buttons at each floor registering on an indicator in the lift. Most other systems can be arranged for dual use so that operators can be employed at times if desired.

Automatic This system operates by having call buttons on each floor and a set of destination buttons in the lift. The first button to be pushed sets the lift in motion and no further calls are accepted until the lift is again at rest. Calls are not stored for further action. This system is only suitable for small, lightly trafficked lifts.

Collective Destination buttons for each floor are provided in the lift and two call buttons on each floor, one for upward and one for downward journeys. Calls

are registered by the control gear and automatically dealt with in sequence. On a downward journey the lift will stop at all floors where downward calls are waiting or where passengers wish to stop. It will then make an upward journey, stopping for upward calls. If further down calls are made it will then return to collect and deliver these passengers. The push buttons, both on landings and in the lift, usually give a visual indication showing that the call and destination have been stored by the control gear for operation. It is possible to have simplified versions of this type of control such as collective control on downward journeys only.

Group collective Up to four lifts may be grouped together in a collective control system so that the closest lift will respond or the first lift of the group moving in the desired direction will stop in response to a call.

Programmed operation The performance of a collective, or group collective, control system may be improved by including in the control gear facilities for functional patterns of lift operation to suit particular circumstances, sometimes described as lift logic. It is possible to have lifts that 'home' to the ground floor during arrival periods and to positions up the building, closer to possible calls, during departure periods. Other lifts may home to ground or to intermediate floors during working periods. If the lift is full it is possible to arrange for calls to be ignored until the load has been reduced, thereby saving the time of wasted stops when no more passengers can be taken. When traffic is light some lifts can be taken out of operation; when it is heavy the journeys of lifts in a group can be timed not to coincide, thereby reducing waiting time.

Motor rooms

Motor rooms are required for all except some special types of lift installation. Normally they are placed over the top of the lift well and require a greater plan area than the lift well. It is possible to place motor rooms at the foot of the well or at an intermediate level. The advantages of motor rooms not at the top of the building are easier sound insulation and the avoidance, for the architect, of a perhaps awkward volume at the top of the building. The disadvantages are the increased loads on the structure; the increased length of ropes required and a constant increase in maintenance costs; there is also the difficulty of achieving adequate ventilation to dissipate the heat generated by the electrical gear. Figure 17.1 shows a

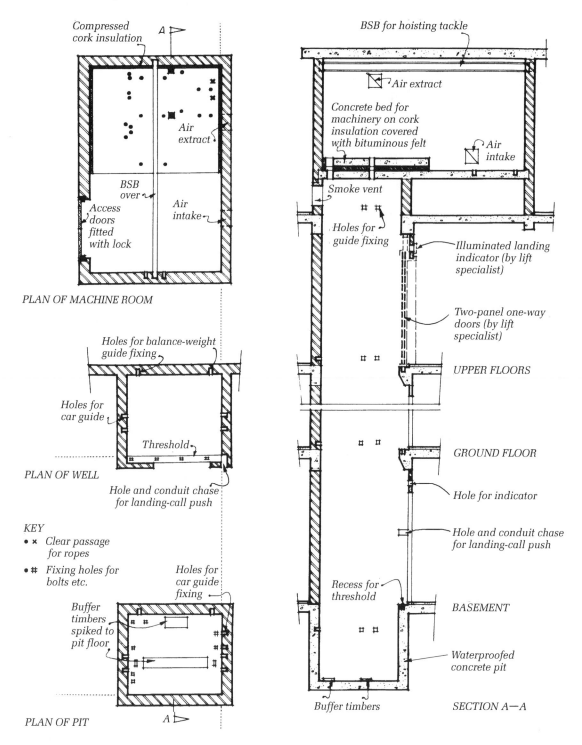

Figure 17.2 Typical construction details of a lift well and motor room.

comparison of the loads imposed on the building and simple roping arrangements for motor rooms at the head and at the foot of a lift well. More complex patterns are used in special circumstances and the ropes for high speed lifts may be wrapped tight round the sheave.

Final size and layout of motor rooms depend on many factors. Lift manufacturers publish tables of typical sizes which should be consulted as early as possible for layout and details of sound precautions. Good ventilation is essential in motor rooms to dissipate the heat generated. Rates of the order of 30 air changes per hour may be needed. Some heating may be needed to prevent undue humidity affecting the electrical gear when it is not in operation. A floor strong enough to sustain the loads of machinery, car and counterweight, and a beam to support hoisting tackle for lifting the machines, both must be provided at an appropriate position in the roof. It is usual for the lift installers to use this beam to raise and install their equipment through the lift well itself.

Machine room equipment

The following items are needed in the machine room (or lift motor room):

- **Winding machine**: comprising the motor, electromechanical brake, worm gear and traction sheave.
- **Controller**: mounted on a frame, usually against a wall and contained in a sheet steel case.
- **Floor selector**: this is operated by a steel tape attached to the car and passing over a sheave top and bottom.
- **Overspeed governor**: operates the safety gear on the car in the event of it exceeding a certain speed.

Sound insulation of the machinery is essential. Variable-voltage motors reduce noise and the gearless traction machine is quieter still. It is best to set the bed plate of the motor on a heavy block of concrete set on a bed of cork. Even so, noise from the lift is difficult to eliminate; the clicking of the contacts is a sharp penetrating noise, so it is best to plan non-living rooms near the lift well. In some cases the lift well forms an independent tower quite independently of the frame or floors of the structure, with separations of strips of cork.

Lift well and pit equipment

The lift car is supported on sets of steel ropes, usually four, clamped to the car sling and to the balance weight. Two guides at the point of balance of the car on each side, are fixed to the lift-well wall or to special steelwork. These fixings are approximately at floor levels. Two other guides are required for the balance weight. These guides are T-section steel accurately machined and finished to 0.05 mm limits.

The lift shaft must extend down below the lowest landing into what is known as the *lift pit*. In this pit are fixed the buffers, spring-type for lower speeds, oil-loaded for high speed lifts. The depth of this pit varies from 1 m for 0.5 m/s lifts to 1.6 m for 1.5 m/s lifts; with gearless machines this will be 2.5 m for 2.0 m/s and 2.8 m for 2.5 m/s. Similarly at the top, head clearance is necessary for overrun. This is a minimum of 4 m up to 4.6 m for 1.5 m/s lifts, but 5.5 m and 5.8 m for gearless lifts.

Plan size of the well depends on the size and shape of car, the type of doors or gates used and can be obtained from manufacturers' tables. Figure 17.3 and Table 17.2 show some typical sizes related to one particular arrangement, having the balance weight at the side and a single-panel sliding door.*

Where two or more lifts are fitted in one well an additional allowance of 100 mm is made between them for the guide supports.

It is very desirable that the inside of the lift well be smooth and free from ledges, recesses, etc. The floor trimming should be very accurate. In a framed building it may be appropriate to surround the lift with light partitioning built off the floor. This should be set true to the lift-well dimensions, so there is no ledge at floor level.

At the front or door side a ledge will, however, be necessary for the landing doors or gates. There is no specific height for this projection, which is often a beam casing, but the underside should be splayed as shown.

The car sizes in Fig. 17.3 are internal sizes, not overall platform sizes, and may have to be varied slightly, depending on the car finish.

Lift doors and gates

Sliding doors of sheet steel are the most satisfactory. They can be made very safe, are easily suited to the closing mechanism and can be made to be fire resistant for long enough to satisfy most authorities.

Generally the doors are 30 mm thick flush, of sheet steel on a light frame, all spot welded, but a finish of special metal, life anodized aluminum can be supplied, or they can be made of wood.

---

* BS 2653, Part 3, gives outline dimensions for 'light traffic passenger lifts' (up to 8 persons, speeds up to 0.75 m/s), 'general purpose passenger lifts' (up to 20 persons, speeds up to 1.5 m/s) and 'general duty goods lifts' (up to 2000 kg load).

Table 17.2 Typical sizes for lift car, well and machine room (see Fig. 17.3)

Number of passengers	Load (kg)	Dimensions (m)										
		A	B	C	D	E[a]			F			CO
						x	y	z	x	y	z	
4	270	1	0.8	1.7	1.2	2.4	–	–	2.1	2.7	–	0.7
5	340	1.1	0.9	1.7	1.3	2.4	–	–	2.7	3	–	0.7
6	400	1.2	0.9	1.9	1.3	2.4	2.7	–	2.7	3	–	0.8
8	540	1.5	1	1.9	1.5	2.7	2.7	3	3	3	4	0.8

[a] E is the width of the machine room.
Dimension x for speed of 0.5 m/s
Dimension y for speed of 0.75–1.0 m/s
Dimension z for speed of 1 m/s
If doors are manually operated, CO is 50 mm less.

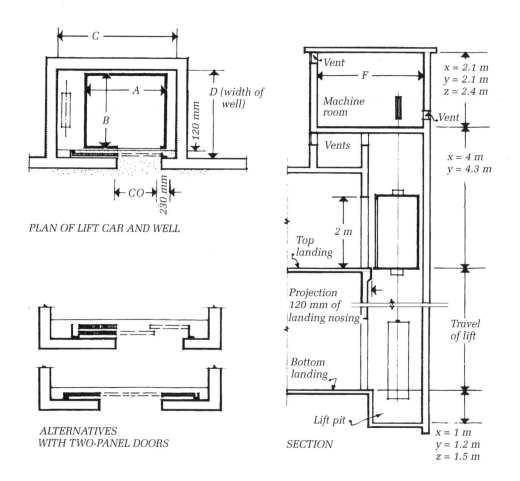

PLAN OF LIFT CAR AND WELL

ALTERNATIVES
WITH TWO-PANEL DOORS

SECTION

Figure 17.3 Typical sizes for a lift car, lift well and machine room (see Table 17.2): geared lift, balance weight at side, single-panel power-operated sliding door.

The door should have a rigid frame of steel angles, so it can be fixed as one complete element to ensure accurate positioning of the tracks.

The hangers for the doors are usually fitted with plastic rollers, of large size, on ball-bearings for reasonably quiet operation.

All landing doors and gates are fitted with electromechanical safety locking gear, with a special emergency lock release operated by a secret key.

Single-panel sliding doors are suitable for small offices, hotels and flats. They are particularly suitable for council flats, as they are simple and safe. Two-panel one-way doors give a wider entrance in proportion to the total width of well. The leading door panel moves at twice the speed of the second panel. Two-panel centre opening is the best arrangement for busy lifts and in openings over 1.5 m four-panel centre opening may be used. Single hinged doors of wood are quite suitable for lifts for residences, small hotels, etc. They should be spring closing and have a check mechanism.

Collapsible steel gates are suitable for low speed attendant-operated lifts and are commonly specified for goods lifts, but they are noisy and provide no smoke check, which the fire authorities usually require. They should not be power operated.

Flush-leaf sliding gates are virtually the same type of gate with flat 1.6 mm metal sheaths fitted by one edge to the outside of the steel pickets so that the sheaths slide over one another and pack away at one side. To get a clear opening of 1 m, this type of gate will need a landing opening of 1.35 m. These gates normally satisfy the fire authorities and look quite neat.

Shutter gates operate rather similarly, but the 150 mm wide 1.6 mm steel plates are hinged together and secured to the pickets alternately so as to open with a concertina effect. They can be made to conform to the 2 h fire test of the Fire Offices Committee. To get a 1 m clear opening the landing opening must be 1.25 m.

The power operation for sliding doors is usually by a small electric motor operating an arm through a worm gear and spur wheel. This is arranged so the door moves fast in the middle of the travel, but slows down to the open and closed position. A sensitive strip can be fitted to the leading edge of a car door so that if an obstruction is met both doors will stop and reopen. Power operation is possible for shutter or flush-leaf sliding gates.

The lift car

This has a basic framework of steel angles and channels, called the sling, incorporating the fixing for the guide shoes and the *safety gear*. There are two kinds of safety gear, the dead-grip type giving instantaneous action for the low and medium speed lifts, and the wedge-clamp type giving gradual action for high speed lifts. When the lifting ropes break or stretch unduly, or when the governor operates, if the lifts exceed a fixed speed, the safety gear is brought into action by an independent steel rope.

In the dead-grip type two pairs of cams with serrated surfaces grip the guides and stop the lift in a very short distance. The wedge-clamp type, controlled by the governor, is operated by wedges forced against the clamps, which steadily close upon the guides and bring the car gradually to rest within a prescribed distance.

The car can be made of wood panels or of sheet metal on light framing. They can have many decorative finishes, such as aluminium sheet with matt, burnished or anodized finish, metal-faced plywood, plywood or blockboard with wood veneer or plastic veneer, or linoleum or rubber sheet. Car surfaces get considerable wear and should be robust, easily cleaned and easily renovated. This particularly applies where luggage or goods are likely to be carried, as in flats, hotels, etc. Floors can be wood, lino, rubber, plastic, cork or with fitted carpet.

Vision panels glazed with wired glass or armour plate are necessary where the lift is manually operated, but large panels in high speed lifts only tend to cause alarm among the passengers, who are otherwise unaware of their speed of travel – if the lift works well. It is important that there is good ventilation in the car, either by simple grilles or by concealed louvres in the roof; these are often combined in the light fitting, which may well be set in a recess in the ceiling with air gaps on all sides. It is important in all passenger lifts, except those in private houses, that the lift is on all the time and operated by a key only.

Planning for lift installations

Separate vertical circulation towers may be justified in very large buildings and separate sets of lifts are essential in staircase access flats. But in most buildings the most critical planning consideration is the grouping of lifts together. This ensures that if any lift is available in the building a person entering may immediately use it. If the lifts are scattered it is possible to have both idle lifts and waiting passengers. The way in which the lifts are grouped is important. Two or three lifts may satisfactorily be placed side by side but if more than this are placed in a row it is not easy for passengers to notice when cars arrive. Even if arrivals are properly announced it may still be difficult for passengers to board before the doors close. Lifts should never be separated by corridors or stairs since this not only worsens the problems of view and access described above but introduces conflicting circulation arrangements and makes control interconnection and maintenance more difficult. However, the stairs and lifts should be close to each other and visible or adequately signposted so that people with relatively short vertical journeys can be encouraged to use the stairs. Figure 17.4 shows how grouping lifts to face each other gives better visibility and quicker access than arranging them in one bank.

Up to three lifts may be placed side by side and give satisfactory service. Four lifts in line may be acceptable but can cause some inconvenience of access particularly if the lift lobby is crowded. Grouping two rows of lifts opposite each other makes access simpler but the practical limit is four lifts on each side. If more lifts are required to deal with the population of the building it will be desirable to provide separate groups of lifts serving different vertical zones. This arrangement may be appropriate for groups of fewer than four lifts.

In very high buildings an intermediate concourse, sometimes described as a sky lobby, can provide an interchange between high speed lifts serving only the ground floor and the concourse and lifts serving individual floors at higher levels. The concourse should also provide an interchange for people wishing to travel between different lift zones. To meet the needs of buildings higher than any yet existing, lift systems have been postulated that enable lift cages to move out of the main shaft when stopping at floors.

Groups of lifts should ideally be placed at the centres of gravity of the circulation systems that they serve. This is not always convenient on the entrance floor since the lifts may then be remote from the entrance and may also constrict the planning of the entrance floor, which often has to serve a different purpose than the floors above. If placed too close to the

entrance, general circulation may be restricted and waiting rendered unpleasant by people passing. In general, therefore, a lift lobby of adequate size to accommodate the numbers likely to be waiting should be provided where it will not be affected by or affect the other movement through the entrance, but where it will be easily seen and readily reached by people coming into the entrance. This arrangement is very compatible with the satisfaction of users who generally appear to be more prepared to walk short distances on upper floors than on the entrance level. The waiting areas for lifts should be clear of other circulation, otherwise unsatisfactory working and general dissatisfaction with both circulation and waiting can result. In very tall buildings it may be economical to divide the building into two from the point of view of lifts. One lift lobby and group of lifts serving ground to mid-height and the other passing from the ground floor directly to mid-height and stopping in the normal way from mid-height to the top. In domestic buildings ease of access for deliveries and adequate manoeuvring space for perambulators must be borne in mind.

Size and number of lifts

Some basic considerations govern the minimum size and number of lifts. Where vertical circulation depends on the lift and walking upstairs is impracticable it is desirable to have two lifts to allow for the possibility of one being out of action due to defect or routine maintenance. Well-maintained lifts are very reliable; in large cities, where routine maintenance can be carried out at night and repair teams can be very rapidly available, thought might be given to relaxing this requirement where it conflicts with economy and where doing so has no critical consequences. In some circumstances minimum size is critical. In multistorey housing lifts must be large enough to take perambulators and furniture. It is not usual, however, for them to be large enough to take coffins, although some local authorities insist on adequate space to take a stretcher. Where firefighters' lifts are required they have to be of eight-person capacity, which is also the standard for perambulator lifts. Particular minimum size considerations may apply in other special cases.

The installation appropriate to a particular building depends on the traffic to be served and the speed, size and number of lifts. Speed is usually considered in terms of round-trip time, which includes the time taken for entering the car, doors closing, acceleration, running at standard speed and deceleration, doors opening and closing, passengers entering and leaving, and acceleration for the number of stops that the lift

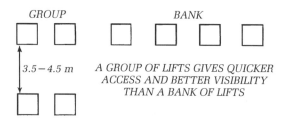

GROUP   *BANK*

3.5 − 4.5 m   *A GROUP OF LIFTS GIVES QUICKER ACCESS AND BETTER VISIBILITY THAN A BANK OF LIFTS*

Figure 17.4   Arranging lifts to give the best service.

will make. The traffic to be served in office buildings is normally assumed to be 75% of the population on the second floor and above, to be taken up from the ground floor in a period of 30 min. In flats the usual standard is 6% of the population on the second floor and above to be moved in 5 min.

If in a particular case the size of the lift is assumed and the round-trip time estimated from the manufacturers' data on lift performance and the probable number of stops (from theoretical lift data) it is easy to determine the number of lifts that will be required.

$$\frac{30 \text{ min} \times \text{lift capacity}}{\text{round-trip time}} = \frac{\text{number of people}}{\text{discharged by one lift}}$$

$$\frac{\text{total number of people}}{\text{number discharged by one lift}} = \frac{\text{number of lifts}}{\text{required}}$$

If very large lifts have been chosen the round-trip time will be long and arrivals may have to wait some time before they can enter. The average waiting interval can be assessed by dividing the round-trip time by the number of lifts. In office buildings this value should not be greater than 30 s, whereas in flats 90 s is appropriate. If the waiting interval is too long a greater number of smaller lifts should be considered.

As an example, consider an office building of 10 floors of lettable area 780 m² each. On the basis of 7 m² per person in offices, this gives 111 persons on each floor, or 888 persons on the eight floors above the first. The transportable population within the half-hour is 75% of this, or 666 persons. If four lifts are used it means that each lift must be capable of taking 166 persons in 30 min. Table 17.3, from a lift manufacturer's handbook, indicates the available choice.

The small advantage of the gearless lift in this case would probably not warrant the considerable extra cost unless an especially smooth service were desired. If three lifts were considered they would all have to be 2.5 m/s gearless, and each would have to take 17 persons. The waiting time works out at 49 s, which is too long.

Table 17.3 Possible four-lift installations

Motor speed (m/s)	Capacity (persons)	Round-trip time (s)	Waiting time (s)
1.25, variable-voltage control	13	133	33
1.5, variable-voltage control	12	124	31
2, gearless	11	109	27

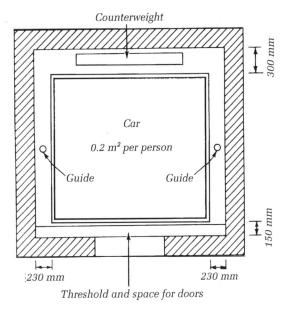

Figure 17.5 Approximate lift car and well dimensions for preliminary planning.

Figure 17.5 shows some dimensions which enable an approximate idea of lift car and well sizes to be obtained. Lift cars should have a square plan, or with the door in the longer side if they are rectangular. Deep narrow lift cars make exits difficult and time-consuming.

Goods lifts and other devices

The type and size of *goods lift* is usually determined by the type of goods transported, their bulk and weight, or more probably the dimensions and loaded weight of the trucks used for their transport. In the absence of any specific figure, 340 kg/m² car area is allowed. Speeds of goods lifts are not usually an important factor, loading and unloading taking up much more time than the travel time; 0.25 and 0.5 m/s are usual.

Service lifts, such as food lifts for restaurants, were mentioned in the section on controls. They can be specially designed to suit any particular set of circumstances, but the following particulars would suit a lift to take 50 kg (balance weight at the back):

Car size	560 mm square × 690 mm high
Well size	780 mm wide × 780 mm deep
Hatch size	560 mm wide × 690 mm high
Headroom	2.7 m

The motor and control gear can usually be accommodated in the same plan area as the well and within the headroom mentioned, provided the doors are set in each wall so there is access to any part. Doors to the hatches can be hinged or, better, sideways parting, in which case more space would be needed in the well than given above, which only allows for vertical sliding or biparting vertical sliding (called rise and fall shutters). Rise and fall shutters give the best service. Lifts for special purposes include *hospital bed lifts*; BS 2655 *Specifications for lifts, escalators and passenger conveyors* gives much information about them. Accurate levelling is important, so low speed is favoured unless it is likely to be very busy. A minimum depth of 2.3 m is advisable to take a wheeled bed; the minimum width is 1.2 m, but this should be increased to 1.6 m wherever possible. Height should be 2.2 m. If the doors are power operated, some means must be provided of keeping them open when required. Open and close buttons in the car are probably the best.

A special lift for *firefighters* may be required by the fire authorities, as in London in buildings over 24 m high. This should have direct access from the street and have a special switch at that entrance which will cancel all calls and return the car to that level, after which the car pushes only are operative. The lift should take 760 kg, or eight firefighters, and have a speed of 0.7 m/s. The car should be made of steel and should be 2.5 m high, with shelf 1.8 m up. Its electricity supply needs to be separate from the main supply to the building.

Hydraulic lifts were used in the latter half of the nineteenth century and several towns had systems of hydraulic power mains that served lifts and other hydraulically operated equipment. They are operated by hydraulic jacks. The most usual arrangement is for the cylinder of the jack to be contained in a borehole and the piston to support and move the lift car. In the early decades of the twentieth century electric power began to be introduced; its speed and convenience, together with the ability to serve higher buildings than could be served by hydraulic systems, led to electrical dominance of the lift field. Although hydraulic mains have disappeared, efficient small pumps are available and there are several applications for which hydraulic lifts are very suitable. Appropriate applications are where large and heavy loads must be carried and where the building cannot sustain the loads that would be imposed by conventional lifts. The main limitations are that the maximum travel is restricted to about 20 m, the speed of operation is slow and the intensity of traffic that can be carried is limited. The simplest arrangement is for the jack to be underneath the centre of the car. This requires a borehole below the bottom of the lift well (Fig. 17.6). Arrangements of jacks at the

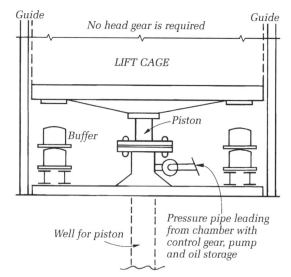

Figure 17.6   Installation at the foot of a hydraulic lift.

side of the lift well can be made, but space requirements are increased and load-carrying capacity is reduced. A small, well-ventilated pump room is required.

Basement hoists for goods only, and to take 1000 kg, consist of a platform with dwarf sides and back running between two channel guides. They are normally operated by lifting ropes, which are attached to the top of one guide and pass under the platform and up to the other guide, then down to a winding drum in a small machine enclosure at the side. A platform 1.4 m² needs a well 1.7 m × 1.5 m and a pit 1 m deep. The machine enclosure at the side may be 2.4 m × 1.5 m. The speed is very slow.

Hand-power lifts are useful for small serveries or food lifts in small restaurants, or even houses, as goods lifts where their use is infrequent, or for basement hoists or ash hoists. They are covered by Code of Practice 407.301.

The lighter types of hand-power lift are made of timber, including guides, the heavier, i.e. over 1000 kg, are usually framed in steel angle with wood panels. Hand-power lifts for over 100 kg are not recommended for frequent use. Dimensions are not much different from those for the electric service lift given below.

The operation of the lift is by hauling rope which passes outside the enclosure. The self-sustaining gear is automatic in action, locking in any position as soon as the hauling rope is not pulled.

The ash hoist is made of steel and is not normally balanced. It is underdriven like the electric basement

Table 17.4 Discharge capacities of escalators

Width of step (m)	Nominal width (m)	Overall width (m)	Discharge capacity (persons per hour)
0.6	0.8	1.3	5 000
0.8	1.0	1.5	7 000
1.0	1.2	1.7	8 000

hoist, but by a chain which is wound through a self-sustaining chain winch.

Escalators

Escalators can move very large numbers of passengers. To achieve a similar service a very large number of lifts, occupying more floor space, would be needed. They can be reversed in their direction of operation, thereby lending themselves to 'tidal flow' situations. Typical applications where large numbers of people have to be moved over limited heights and where tidal flow is marked are department stores, underground railway stations and exhibitions. Where numbers of passengers justify it two or even three escalators may be used, offering various possibilities of inward, outward and two-way flow. The horizontal speed of escalators is normally 0.50–0.65 m/s; faster

travel is possible but apparently produces little increase in the capacity of the escalator. Typical widths and discharge capacities are shown in Table 17.4.

Figure 17.7 shows other typical dimensions for small escalators. Steps 0.6 m wide will permit one laden adult or an adult and child to stand on the step, whereas steps 1.0 m wide permit passing even when carrying parcels. Figure 17.8 shows a system with continuous upward and downward flow.

It is possible to use a series of escalators to serve several floors. A corkscrew or scissors arrangement gives almost continuous travel for passengers and a double corkscrew may be arranged to give a choice of both up and down directions from each floor.

The large floor openings required by escalators pose a problem of fire spread and it may be necessary to provide a separate hall on each floor separated from the rest of building by walls with openings protected by automatic fire doors. Automatic shutters closing the openings and water curtains may be considered in some cases.

The need to plan for escalators is clear and also the need to carry the loads that they may impose on the structure. It is also important to remember that escalators are normally delivered to the buildings as

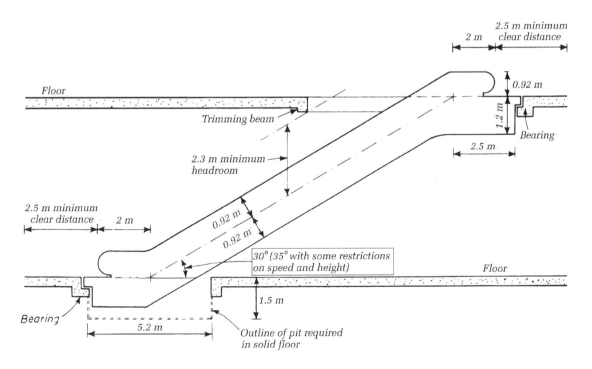

Figure 17.7 Typical dimensions for a small escalator.

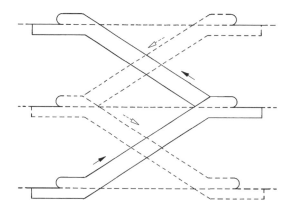

Figure 17.8   Escalator system with continuous upward and downward flow.

a unit ready for installation, and adequate access is important.

Paternosters

A paternoster consists of several open-fronted cars supported on endless chains running over sprocket wheels at head and foot. The cars move continuously, providing both up and down movement. Support is only at the top of the car, so the cars remain vertical and the same way up at all times. No doors are fitted to the car or opening, although aprons are fitted above and below the car to prevent anyone entering the wrong area. Safety devices to stop the motor are fitted and leading edges are hinged to prevent injury if improper use is made. The safety record of paternosters is good. The cars are shallow and capable of taking two people side by side. Passengers have to judge the motion of the car, usually about 0.35 m/s, and step in as it passes. The maximum discharge capacity of a single paternoster is stated by manufacturers as being about 600 persons per hour, but this type of apparatus is normally selected for the type of service it offers rather than a specific capacity.

The main advantage of the paternoster is that passengers can arrive at any floor, intending to travel up or down, and immediately begin the journey. Normal lift installations would inevitably cause some waiting. Paternosters are therefore particularly suited to serve the internal traffic of the building. They do not deal so effectively with arrival and departure situations nor with members of the general public who, in this country, are not familiar with them. They are therefore appropriate in buildings of medium rise, with a substantial amount of interfloor circulation. Another

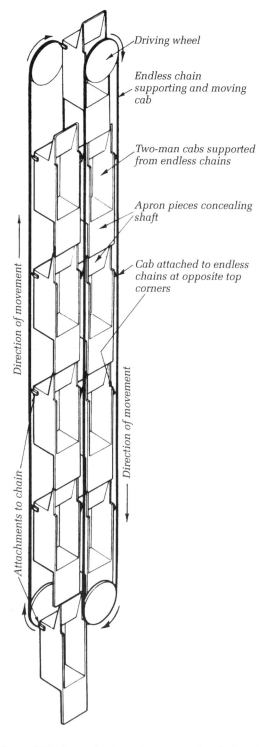

Figure 17.9   Isometric view of paternoster installation: the front and back driving wheels are offset (Fig. 17.10).

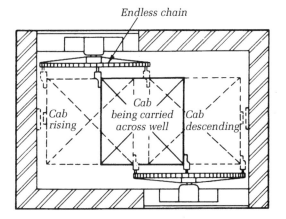

Endless chain

Cab rising

Cab being carried across well

Cab descending

Figure 17.10 Plan view of paternoster installation: the cab is supported from the two endless chains using attachments at opposite top corners.

advantage of the paternoster is its reliability. The absence of control gear and the continuous motion as compared with the start, accelerate, travel, decelerate, stop movement of lifts renders them remarkably free from breakdown. Figure 17.9 shows a view of a paternoster installation free of the surrounding building and Fig. 17.10 gives a plan of the well.

It will usually be desirable to provide a lift in association with paternosters for the use of handicapped people, tea trolleys and goods.

17.2 Tube conveyors

Containers fitted with air seals may be moved along tubes by exhausting air from ahead of the container. At their simplest a pair of tubes link two stations. Air is exhausted from the end of one tube and containers can therefore be despatched in one direction. Excess airflow is prevented by flaps at entry and exit points, a bypass connection between the ends of the tubes remote from the point of exhaust will produce an airflow away from the exhaust end; carriers can then be sent in this direction. This arrangement is very simple and robust; no sophisticated controls are required. No choice of destination is offered unless further pairs of tubes are installed. The system is therefore particularly suitable for cases where all communication is from one central point to several outlying stations and where the outlying stations do not communicate with one another. This is the case in department stores with a central cashier's office and this type of installation has been used for many years. The convenience of the system for conveying papers and a variety of small articles has led to its use in other circumstances where outlying stations do communicate. At first this necessitated a central exchange where containers were manually transferred from one tube to another. This arrangement required a great deal of tube and substantial space was required for this near the main station. Improved efficiency and economy can be achieved by a ring-main type of installation with automatic selection of destination and several

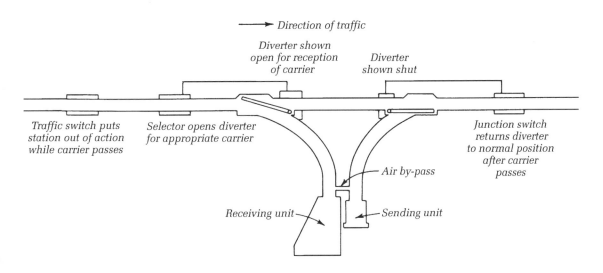

Direction of traffic

Diverter shown open for reception of carrier

Diverter shown shut

Traffic switch puts station out of action while carrier passes

Selector opens diverter for appropriate carrier

Junction switch returns diverter to normal position after carrier passes

Air by-pass

Receiving unit

Sending unit

Figure 17.11 Receiving and sending station on a pneumatic tube ring main.

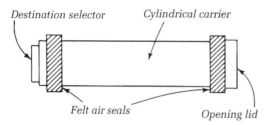

Figure 17.12   Typical carrier unit for a pneumatic tube ring main.

systems have been devised to give this sort of service. Table 17.3 shows typical layout patterns of tube systems. Figure 17.11 shows a station on a ring main, demonstrating how containers may be received and sent from a point on a ring main. Figure 17.12 shows a typical carrier unit.

Both round and rectangular tubes are used for tube installations, sizes varying from just over 50 mm to 150 mm diameter for round tubes and up to 380 mm × 130 mm for rectangular tubes. As well as cash, bulky documents, drawings, films, diet sheets, drugs, tools and small parts may be carried. Movement is possible in all directions, including the vertical direction, and there is no close limitation on length of run. Provision should be made in planning so that easy bends can be accommodated. Electrical and acoustic devices are used for destination selection and a large number of stations (80–120 depending on the system employed) can be used on one ring main. Steel tube is most commonly used but PVC is now proving successful. Some noise at stations is inevitable as carriers are sent or received and a plant room for the exhauster unit must be allowed for in planning and insulated if necessary. In central installations it should be near the central station but in ring-main systems discussion with the manufacturers will be necessary to establish the right position. In all cases exhaust to the fresh air will be needed.

18 Firefighting equipment

18.1 General considerations

Precautions for fire safety in buildings involve not only alarms and firefighting equipment but also access, building form, materials of construction, internal planning and sometimes mechanical ventilation. Many of these aspects are interdependent and fire safety should be part of an integrated design process from the outset. The risks associated with fire and the complex technical considerations involved in the choice of suitable precautions mean that, in substantial buildings, many parties will be concerned in the conception and development of the design. In addition to the building owner, architect, structural and services engineers, the local authority and fire service, the Health and Safety Executive (HSE), the insurers of the building, and manufacturers and installers of equipment may have essential contributions to make. Even in the case of small buildings local fire services are normally very pleased to be asked for advice about fire precautions and it is sensible to take advantage of this at the early stages of design. All new buildings must comply with the Building Regulations and may also be subject to health and safety legislation.

Approved documents made from time to time in relation to Part B of the Building Regulations 1991 give increasingly more detailed standards and guidance. The Building Regulations are concerned with the safety of the people in the building, not with the preservation of the building or its contents. Where these are important considerations precautions beyond those in the Building Regulations may be required. The philosophy which underlies the regulations is that the possibility of fire cannot be eliminated but that occupants can be alarmed so that small fires can be extinguished or, if this cannot be achieved, occupants can use safe escape routes and the fire service can have access and facilities to control established fires. Building services play an important part in all these aspects.

Fire precautions are simple in one- and two-storey dwellings. Each room must open directly into a hall or landing giving access to an external door, or escape must be possible through a window. The only building services requirement is the provision of alarms. The minimum requirement for dwellings is that there should be at least one self-contained, mains-operated smoke alarm on each floor. There should be an alarm mounted on the ceiling of the corridor forming the escape route within 7 m of kitchen and living-room doors and within 3 m of bedroom doors. Where there is more than one alarm, interconnection should be provided so that all alarms will be activated if any one detects smoke. Progressively more elaborate precautions are required for larger and more heavily populated buildings. Protected escape routes to provide safe exits from the building may be needed and may require pressurization as well as escape lighting. Fire detectors will be connected to central alarm systems so that all the occupants can be warned when a fire has started. Occupants may be exercised so they can leave the building quickly and some may be trained to use firefighting equipment. Extinguishers will be provided and, where appropriate, automatic sprinklers, permanent piped installations of hose reels and fire mains may be installed. The escape routes will be signposted and have emergency lighting; they will be protected against fire and especially against the entry of smoke. Special mechanical ventilation may be provided to pressurize escape routes and ensure that smoke cannot penetrate. In very large buildings 'firefighting shafts' may be provided with protection from fire and smoke; they contain stairs, a firefighters' lift, normally part of the circulation system but able to be taken over and controlled by the fire services, and fire mains and hydrants for fighting the fire. The building may be divided into separate compartments that can be sealed to prevent fire spread. Building services often penetrate walls and floors which would otherwise be able to resist the spread of fire. In these cases special precautions will be needed as part of the services installation, in order to prevent fire penetration. The strategy and layout of the services installations as a whole may be affected by these planning constraints.

18.2 Types of firefighting equipment

The following notes give a general description of the principles and the types of service equipment which may be required for firefighting. They include fire detectors, fire alarms, hand fire extinguishers, hose reels, sprinkler systems, emergency lighting and signs, fire hydrants, firefighters' lifts and mechanical ventilation to pressurized escape routes. Emergency power supplies and special water storage for firefighting may be needed. The Building Regulations and relevant British Standards must be consulted for full details. In all but the simplest buildings special skill will be required to balance and implement the planning, structural and service aspects of preventing, fighting and providing escape from fire.

Fire detectors

It is important to have automatic warning of fire. Two main systems are employed. They involve the detection of smoke or heat. Smoke is the preferred and most widely used method. Smoke gives an early indication of fire and smoke detectors are very sensitive. In some cases the normal activities in the area will produce dust or smoke without any risk of fire. This precludes the use of smoke detectors. Heat-detecting devices activated by air temperature or radiation levels can be employed in these cases.

The location of detectors depends on the building layout and the risks involved. In general detectors should be located at ceiling level near points of fire risk and in the approaches to escape routes. Care must be taken to ensure that detectors are clear of obstructions and local heat sources; the ceiling configuration must not shield the detector from smoke or hot gases.

Smoke detectors

Smoke detectors either use a small radioactive source that emits ions attracted to charged electrodes, or they use a beam of light and a photocell. Smoke will interrupt the flow of ions or the passage of light and thereby activate the detector. The light-operated detector can be self-contained or it can operate by projecting a beam of light across the area being protected to impinge upon the photoelectric cell. Small ionization detectors can be self-contained with their own battery and alarm. They are widely used in existing domestic premises. Mains supply with battery backup is preferred for new dwellings and larger buildings.

Heat detectors

Air temperature or radiation levels may be used for heat-based fire detection. Possible sensors include fusible alloys, bimetallic strips, infrared or ultraviolet sensors and frangible bulbs. All these sensors can close an electrical circuit to provide alarm. In the case of fusible alloys and frangible bulbs they can also set a mechanical process in action without the need for electrical circuitry.

Mechanical detectors are based on the melting of solder breaking the coupling between two pieces of brass (fusible link), or the breaking of a glass bulb by expansion of the liquid contained in it (frangible bulb). The fusible link can be used in a wide variety of applications, often to activate firefighting measures as well as sound the alarm. The fusible link can be used to couple wires strained across items of high fire risk. Boiler rooms and cinema projection rooms are examples where the fusible links can be positioned above the projectors or oil burners. The parting of a fusible link can both sound the alarm and cut off the oil or electrical supply. Automatic firedoors, fly-tower rooflights, firedampers in ventilation trunkings and many other devices may be operated in the same way. With suitable holders frangible bulbs can achieve the same results. A more common use of the bulb is in sprinkler heads, where the head is sealed by the bulb until it breaks and allows the sprinkler to operate.

Heat detectors may also be used for purposes other than fire. Machinery which might be damaged at high temperature can be protected, and line detectors can be run beside electric cables to ensure that working temperatures are not exceeded.

Fire alarms

An alarm system capable of alerting all the occupants of the building is required. In dwellings and very small buildings self-contained smoke detectors incorporating their own alarm can be adequate. In larger buildings it will be necessary for any single fire detector to activate audible alarms throughout the building. Manual alarm pushes as well as automatic detectors should be incorporated into the system. Special circuitry with an independent power supply will be required.

In very large buildings sounding the alarm system may operate on a phased basis to avoid congestion in the escape route. Those nearest the fire will normally be alerted first. In buildings which are subdivided into separate fire compartments it may be appropriate to regard each compartment separately from the point of view of alarm and escape.

18.3 Media for fire extinguishment

Fires are classified into four types which present particular types of hazards and requirements for extinguishment:

- Class A: solid materials usually of organic origin (e.g. wood, textiles and paper)
- Class B: inflammable liquids
- Class C: inflammable gases
- Class D: inflammable metals

Fires involving electrical equipment represent a special hazard. Types C and D involve special risks outside the usual scope of provision in buildings. Fires may be extinguished by removing combustible material, cooling, excluding oxygen or inhibiting the process of combustion. Fire extinguishers mainly act by cooling or excluding oxygen. Here are the principal media used to extinguish fires in buildings.

Water

Water is cheap, readily available and very effective as a cooling agent. It is particularly appropriate to class A fires. If oil is present the use of water might tend to spread the fire rapidly, but in spray form it can be useful against some types of class B fires. Water in spray form can also be used to prevent the spread of fire through openings or from building to building. There are limitations to the use of water. Its use can cause damage to buildings, equipment, documents and furnishings. It cannot be used on live electrical equipment. Unprotected pipework or containers may become frozen.

Foams

Foams consist of water together with a foaming agent. They act by excluding oxygen. Different foaming agents can produce different densities of foam and different chemical properties. Low expansion (up to 20 times the volume of water) and medium expansion foams (20–200 times the volume of water) are appropriate for class B, flammable liquid fires where they will float on the surface and exclude oxygen and also provide some cooling. They can also be used for class A fires and mixed class A and class B. Medium expansion foams are appropriate where a depth of several metres of foam is required. High expansion foams (200–1000 times the volume of water) require a special generator and are used where a great depth of foam is required or where a space in a building is to be 'flooded' with foam. Foams are not toxic but care must be taken to ensure compatibility if powder extinguishers are used in conjunction with foam. Care must be taken that foam does not cover objects which might trip or injure occupants. In large installations where volumes are flooded, occupants should be evacuated before discharge begins. Foams conduct electricity but their conductivity decreases in proportion to their density.

Gaseous media

Halogens, nitrogen and carbon dioxide

These gases are stored at high pressure, usually in liquid form; on discharge they vaporize and will extinguish fires when they reach an appropriate concentration. Carbon dioxide has some cooling effect. All these media are clean and non-conductive. They do not damage materials or equipment and can be used safely on electrical equipment. A particular advantage is their ability to fill enclosed spaces which other media might not be able to penetrate. Machinery casings or interfloor spaces are typical examples. Rooms containing equipment can be flooded. The concentration required for extinguishment might be dangerous to health, so occupants should be evacuated before the extinguisher is applied. Small quantities of nitrogen or carbon dioxide used in large spaces may be safe but halogens may produce toxic fumes when exposed to fire, so ideally they should be used externally.

Dry powder

Dry powder usually means powdered bicarbonate of soda but other chemicals are used for particular applications. Powders are available for all classes of fire. Most powders are non-toxic. They can affect visibility. If used for flooding an enclosure the occupants should be evacuated before the powder is released. Powders can set on hard surfaces and corrosion may follow in damp conditions. Equipment should be cleaned as soon as possible.

18.4 Portable fire extinguishers

The Building Regulations do not address firefighting equipment for the use of occupants. However, the Fire Precautions Act and Health and Safety at Work Act do cover this issue for many building types and the Housing Act requires local authorities to consider houses in multiple occupancy.

All the media described above can be used in portable extinguishers. All portable extinguishers are prominently labelled with the classes and sizes of fires to which they are appropriate. Water and foam extinguishers, which may present a risk of electrocution, are labelled: Do not use on live electrical equipment. All buildings are subject to the risk of class A fires, so appropriate extinguishers should be sited in prominent positions in corridors and landings near to room exits. There should be an extinguisher within 30 m of any possible fire and appropriate extinguishers should be provided for any special risks. The weight of portable extinguishers should not exceed 23 kg to ensure ease of use by occupants.

Water

The earliest form of portable water extinguisher was a row of red-painted hanging water buckets. The bottoms were rounded so they could not be employed for other activities. They were superseded many years ago by the soda–acid extinguisher (Fig. 18.1). This type of extinguisher was sealed so that evaporation was overcome and rubbish could not be deposited. Instead of a single inaccurate short-range flush of water from a bucket, a jet of water could be accurately directed. Bicarbonate of soda was dissolved in the water and the power for the jet provided by mixing sulphuric acid with the bicarbonate solution to give large volumes of carbon dioxide gas, the pressure of which provided the jet. One type of extinguisher released the sulphuric acid by breaking a phial, another type emptied a bottle of acid when the extinguisher was inverted. Once started the whole of the contents were discharged. The soda–acid propellent system is no longer recommended and the power for the water jet is now by a cartridge of high pressure carbon dioxide, broken by operating a plunger. Water extinguishers are only appropriate for class A fires. Once the extinguisher has been activated it will continue until empty.

Foam

Foam extinguishers may operate by inversion and the mixing of two chemicals to produce and deliver a jet of foam (Fig. 18.2) or by breaking a compressed carbon dioxide cartridge. Similar to water extinguishers the delivery of foam will continue until the extinguisher is empty.

Gaseous media

Halogens, nitrogen and carbon dioxide can be compressed and stored in liquid form within pressurized cylinders. A short hose and nozzle are provided to direct the gas to the seat of the fire. The delivery is controlled by a valve. Figure 18.2 shows a typical carbon dioxide extinguisher.

Powder

The powder is contained in a pressure vessel and expelled by compressed carbon dioxide or nitrogen.

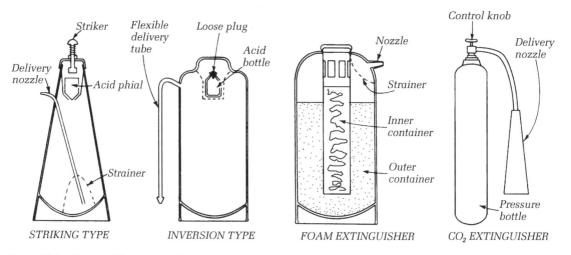

STRIKING TYPE   INVERSION TYPE

Figure 18.1 Soda-acid fire extinguishers.

FOAM EXTINGUISHER   CO₂ EXTINGUISHER

Figure 18.2 Foam and carbon dioxide extinguishers.

Asbestos blanket

Asbestos blankets may look different from the other extinguishers but they work in a similar way – they exclude oxygen from the fire. They are effective against small fires and liquids in open containers. They can be used to wrap people whose clothes are on fire. Widely used in kitchens and welding shops, asbestos blankets cause no damage or contamination.

18.5 Fixed firefighting apparatus

Hose reels

Small-diameter rubber hoses for use by the occupants of buildings provide a very much more effective measure than buckets or soda–acid extinguishers and in many circumstances will be competitive in price. Rubber hose is used, usually of 18 or 25 mm internal diameter. Up to 22 mm of unreinforced hose or up to 36 m of reinforced hose is wound on to a drum. The hose is connected to a water supply serving the spindle of the drum and fitted with a small-diameter nozzle with control cock. Figure 18.3 shows such a reel installed, as is usual, in a recess in corridor or landing wall. In case of need the control valve is turned on, then the person using the hose takes the nozzle end of the hose and moves towards the fire, pulling out the hose as he or she goes. Figure 18.3 shows the guide which is used to make this possible with fixed-drum hose reels. A superior arrangement is the use of a swinging-arm hose reel, where the drum automatically swings to follow the line of movement of the hose.

The hydraulic requirements for hose reels is that they should be able to delivery 22 l/min at a distance of 6 m from the nozzle and that three should be capable of operating at any one time. A pressure of $207\,kN/m^2$ (21 m head) at the connection is usually considered appropriate for each reel. Many water authorities will allow connection of hose reels direct to the main, and when the mains pressure is adequate, this provides a simple supply arrangement. Some cases call for break tanks, usually of 1150 l minimum capacity. If the break tanks can be high enough to supply the hose reels by gravity, no other special provision is needed. If they are not or if in the case of direct connection the building is too high for the hose reels to operate, duplicate pumps capable of appropriate water delivery must be provided. They are usually installed in the basement and means for bringing them into operation must be provided, such as automatic flow or pipeline switches, or by manual operation of the reel.

Hose-reel installations must be taken into account when buildings are being planned. They should be sited near means of escape so they are readily available to people leaving the building and, if the floor is full of smoke after the hose reel has been in use, the hose itself can guide its users to safety. There should be sufficient hoses, of appropriate lengths, so positioned that the hose can enter every room, and when fully extended reach to within 6 m of every part of the floor.

Sprinklers

There is clearly great advantage, particularly for buildings involving special fire risks or only intermittently supervised, such as very large or high buildings, underground car parks, warehouses or stores, for an automatic firefighting installation that does not depend for its operation on the presence of occupants. Sprinkler installations fulfil such a purpose. They consist of a grid of water pipes fixed under the ceiling with delivery heads normally on a 3 m square grid. Water is prevented from emerging by a glass or quartzoid bulb containing liquid. Figure 18.4 shows a typical sprinkler head. When the temperature rises the liquid expands, breaking the bulb that is preventing the water passing. Bulbs are available that break at temperatures in the range 68–180 °C. A jet of water then impinges on the shaped deflector plate, delivering a spray of water over an area of about $10\,m^2$. The flow of water is arranged so that it works a turbine-operated fire alarm. The sprinkler installation therefore sounds the alarm and helps to minimize and contain the fire. In an unheated building there would be danger of freezing in the winter. It is possible to overcome this by using the dry pipe system, where special valve arrangements enable the delivery pipes to be filled with air under pressure. When this pressure drops, due to the opening

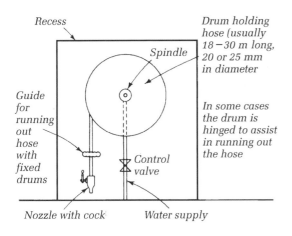

Recess

Drum holding hose (usually 18 – 30 m long, 20 or 25 mm in diameter

Spindle

In some cases the drum is hinged to assist in running out the hose

Guide for running out hose with fixed drums

Control valve

Nozzle with cock   *Water supply*

Figure 18.3   Hose reel in recess.

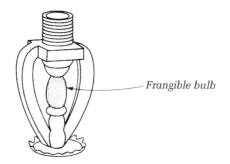

Figure 18.4   Typical sprinkler head.

— *Frangible bulb*

of a sprinkler head, the water is admitted to the delivery pipes.

Sprinkler systems require central control and test gear. This is often arranged in the basement, but there is much to be said for having this gear, or at least the alarm and stop valve, in a prominent position near the entrance to the building. This will enable the water to but shut off quickly before too much damage is done as soon as the fire is under control. Figure 18.5 shows the parts and function of a sprinkler control. Two independent water supplies are generally thought desirable for sprinkler installations. In towns it is sometimes possible to have supplies from water mains

served by two independent trunk mains. When this is not the case storage on the site is necessary. High level cisterns capable of holding 22.5–34.0 m^3 used to be the standard provision. Pressurized cylinders housed in the basement are becoming universal; 22.5–50.0 m^3 capacity is usual.

A regular grid is required for sprinklers and is covered by complex standards that depend on the fire protection of the building and the nature of the activities. Coupled with the need for control gear and storage space, the importance of the grid means that installations must be planned for in the early stages of building design. Rules are laid down for sprinkler supply pipes. Their minimum diameter is 25 mm, rising in stages until a 50 mm diameter pipe can serve 18 sprinklers (or 10 in the case of some special risks). Installations with more than 150 sprinklers will require a 150 mm diameter pipe.

A special type of sprinkler can be used for oil fires. Although a jet of water would do more harm than good, a fine spray falling on the surface of burning liquid cools the surface, and by turning into steam it excludes oxygen.

It is also possible to obtain valves which open on the bursting of a sprinkler bulb. They then admit water to a series of sprinklers. This technique is sometimes used in the case of rooms containing electrical equipment.

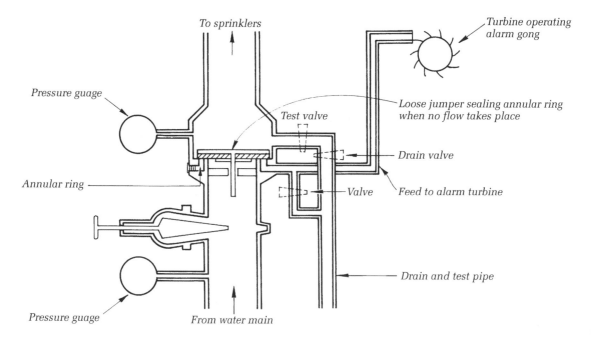

Figure 18.5   Sprinkler control gear.

Sprinklers cannot discharge over the apparatus but they can be used to protect door and window openings to prevent the spread of fire if the water supply is controlled by the temperature within the electrical room itself.

Drenchers

Drenchers are similar to sprinklers but they deliver a curtain spray and are normally used to protect the external face of a building from some adjacent fire risk. Since they would inevitably be subject to freezing they are not normally provided with frangible bulbs but have open waterways. The water supply is turned on manually when required. As in the case of sprinklers, 25 mm is the minimum pipe diameter for serving two drenchers. Ten drenchers require a 50 mm diameter pipe, 36 require a 75 mm diameter pipe and over 100 require a 150 mm diameter pipe. Not more than 12 drenchers may be fixed in any one horizontal branch. Periodical flushing of the pipework is required.

Total flooding systems

There are many rooms in buildings where there is a particularly high risk of fire or contents which are particularly valuable. Examples are boiler rooms, oil stores, archives and computer installations. In the case of fire it would be possible to completely fill (flood) the room with extinguishing medium. Foam would be the appropriate medium for boiler rooms and oil stores. In the case of archives or computer rooms, where the contents must not be damaged, gaseous media can be used. Installations of this type have a central pressure vessel to contain the extinguishing medium; pipework leads to delivery points at high level in the protected room and there is also a control system. It would be dangerous to flood the room without evacuating any occupants. Realistic systems wait for the fire detector to sound an alarm then allow a delay before the extinguishing medium is released, or they sound the alarm and leave the occupants to choose the appropriate action.

18.6 Facilities for the fire service

In buildings of modest size hard access for fire appliances to a reasonable proportion of perimeter is required but no special installation for use by the fire services is normally needed. Water for firefighting purposes is taken from hydrants in the highway.

Fire service hydrants

If the building is extensive or sited a long way from the road, one or more fire service hydrants may be needed. Fire service hydrants are 65 mm in diameter with a valve and a special type of coupling for hoses. If the water serving the hydrants can flow only in one direction, as in communication pipes to buildings, a minimum 100 mm diameter is needed, so the section of the service pipe up the hydrant will have to be increased to this diameter. If the water supply to the building is metered a problem arises since it might often be inappropriate to install a 100 mm diameter for comparatively small flow. Figure 7.34 shows a way in which the difficulty may be overcome by interrupting the large-diameter pipe with a valve under the control of the fire service to be turned on only in case of fire; the supply to the building is maintained by a small diameter bypass including the meter.

Access by fire service appliances

For firefighting and emergency escape to take place effectively fire service appliances such as pumps and escape ladders must be able to be positioned close to buildings. It is usual for the local authority to insist that an appropriate proportion of the perimeter of the building is accessible in this way; see the MBS title *Structure and Fabric Part 1*. In buildings remote from the highway with only access roads on the side it may be necessary to construct special roadways or hardstanding to meet this requirement.

High buildings

Until comparatively recently the height of buildings in England was limited to 24 m with a further 6 m in the roof space. This requirement was based on firefighting taking place from the street. The move towards high buildings depended on providing adequate means of vertical access for firefighting even when several floors were on fire. Access is required not only for firefighters but also for apparatus, which would be difficult to carry up stairs, and for a supply of water well beyond the normal mains head. The *firefighting shaft* has been developed to meet this need. It consists of a protected enclosure containing a firefighting stair, an eight-person firefighters' lift and a firefighting lobby. The doors leading from the firefighting lobby must be self-closing and fire resisting. The lift can be part of the normal building circulation but it must be separately powered and controlled in such a way that the fire service can bring it under their control in case of fire. At ground level the firefighting shaft must be readily accessible to the fire service and pumping appliances must be able to park within 18 m. The firefighting shaft will be equipped with a fire main serving a landing valve at each floor to allow hose connections. The rising main will be 100 mm in diameter unless more than one landing valve is served at any level, in which case 150 mm is required. In buildings with no floor higher than 60 m above ground level access to the fire main can be dry. In this case an inlet at access level allows a pumping appliance to provide the water supply. At heights of over 60 m a stored water supply and pumping facilities must be supplied within the building, although access for the pumping appliance must still be provided to replenish the water supply if necessary. Lower buildings with large floor areas and basements with two or more levels having large floor areas may be required to have firefighting shafts but not lifts. The location and number of firefighting shafts depends on floor area, location of the shafts and whether or not sprinklers are provided. These planning requirements are set out in Approved Document B5 of the Building Regulations.

The firefighting shaft must be kept clear of smoke. This may be difficult to achieve by natural means, particularly if the shaft is not on the external wall. A system of pressurized mechanical ventilation can overcome the problem. A secure air supply free from smoke must be provided to maintain a pressure of 50 Pa in the shaft. The design of the system must take into account air leakage from the construction and door opening. Pressure should be controlled to between 40 and 60 Pa. The areas surrounding the shaft must be able to allow free leakage of air, otherwise it would be difficult to maintain the pressure differential. In the case of an internal shaft having no natural ventilation it is possible to have a two-stage system, where a basic ventilation level is maintained at all times and increased in the case of fire. The smoke pressurization level should be activated by the smoke alarms in the building.

Emergency control of services

It should be possible for the fire services to turn off the main electrical and gas intakes at a readily accessible spot so that firefighting may be easier and safer.

Escape routes

Ventilation

In buildings with mechanical ventilation and escape routes that are naturally vented the ventilation system should not promote a flow of smoke into the escape routes. In many cases it may not be practicable to prevent smoke ingress into escape routes by natural venting. Pressurization may be used on escape stairs to overcome the problem in a similar way to firefighting shafts. Access from the accommodation on each floor will normally be via an unventilated lobby, automatically pressurized by air from the stairwell. If the layout of the floors requires protected corridors as part of the escape route, they may also be pressurized to protect against smoke ingress. Protected corridors and protected stairwells should have separate air supplies and the pressure in the corridor should be 5 Pa lower than the pressure in the protected stairwell to ensure no smoke penetration to the stairwell.

Lighting

All escape routes must be provided with escape lighting served by separate protected electrical circuits; Chapter 8 describes the types of installations required. Besides the actual escape routes, escape lighting is required in several other areas: in electricity generator rooms, in battery rooms and switchrooms for emergency lighting and in emergency control rooms. In underground office buildings or windowless accommodation escape lighting is required for stairways in central cores or stairways extending more than 20 m above ground, internal corridors longer than 30 m and open offices of area larger than 60 m^2. There are also special requirements for shops, commercial buildings, car parks and assembly and recreation buildings.

Large open volumes

Buildings with atria where virtually the whole floor area of the building forms part of a single volume present a special fire hazard. The definitions to Approved Document B distinguish between atria and open spatial planning where more than one storey is contained in one undivided volume, but the nature of the distinction is not made clear. The approach so far adopted in buildings with atria is to raise the ceiling of the atrium substantially above the level of the top-floor ceiling, so that smoke rising is not immediately spread to the top office level, and to provide substantial smoke venting at the top of the atrium to ensure the smoke rises through it rather than spreading at each floor level.

Precautions where pipes and ducts pass through fire resistant walls and floors

Walls and floors which have to provide resistance to the spread of fire are described in the Building

Regulations as 'compartment' walls or floors. There are special requirements for ensuring the fire resistance is not compromised at locations where pipes and ducts must pass through. The problem should be considered in the early stages of planning and it may be advantageous to group verticals runs of pipes and ducts into protected shafts, effectively vertical service ducts with walls that provide the same fire resistance as compartment walls. Horizontal branches coming from the main vertical runs are likely to be of smaller diameter, so they should present fewer problems when they penetrate the walls of the protected shaft.

The requirements for maintaining fire resistance where pipes and ducts pass through compartment structures are complex. They are set out in Approved Document B3 and Part 9 of BS 5588. They include some general principles which affect planning as well as detailed fire-stopping provisions:

- Openings should be as few as possible.
- Openings should be as small as possible.
- All gaps should be fire-stopped.

Precautions for pipes
Each situation must be considered in detail and the requirements of the Building Regulations must be satisfied. The following methods may be considered:

- For pipes of any material or diameter use a proprietary pipe seal which has satisfied approved tests.
- For pipes of internal diameters up to 160 mm use a non-combustible material such as cast iron or steel; or for pipes of lead, aluminium, aluminium alloy, PVC or fibre–cement use a close sleeve of non-combustible material extending for 1 m on each side of the wall or floor.
- For pipes of lead, aluminium, aluminium alloy, PVC or fibre–cement which form part of an above-ground drainage system and are in a sealed enclosure with internal surfaces of class 0 construction, use a maximum internal diameter of 160 mm for stacks or 110 mm for branches.
- For any other circumstance or material use 40 mm internal diameter pipes.

Precautions for ventilation systems
Ventilation systems present a particularly acute risk of fire spread. In addition to the relatively large holes in compartment structures needed for ventilation trunkings and requiring special precautions, the system itself is capable of spreading smoke and sparks.

The expanding gases which can result from fires and the strong buoyancy effects may overcome the normal pattern of air movement and spread smoke in directions other than the normal airflow. Fire adjacent to trunking may ignite the material of the trunking or internal linings if they are combustible. Fire established in the linings of trunking can spread against the direction of airflow.

In many mechanically ventilated or air-conditioned buildings air in excess of that required for ventilation is recirculated as part of the heating or cooling cycle. This can clearly present an acute problem of smoke distribution. Types of precautions which may be employed include:

- Separate trunking systems that service high risk areas such as large kitchens, car parks, boiler rooms, etc.
- In domestic buildings extract systems to bathrooms and WCs, and sometimes to kitchens, can be provided with shunt connections that join the main trunking 900 mm above bathroom or WC exit grilles, or 1800 mm above kitchen extract grilles. This arrangement is also adopted to minimize noise penetration.
- Smoke detectors may be provided in ducts to shut off fans and close dampers. Smoke detectors in rooms may also control the ventilation system. They have the advantage of giving earlier indication of fire. This precaution is particularly important in systems which recirculate air.
- Fire service control of the ventilation system may be needed in large buildings, both to stop the operation of the system and, if appropriate, to start the fans after the fire has been extinguished to clear smoke from the interior.
- Firedampers that close automatically in a fire will be required where trunkings pass through compartment walls or floors. Frangible bulbs or fusible links may be used to hold the dampers open against gravity or spring pressure. Some time may be required before detectors in the duct can respond to any temperature rise, so smoke detectors can give an earlier response, either in the duct or in the rooms served.
- Protected shafts, as described for pipe systems, may be used to enclose main runs of ventilation trunking. Firedampers can be provided where branches serving rooms penetrate the protected shaft. Protected shafts may run vertically or horizontally and may include other services.

19 Ducted distribution of services

The space required to accommodate the distribution services is a significant proportion of the total building volume. Together with the special requirements for location of some of the plant required, it means that accommodation for and the layout of services are two essential factors which must be taken into account from the earliest stages of design, long before detailed calculations of plant, pipe and trunking sizes can take place.

In houses pipe runs are short, pipes are relatively small and cupboards and joisted floors provide convenient routes for service runs. But even in dwellings the cold-water cisterns, hot-water cylinders, fuel stores, boilers and meters must be considered at the outset of design, and the runs of waste and soil pipes must be taken into account when planning the sanitary accommodation if economic and efficient installations are to be achieved. In larger buildings it is usual to need some 7–10% of the total floor area for plant spaces and ducts and a substantial extra volume of ceiling voids and below-ground subways is often required.

It is essential that building designers should be aware from the outset of the nature and the scale of the plant and storage spaces required, and the factors which govern their location. They should then satisfy themselves that feasible routes exist which will be adequate to accommodate service runs when they are finally determined and sized at the detailed design stage.

Properly designed facilities for services have to do more than simply contain the installations. Access for ease of assembly, a layout and duct spaces that enable installations to be completed in one stage of work rather than in sections which depend on other work,

access for efficient testing and commissioning and feasibility of extension, alteration and replacement, all these are critical factors in ensuring the satisfactory and economical performance of the installations.

The terminology associated with the accommodation of services is confused at present and the same terms are used for different aspects.

Plant space
Area required for the accommodation of mechanical or electrical equipment or control gear required for the operation of services.

Storage space
Area required for the accommodation of storage containers required for particular services.

Duct
A space within a building specially enclosed for the accommodation of services and allowing facilities for working and inspection.

Subway
A horizontal passage for the conveyance of services underground or below the bottom floor of the building which allows walking headroom for access. Also termed **walkway**.

Crawlway
A passage for services similar to a subway, but where there is insufficient headroom to stand upright.

Trench
A horizontal passage for services below floor level where the access is by removable covers in the floor.

Wells

Vertical space used for the accommodation of stairs or lifts or to allow natural light or ventilation.

Casing

An enclosure formed over pipes or cables running on the surface of a wall or ceiling. Casings are usually for decoration but can also provide protection from impact or corrosion.

Chase

A recess cut in a wall or floor when building is over; it accommodates pipes or cables and is screeded or plastered over.

Void

A space which may be used for the accommodation of services but which is not primarily intended for this purpose.

Sump

A pit for seepage, leakage and draining down of pipework that cannot be discharged by gravity to the drain and must be collected and pumped.

Trunking

Lightweight, usually sheet metal enclosure for the passage of air or cables.

Flue

Builders' work or metal passage to convey the products of combustion to an acceptable point for discharge to the atmosphere.

Service core

Zone extending vertically through a high-rise building containing vertical circulation, service ducts and other utility and sanitary provisions.

Importance of a unified system of service distribution

If each service were to be considered in isolation a number of different layout patterns would emerge and the accommodation of the services would become expensive and confused. With the exception of flammable liquids and gases, and subject to satisfactory location, most services can be run in common ducts. This enables the same space to be used for access and installation to all the services and greatly simplifies inspection and maintenance. It is important, therefore, that the pattern of distribution of services is considered as a whole in relation to the building planning. The increase in efficiency which results from well-maintained services cannot be costed with present techniques but is undoubtedly significant.

It is possible to distinguish three different categories of service run having different requirements for patterns of distribution: wells and flues, pipes and ventilation trunkings, and electrical cables. Stair and lift wells run vertically through buildings. They do not require linkage with services at each floor level but prevent an obstacle to horizontal distribution of other services. On plan it often appears convenient to group vertical ducts with wells and flues, but care must be taken to ensure the horizontal service branches from the vertical ducts can be run satisfactorily. Electrical cables are relatively small and can often be contained in very small trunkings or conduit, even within the thickness of plaster or floor screed. Since cables can be threaded into buried conduct the problems that arise from burying normal pipework do not arise. With suitable precautions to ensure that leakages from other services do not fall on them, electrical cables may be run in normal ducts with other services. But the very different space requirements mean that electrical services can often be partly or completely separate from the pattern of pipework.

Location and approximate sizing of plant spaces and ducts

There can be no absolute rules governing the location of plants and ducts. Different building functions will impose different priorities on the service. Electric cables do not normally impose serious limitations on planning. In the case of large computers or very heavy electrical load, however, the length of cables may be a dominant factor. Other functions or site problems may mean that a less than ideal location for services will be the best overall solution for building performance. It is important that designers are conscious of the factors involved and take them into account from the earliest stages of design. In addition it is vital that sensible approximate sizes for services should be used as soon as the remainder of the design begins to take dimensioned form. There is insufficient information in the early stages to design the services in full engineering detail but it is important to incorporate service sizes that will not be so inaccurate as to make the final design impracticable. This does not require great precision since space for access forms a major part of the service spatial requirement and this is not very sensitive to variations in load. Here are some planning considerations for service distribution:

- *Air trunkings* are large and it is essential to consider them at the outset of any design. The space which they require will mean that in buildings where they are present they must be given considerable priority.
- *Waste and soil installations* are not as large as air trunking but because of the continuous fall which must take place they exercise a limiting influence on planning and cannot be ranked too far behind air trunkings in conceiving appropriate layouts in the early stages of design.
- *Heating pipes* may be larger than waste and soil pipes in large buildings, particularly if the insulation is included in the diameter. For much of their length, however, they will be smaller and they can be more freely varied than waste and soil pipes.
- *Water supply* pipes are very often smaller than heating supply pipes and do not require such heavy, if any, insulation. They are equally flexible in their detailed layout and may be taken up and over or under obstructions without any major problems.
- *Electrical cables* are very small in most buildings and they can very easily be accommodated in comparatively small trunkings; often they are disposed within the thickness of a floor screed. It would rarely be appropriate to attempt to save runs of cables at the expense of increased lengths of circulation area or other service provision.

It is sometimes possible to take advantage of the reverse pattern of diminishing sizes to economize in the provision of ducts and trunkings. If electrical intake and telephone intake are at opposite ends of the building, each with a primary vertical distribution leading to secondary floor trunkings, then the density of cables from each type of installation will diminish as the opposite end of the building is approached. If suitable arrangements can be made to transfer the use of the trunking from one service to the other, a regular layout of trunking can accommodate two services with considerable economy.

Table 19.1 lists common types of plant and storage together with a summary of the considerations affecting their location. It gives approximate sizing techniques and planning considerations for individual services. Approximate sizing methods for ventilation trunkings have already been described; comparative sizes of heating pipes and trunkings are given in Table 7.6.

Approximate sizing of ducts

In establishing the sizes for ducts it is not merely the sizes of pipes and trunkings that must be considered.

Access for installation, space for fixing, allowance for valves and expansion provisions and space to allow branches to be connected are all important factors, as is the space required for operative access during inspection and maintenance. Several factors must be considered in deciding the size of ducts and recesses:

- The number and size of pipes, cables and trunkings to be accommodated.
- Any critical spacing or fixing position which must be followed for certain pipe types.
- Clearance required for placing the pipes in position, which must also allow for ease of removal should it become necessary during maintenance.
- Clearance to allow for position of fixings and to permit jointing – particularly where manipulation of joints or tools is required – as with push-fit plumbing assemblies.
- Allowance for additional services which may be needed.
- Space for access in the case of ducts.
- Space for valves, dampers, etc.
- Space for expansion bands in long, straight horizontal or vertical ducts.
- Space for branching and service junctions, and to carry these branches past adjoining services.

Simple rules can be laid down to cover basic pipe spacing, but it is not so easy to deal with the problem of branches because of the many variations in the sizes of pipes, branches and adjacent pipes, and the different directions required. It is critical that the general layout of the various services be borne in mind during the design period when duct sizes are being determined. In planning it is not possible to detail all parts of the installation and it is important to identify and distinguish those which can be varied in their dimensions during the design stage without major consequences and also those where dimensioning is critical and careful studies should be made.

Vertical ducts

With vertical ducts running down through a building it is often found that dimensions can be varied without disastrous planning consequences until quite late in the design process.

Underground ducts

To a considerable degree the same thing applies to underground ducts except where they would conflict with columns and foundations.

Table 19.1   Planning considerations for typical plant spaces

Plant or store	Size	Weight	Vibration and noise	Access considerations	Location	Information on sizing
Fuel store	–	Heavy	None	Fuel delivery	Ground level or basement	Page 137 Table 7.4
Boiler room	–	–	Medium	Fuel Flue Air for combustion heating flow and return	Ground level ideal Roof possible with gas and oil Basement possible	Page 172 Fig. 7.47(a)
Refrigeration plant	Small	Heavy	High	Heat dissipation from plant Linkage to costing towers	Should be based on ground independent of building structure	Page 172 Fig. 7.47(c)
Air-handling plant	Large	Light	Low	Air intake and exhaust Delivery and return	Should be near area served to minimize trunking runs Roof level convenient for intake and exhaust	Page 172 Fig. 7.47(b)
Cooling tower	Large	Light	Low	Requires free airflow for heat dissipation Linkage to refrigeration plant	Ground level near building or roof	
Calorifiers and hot-water cylinders		Heavy	Low		Not critical Near load which is served	Fig. 9.27 Fig. 11.11 Tables 11.2, 11.3
Cold-water cisterns		Heavy	None	Service, distribution and warning pipes	At high level to supply building by gravity In high-rise buildings 75% of storage may be at low level	Domestic: Fig. 9.10 Non-domestic: Tables 11.1, 11.2
Lift and motor rooms	Small	Heavy	Medium	Ventilation to dissipate heat (required at top of shaft)	At top of lift well; other positions are theoretically possible but expensive in initial and running costs	Passenger: Fig. 17.5 Goods: page 343
Electrical substation	Small	Heavy	Low	Ventilation to dissipate heat from transformer Access for electricity company Gentle entry bends for large cables	Separately located on site but may be incorporated into building if adequate fire precautions are taken	
Electrical switchroom	Small	Light	None	Gentle entry bends for large cables	Near electrical entry Some large oil-quenched switches present fire risk	

Ceiling voids

The points at which branches leave the primary vertical ducts to enter secondary ceiling recesses present a very different problem. The headroom on the various floors will normally have been reduced to the acceptable minimum, and the space available above a suspended ceiling will also normally be kept as low as possible in order to avoid increasing the overall height of the building. Generally there will be adequate space to accommodate services within the ceiling void. However, near the point at which the secondary recess joins the primary vertical duct, the trunkings and pipes will be at their maximum diameter, and the necessity for crossings will be relatively high. A dimensional problem at this point will be particularly difficult. It is an important area that should be investigated reasonably thoroughly at an early stage in every design where it occurs. Replanning will often provide a better solution to any difficulties that may arise than would be possible by technical manipulation or by adding to the height of the building.

Floor voids

Increasing use of electrical, telecommunications and computer equipment in offices has led to the use of suspended floors to contain the wiring and provide flexibility for additions and alterations.

The recent development of natural ventilation for deep offices which would previously have been air-conditioned has led to the need for the structural floor to be used as a heat sink to control extremes of temperature. To achieve this the floor slab must be exposed to form the ceiling of the office below. This precludes the use of a suspended ceiling containing light fittings and air trunkings which had become a standard feature in large offices. Fortunately natural ventilation eliminates the need for large air trunkings at high level. There may still be a requirement for small amounts of air input to assist in the control of temperature peaks. The amounts will be small in comparison with full mechanical ventilation and can be provided by flexible circular trunkings running in the floor voids. Lighting fittings which would have been located in the suspended ceiling must be replaced by surface-mounted ceiling fittings or uplighters. In both cases the cables serving the fittings can be run in the floor voids.

Crossing services

Where services cross it is possible to have a great deal of wasted space in ducts, but it is equally possible that critical restrictions may occur which may impair both the planning arrangements for the building and the efficiency of operation of the services themselves. Extract and input air trunkings distributed in ceiling recesses form a typical example. In a ceiling recess very little of the space may be occupied, but the headroom available may be inadequate at the point where trunkings cross. This may lead to decisions to restrict headroom in the circulation space or to reduce the dimensions of the ducts or the recesses with consequent impairment of performance, alternatively the height of the ceiling space could be increased with a consequent increase in total building height. A better solution would obviously be a planning arrangement that eliminated the necessity for the two trunkings to cross.

Clear space for access in ducts

Table 19.2 compares the recommendations for clear space in walk ducts and crawl ducts required by BS 8313 *Ducts for building services* and DHSS Memorandum 23. Although there is no reason to suppose the requirements for access in hospital ducts are significantly different to ducts in other buildings, notice how the recommendations vary. Designers must decide for themselves about the requirements of the particular situation for which they are designing. The code of practice requirement is less generous than the DHSS requirement and it will be more important to ensure that clear space is not reduced significantly by projecting valves or by branches crossing other pipes. Headroom should not be reduced by light fittings.

Table 19.2   Recommended clear space: comparison of walk ducts and crawl ducts[a]

Duct type	BS 8313 : 1988 requirement[b]	Access and accommodation for engineering services[c]
Walk duct		
Height (m)	2.0	2.0
Width (m)	0.7	0.9[d]
Crawl duct		
Height (m)	1.0	1.1
Width (m)	0.7	0.9

[a] The amount of clear space in a duct is the area $w \times h$. Values of width w and height h can be obtained from the table. The area outside $w \times h$ is occupied by pipes, etc.
[b] BS 8313 : 1988 *Ducts for building services*.
[c] DHSS, Welsh Office Hospital Technical Memorandum 23, *Access and accommodation for engineering services*, March 1972.
[d] Width of 1.2 m recommended where trolleys are used and 1.4 m for powered vehicles.

Table 19.3 Dimensions of square space to be allowed in the cross-section of ducts

Pipe diameter (mm)	Size of square (mm)	Pipe diameter (mm)	Size of square (mm)
20	100	65	200
25	120	75	225
32	130	100	300
40	140	150	450
50	150	200	600

Space around pipes

At the early stages of planning, the best way to decide on spatial allocation is to consider the duct in cross-section, then to regard each pipe as occupying a square of the dimensions given in Table 19.3. Where two pipes are adjacent they allow more easy access than would be the case if a trunking were adjacent and the spacing between could be reduced to approximately two-thirds of that shown in the table. Generally pipes should not be situated behind other pipes. Where this arrangement cannot be avoided the space allocation should be more generous than shown in the table. Table 19.3 takes rough account of insulation but not of valves or branches.

Space around trunking

Electrical trunkings are usually relatively small, mounted directly on to walls and require only nominal clearance. Cable trays suspended from ceilings should have a minimum clearance of 250 mm at each side and a similar distance above each edge of the tray.

Ventilation trunkings

Ventilation trunkings should also have 250 mm clearance at the sides and at least 300 mm clearance above the trunking. The clearance above the trunking should be increased up to 500 mm for 2 m wide trunkings. Space limitations often preclude this and trunkings may be suspended close to structural floors which usually support them. Fabrication and insulation is more difficult and, although galvanized steel trunkings fitted internally require little maintenance, access for any purpose would be difficult. One particular problem is that joints in ductwork frequently require further sealing after installation in order to limit air losses and additional stiffening may be required to prevent drumming. Both of these repairs would be difficult to make where there are small clearances.

Fire hazards

Ducts and trunkings which run through fire-resisting

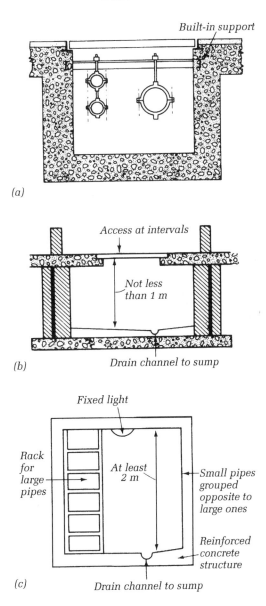

(a)

(b)

(c)

Figure 19.1 Service access: (a) floor trench; (b) crawlway (socket outlet every 10 m); (c) walkway (socket outlet every 10 m).

constructions present a potential fire hazard, and careful precautions must be taken. Building regulations and BS 8313 should be carefully consulted. In general vertical ducts should have vents at the top so that smoke and flame can be dissipated rather than forcing its way out into the building at lower levels. Ducts must either be sealed where they pass through fire-resisting constructions or the walls and doors of the ducts must have adequate fire resistance in themselves. In the case of some air trunkings it is not possible to comply with these requirements, so automatic shutters and dampers which close in the case of fire have to be provided. Mechanical and chemical intumescent types are available in a range of shapes and sizes.

Where pipe runs are forced to cut through fire-resisting walls, it is equally important to ensure that any gap left around the pipe is tightly packed with fire-stopping material. Materials which can be used include exfoliated vermiculite, bound with a non-combustible water-soluble binder, intumescent mastics or mineral wool packing. It will be important to judge how much movement there is likely to be in the service before selecting one of these materials; mastics are able to stop small gaps where only small movement is likely and mineral wool packing can stop large gaps with large movements; for a small gap with large movement a mineral wool sleeve will be sufficient.

Duct details

Typical construction details for floor trenches, crawlways and walkways are shown in Fig. 19.1. In main ducts (Fig. 19.2) considerable care must be taken with the means of suspending pipes. Some have to be laid to fall, so they require adjustable hangers. Heating pipes, especially if laid in long straight runs, will

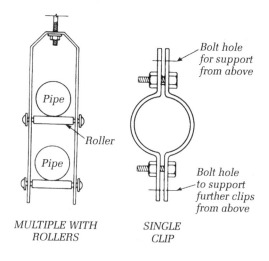

Figure 19.3 Details of pipe supports.

exhibit significant thermal movement and must be firmly supported but not restrained for longitudinal movement. Figure 19.3 gives details of the roller pipe support and the adjustable-level pipe clip. Where long straight lengths of pipe occur it may be necessary to have loop- or bellows-type expansion joints for pipes other than heating and hot water. In underground ducts care must be taken to avoid damp penetration. The MBS title *Structure and Fabric Part 1* gives details of their construction.

All major underground ducts should be provided with channels to carry away seepage and leakage. It is usual to arrange these channels to drain to the sump in the boiler room.

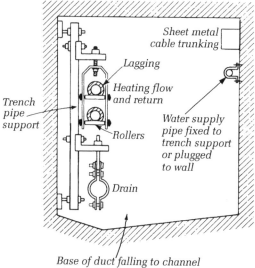

Figure 19.2 Typical main duct running below groundfloor level with access for inspection and maintenance.

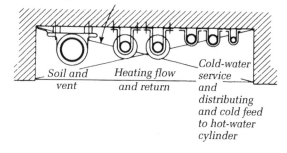

Figure 19.4 Typical vertical duct: pipes are plugged to masonry wall; where this is not practical, angle-iron supports could be provided and pipes fixed by U-bolts or saddlebands.

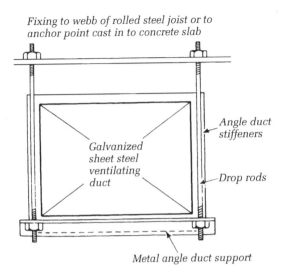

Fixing to webb of rolled steel joist or to anchor point cast in to concrete slab

Angle duct stiffeners

Galvanized sheet steel ventilating duct

Drop rods

Metal angle duct support

Figure 19.5 Ventilation trunking suspended from structural flooor and concealed by suspended ceiling.

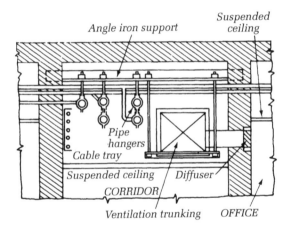

Angle iron support

Suspended ceiling

Pipe hangers

Cable tray

Suspended ceiling

Diffuser

CORRIDOR

Ventilation trunking

OFFICE

Figure 19.6 Typical corridor suspended ceiling: service distribution takes place above the ceiling.

Figure 19.4 shows a vertical duct. In this case the duct is simply a recess in a corridor wall covered by a ply facing. A hole is left in the concrete floor to allow the services to be installed and this is subsequently filled in as a fire precaution. Access to the service runs is very easy and on upper floors the interference to circulation while work is going on is not likely to be critical. If this arrangement is not acceptable and work on the services must be able to be carried out without access from the corridor, a walk-in type of duct can be provided, running continuously up the building with

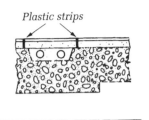

Plastic strips

Figure 19.7 Floor chases.

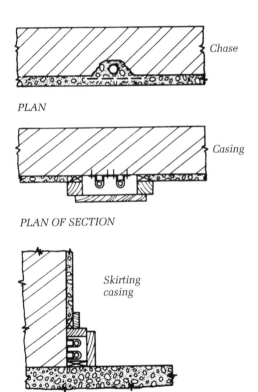

Chase

PLAN

Casing

PLAN OF SECTION

Skirting casing

SECTION

Figure 19.8 Wall chase and pipe casings.

open gratings at each working stage and ladders for access. It is very much more expensive and space-consuming than the type previously described, and fire precautions may be necessary to prevent spread of flames from one floor to another.

Positions for vertical ducts should be selected with care. Connection to lateral ducts may often be difficult to contain within ceiling spaces as they leave the vertical duct and distribution in that direction.

Lateral ducts are very often formed in ceiling spaces. Figure 19.5 shows a basic method of suspending a ventilating trunking at high level in a corridor, and Fig. 19.6 shows a typical corridor ceiling space used for services. The construction of floor and wall chases and pipe casings is shown in Figs 19.7 and 19.8.

Special references

BS 8313 : 1988 *Ducts for building services* does not deal with early design consideration, although it gives useful and detailed information on fire precautions. Two documents do give particular assistance for design. They are Hospital Technical Memorandum 23, *Access and accommodation for engineering services* and BSRIA Technical Notes TN 3 and 4/79, *Space allowances for building services, outline and detailed design stages.*

Appendix I

Thermal and vapour properties of
materials

Material	Density (kg/m^3)	Conductivity, k or λ (W m^{-1} K^{-1})	Resistivity, r (m K W^{-1})	Volume specific heat capacity (MJ K^{-1} m^{-3})	Diffusivity/10^{-6} (m^2/s)	Vapour resistivity (MN s g^{-1} m^{-1})
Masonry						
Brickwork (1% mc)[a]	1 200	0.31	3.23	0.9	0.34	
	1 400	0.42	2.38	1.05	0.4	
	1 600	0.54	1.85	1.20	0.45	25–150
	1 800	0.71	1.41	1.35	0.53	
	2 000	0.92	1.09	1.5	0.61	
Concrete (3% mc)[a]	400	0.15	6.67	0.37	0.41	
	600	0.20	5.00	0.55	0.36	
	800	0.23	4.35	0.74	0.31	
	1 000	0.30	3.33	0.92	0.33	
	1 200	0.38	2.63	1.10	0.35	30–100
	1 400	0.51	1.96	1.29	0.40	
	1 600	0.66	1.52	1.47	0.45	
	1 800	0.87	1.15	1.66	0.52	
	2 000	1.13	0.88	1.84	0.61	
	2 200	1.45	0.69	2.02	0.72	
	2 400	1.83	0.55	2.21	0.83	
Granite	2 650	2.94	0.34	2.22	1.32	400
Limestone	2 000	1.54	0.65	1.85	0.83	150
Marble	2 720	2.50	0.40	2.30	1.09	200
Sandstone	2 400	1.30	0.77	2.02	0.64	150
Slate	2 800	1.89	0.53	2.20	0.86	250
Cement mortar	1 300	0.50	2.00	1.00	0.50	100
	1 800	0.53	1.88	1.72	0.31	
Plaster, dense	1 300	0.50	2.00	1.37	0.36	60
Plaster, lightweight	600	0.17	5.88	0.55	0.31	50
Tiles						
Clay	1 300	0.83	1.2	1.8	0.46	270
Terrazzo	2 400	1.75	0.57	2.04	0.86	250
Metals						
Aluminium	2 700	222	0.0045	2.72	82	∞
Copper	8 850	400	0.0025	3.34	120	∞
Lead	11 325	36	0.028	1.80	20	∞
Steel	7 830	50	0.02	3.94	13	∞
Zinc	7 100	118	0.0085	2.68	44	∞
Timber						
Softwood	500	0.14	7.2	0.88	0.16	60
Hardwood	650	0.16	6.25	1.5	0.11	60
Sheet materials						
Plasterboard	950	0.16	6.25	0.80	0.20	55
Asbestos cement	750	0.36	2.80	1.60	0.49	1 000
Plywood	650	0.14	7.30	1.50	0.09	3 500
Chipboard	700	0.11	9.20	1.10	0.10	500
Hardboard	900	0.20	5.00	2.20	0.09	600
Fibreboard	300	0.05	20.00	0.70	0.07	40
Wood wool	500	0.10	11.00	0.60	0.60	15
Glass	2 500	1.02	0.98	2.10	0.04	∞
Miscellaneous						
Asphalt or bituminous felt roof		0.20	25.00	1.7	0.29	100 000
Lino		0.18	5.50	2.00	0.09	1 000
Thatch		0.12	8.60	0.28	0.43	15

Material	Density (kg/m³)	Thermal properties				Vapour resistivity (MN s g⁻¹ m⁻¹)
		Conductivity, k or λ (W m⁻¹ K⁻¹)	Resistivity, r (m K W⁻¹)	Volume specific heat capacity (MJ K⁻¹ m⁻³)	Diffusivity/10⁻⁶ (m²/s)	
Insulation slabs						
Mineral wool	50	0.033	30	0.05	0.66	6
Fibreglass	25	0.035	30	0.025	1.32	15
Cork	145	0.04	25	0.32	0.13	75
Expanded polystyrene	25	0.033	30	0.03	1.10	350 open cell / 1 000 closed cell
Polyisocyanurate	40	0.23				40
Loose insulation						
Fibreglass	12	0.04	25	0.01	4.00	10
Vermiculite	80	0.06	16	0.07	0.86	15
Cork granules	95	0.04	25	0.18	0.22	10
Cellulose fibre	50	0.032	32			

Appendix II

Thermal factors for standard
construction

External walls: surface resistance 0.055 m^2 K W^{-1} external, 0.123 m^2 K W^{-1} internal

Description	Density (kg/m^3)	Conductivity (W m^{-1} K^{-1})	Specific heat (J kg K^{-1})	U (W m^{-2} K^{-1})	Admittance (W m^{-2} K^{-1})	Decrement factor	Time lag (h)
Brickwork							
Solid brickwork, unplastered							
Brick 105 mm	1 700	0.84	800	3.3	4.2	0.88	2.5
Solid brickwork, unplastered							
Brick 220 mm	1 700	0.84	800	2.3	4.6	0.54	6.0
Solid brickwork, unplastered							
Brick 335 mm	1 700	0.84	800	1.7	4.7	0.29	9.4
Solid brickwork with dense plaster							
Brick 105 mm	1 700	0.84	800 }	3.0	4.1	0.83	3.0
Dense plaster 16 mm	1 300	0.50	1 000				
Solid brickwork with dense plaster							
Brick 220 mm	1 700	0.84	800 }	2.1	4.4	0.49	6.5
Dense plaster 16 mm	1 300	0.50	1 000				
Solid brickwork with dense plaster							
Brick 335 mm	1 700	0.84	800 }	1.7	4.4	0.26	9.9
Dense plaster 16 mm	1 300	0.50	1 000				
Solid brickwork with lightweight plaster							
Brick 105 mm	1 700	0.84	800 }	2.5	3.1	0.82	3.1
Lightweight plaster 16 mm	600	0.16	1 000				
Solid brickwork, with lightweight plaster							
Brick 220 mm	1 700	0.84	800 }	1.9	3.4	0.45	6.7
Lightweight plaster 16 mm	600	0.16	1 000				
Solid brickwork with lightweight plaster							
Brick 335 mm	1 700	0.84	800 }	1.5	3.4	0.23	10.0
Lightweight plaster 16 mm	600	0.16	1 000				
Solid brickwork with plasterboard							
Brick 220 mm	1 700	0.84	800 }	1.9	3.4	0.45	6.7
Plasterboard 10 mm	950	0.16	840				
Cavity walls (unventilated)							
Cavity wall with 105 mm inner and outer brick leaves with dense plaster on inner							
Brick 105 mm	1 700	0.84	800				
Cavity >20 mm	resistance = 0.18 m^2 K W^{-1}			1.5	4.3	0.43	7.8
Brick 105 mm	1 700	0.62	800				
Dense plaster 16 mm	1 300	0.50	1 000				
Cavity wall as above but with lightweight plaster							
Brick 105 mm	1 700	0.84	800				
Cavity >20 mm	resistance = 0.18 m^2 K W^{-1}			1.3	3.3	0.39	8.0
Brick 105 mm	1 700	0.62	800				
Lightweight plaster 16 mm	600	0.50	1 000				
Cavity wall as above but with 230 mm outer leaf							
Brick 230 mm	1 700	0.84	800				
Cavity >20 mm	resistance = 0.18 m^2 K W^{-1}			1.2	4.3	0.20	11.7
Brick 105 mm	1 700	0.62	800				
Dense plaster 16 mm	1 300	0.50	1 000				
Cavity wall as above but with lightweight plaster							
Brick 230 mm	1 700	0.84	800				
Cavity >20 mm	resistance = 0.18 m^2 K W^{-1}			1.1	3.3	0.18	11.8
Brick 105 mm	1 700	0.62	800				
Lightweight plaster 16 mm	600	0.16	1 000				
Cavity wall with brick outer and lightweight concrete block inner leaf with dense plaster on inner							
Brick 105 mm	1 700	0.84	800				
Cavity >20 mm	resistance = 0.18 m^2 K W^{-1}			0.96	2.9	0.56	7.1
Lightweight concrete block 100 mm	600	0.19	1 000				
Dense plaster 16 mm	1 300	0.50	1 000				
Cavity wall as above but with 13 mm expanded polystyrene in cavity							
Brick 105 mm	1 200	0.84	800				
Cavity >20 mm	resistance = 0.18 m^2 K W^{-1}						
Polystyrene 13 mm	25	0.033	1 380	0.70	3.0	0.49	8.0
Lightweight concrete block 100 mm	600	0.19	1 000				
Dense plaster 16 mm	1 300	0.50	1 000				
Cavity wall, rendered externally with 75 mm aerated concrete block outer and 100 mm aerated concrete inner with dense plaster on inner							
Rendering 10 mm	1 300	0.50	1 000				
Aerated concrete block 75 mm	750	0.24	1 000				
Cavity >20 mm	resistance = 0.18 m^2 K W^{-1}			0.85	3.2	0.54	7.2
Aerated concrete block 100 mm	750	0.22	1 000				
Dense plaster 16 mm	1 300	0.50	1 000				
Concrete							
Solid cast concrete 150 mm thick							
Concrete 150 mm	2 100	1.40	840	3.5	5.2	0.71	3.9
Solid cast concrete 200 mm thick							
Concrete 200 mm	2 100	1.40	840	3.1	5.4	0.57	5.3
Precast panel 25 mm thick							
Concrete 75 mm	2 100	1.40	840	4.3	4.9	0.92	1.8

Description	Density (kg/m³)	Conductivity (W m⁻¹ K⁻¹)	Specific heat (J kg K⁻¹)	U (W m⁻² K⁻¹)	Admittance (W m⁻² K⁻¹)	Decrement factor	Time lag (h)
Cast concrete (150 mm) with 50 mm wood wool slab on inner surface finished with 16 mm dense plaster							
Concrete 150 mm	2 100	1.40	840				
Wood wool 50 mm	500	0.10	1 000	1.2	2.3	0.50	6.5
Dense plaster 16 mm	1 300	0.50	1 000				
As above but with 200 mm concrete							
Concrete 200 mm	2 100	1.40	840				
Wood wool 50 mm	500	0.10	1 000	1.2	2.3	0.36	7.8
Dense plaster 16 mm	1 300	0.50	1 000				
75 mm concrete panel with inner sandwich panel of 5 mm asbestos cement sheet, 25 mm expanded polystyrene and 10 mm plasterboard							
Concrete 75 mm	2 100	1.40	840				
Cavity >20 mm	resistance = 0.18 m² K W⁻¹			0.80	1.0	0.82	3.1
Asbestos cement sheet 5 mm	700	0.36	1 050				
Expanded polystyrene 25 mm	25	0.033	1 380				
Plasterboard 10 mm	950	0.16	840				
Precast sandwich consisting of 75 mm dense concrete, 25 mm expanded polystyrene and 150 mm lightweight concrete							
Concrete 75 mm	2 100	1.40	840				
Expanded polystyrene 25 mm	25	0.033	1 380	0.72	3.8	0.28	9.8
Lightweight concrete 150 mm	1 200	0.38	1 000				
Roofs: surface resistance (0.045 m² K W⁻¹ external, 0.123 m² K W⁻¹ internal)							
Asphalt 15 mm on lightweight concrete screed 75 mm on dense concrete 150 mm							
Asphalt 15 mm	1 700	0.50	1 000				
Screed 75 mm	1 200	0.41	840	1.9	5.1	0.36	7.4
Dense concrete 150 mm	2 100	1.40	840				
Dense plaster 15 mm	1 300	0.50	1 000				
Asphalt 19 mm on 150 mm autoclaved aerated concrete roof slabs with dense plaster internally							
Asphalt 19 mm	1 700	0.50	1 000				
Aerated concrete 150 mm	500	0.16	840	0.86	2.5	0.78	4.7
Dense plaster 15 mm	1 300	0.50	1 000				
Asphalt 19 mm on fibre insulation board 13 mm on hollow or cavity asbestos cement decking							
Asphalt 19 mm	1 700	0.50	1 000				
Fibre board 13 mm	300	0.057	1 000				
Asbestos cement 10 mm	1 500	0.36	1 000	1.5	1.9	0.96	1.8
Cavity >55 mm	resistance = 0.18 m² K W⁻¹						
Asbestos cement 10 mm	1 500	0.36	1 000				
Asphalt 19 mm on 13 mm cement and sand screed on 50 mm wood wool slabs on steel framing with cavity and 10 mm plasterboard ceiling							
Asphalt 19 mm	1 700	0.50	1 000				
Cement and sand 13 mm	2 100	1.28	1 000				
Wood wool slab 50 mm	560	0.10	1 000	1.03	1.45	0.89	3.0
Cavity >100 mm	resistance = 0.18 m² K W⁻¹						
Plasterboard 10 mm	950	0.16	1 000				
Felt/bitumen layers on 25 mm expanded polystyrene on metal decking							
Felt/bitumen 19 mm	1 700	0.50	1 000	1.03	1.0	0.99	0.6
Expanded polystyrene 25 mm	25	0.033	1 000				

Description				U (W m⁻² K⁻¹)	Admittance (W m⁻² K⁻¹)	Density (kg/m³)	
Internal walls							
105 mm brick, 15 mm dense plaster each side					4.5	1 650	
75 mm lightweight concrete block 15 mm dense plaster both sides					2.6	750	
Two fibreboard sheets with cavity between					0.3	300	
Stud partition, plasterboard both sides					1.0	<1 200	
Windows							
Wood frame							
single-glazed					4.3		
double-glazed					2.5		
Metal frame							
single-glazed					5.6		
double-glazed (with thermal break in frame)					3.2		

Floors and ceilings				*Floor*	*Ceiling*		Time lag (h)
Timber							
10 mm timber cavity / 16 mm plasterboard ceiling				0.1	0.3		800
Concrete							
50 mm screed / 150 mm concrete				5.6	5.6		2 100
As above but with wood block or carpet floor finish				3.1	5.8		2 000

(*Source*: BS CP 61/74, N.O. Millbank and J. Harrington-Lynn, *Thermal response and admittance procedure*; reproduced by permission of the Controller, HMSO, Crown copyright)

Appendix III

SI units

All quantities in this volume are given in SI units, which have been adopted by the United Kingdom for use throughout the construction industry from 1971.

Traditionally, in this and other countries, systems of measurement have grown up employing many different units not rationally related and indeed often in numerical conflict when measuring the same thing. The use of bushels and pecks for volume measurement has declined in this country but pints and gallons, and cubic feet and cubic yards are still both simultaneously in use as systems of volume measurement, and conversions between the two must often be made. The subdivision of the traditional units vary widely: 8 pints equal 1 gallon; 27 cubic feet equal 1 cubic yard; 12 inches equal 1 foot; 16 ounces equal 1 pound; 14 pounds equal 1 stone; 8 stones equal 1 hundredweight. In more sophisticated fields the same problem existed. Energy could be measured in terms of foot pounds, British thermal units, horsepower, kilowatt-hours, etc. Conversion between various units of national systems were necessary and complex, and between national systems even more so. Attempts to rationalize units have been made for several centuries. The most significant stages are as follows:

- Establishment of the decimal metric system during the French Revolution.
- Adoption of the centimetre and gram as basic units by the British Association for the Advancement of Science in 1873, which led to the CGS system (centimetre, gram, second).
- Use after approximately 1900 of metres, kilograms and seconds as basic units (MKS system).
- Incorporation of electrical units between 1933 and 1950 giving metres, kilograms, seconds and amperes as basic units (MKSA system).
- Establishment in 1954 of a rationalized and coherent system of units based on MKSA but also including temperature and light. This was given the title *Système International d'Unités*, abbreviated to SI units.

The international discussions which have led to the development of the SI system take place under the auspices of the Conference Général des Poids et Mésures (CGPM) which meets in Paris. Eleven meetings have been held since its constitution in 1875.

The United Kingdom has formally adopted the SI system and it will become, as in some 25 countries, the only legal system of measurement. Several European countries, while adopting the SI system, will also retain the old metric system as a legal alternative. The United States has not adopted the SI system. The SI system has six basic units.

Quantity	Unit	Symbol
Length	metre	m
Mass	kilogram	kg
Time	second	s
Electric current	ampere	A
Temperature	kelvin	K
Luminous intensity	candela	cd

The kelvin (K) is the basic SI unit. Celsius (°C) will still be used for customary values and particular temperatures. One kelvin and one degree Celsius measure an identical temperature interval, but the kelvin scale commences at absolute zero, which is taken as −273.15 °C.

Quantity	Unit	Symbol
Plane angle	radian	rad
Solid angle	steradian	sr

Degrees °, minutes ′ and seconds ″ will also be used as part of the system.

From these basic and supplementary units are derived the other units necessary for measurement, e.g.

Area is derived from length	m^2
Volume is derived from length	m^3
Velocity is derived from length and time	m/s

Some derived units have special symbols.

Quantity	Unit	Symbol	Basic units involved
Frequency	hertz	Hz	$1\,Hz = 1/s$ (1 cycle per second)
Force, energy	newton	N	$1\,N = 1\,kg\,m\,s^{-2}$
Work, quantity of heat	joule	J	$1\,J = 1\,N\,m$
Power	watt	W	$1\,W = 1\,J/s$
Luminous flux	lumen	lm	$1\,lm = 1\,cd/sr$
Illumination	lux	lx	$1\,lx = 1\,lm/m^2$

Multiples and submultiples of SI are all formed in the same way and all are decimally related to the basic units. It is recommended that the factor 1000 should be consistently employed as the change point from unit to multiple or from one multiple to another. The following table gives the names and symbols of the multiples. When using multiples the description or the symbol is combined with the basic SI unit, e.g. kilojoule kJ.

Factor		Prefix	
		Name	Symbol
one million million	10^{12}	tera	T
one thousand million	10^{9}	giga	G
one million	10^{6}	mega	M
one thousand	10^{3}	kilo	k
one thousandth	10^{-3}	milli	m
one millionth	10^{-6}	micro	μ
one thousand millionth	10^{-9}	nano	n
one million millionth	10^{-12}	pico	p

Note that the kilogram departs from the general SI rule with respect to multiples; it is already 1000 g. Where more than three significant figures are used it has been United Kingdom practice to group the digits into three and separate the groups with commas.

This could lead to confusion with calculation from other countries where the comma is used as a decimal point. It is recommended therefore that groups of three digits should be separated by spaces, although for clarity this book has adopted the comma convention for numbers greater than 9999. In the United Kingdom the decimal point can still, however, be presented by a point either on or above the bottom line.

A3.1 Definitions

Basic SI units

Metre
The metre is the length equal to 1,650,763.73 wavelengths in vacuum of the radiation corresponding to the transition between the levels $2p_{10}$ and $5d_5$ of the krypton-86 atom (Eleventh General Conference of Weights and Measures, 1960; XI CGPM, 1960).

Kilogram
The kilogram is equal to the mass of the interational prototype of the kilogram (III CGPM, 1901). The international prototype is in the custody of the Bureau International des Poids et Mésures (BIPM), Sèvres, near Paris.

Second
The second is the duration of 9,192,631,770 periods of the radiation corresponding to the transition between the two hyperfine levels of the ground state of the caesium-133 atom (XII CGPM, 1967).

Ampere
The ampere is that constant current which, if maintained in two straight parallel conductors of infinite length, of negligible circular cross-section, and placed 1 metre apart in a vacuum, would produce between these conductors a force equal to 2×10^{-7} newton per metre of length (IX CGPM, 1948).

Kelvin
The kelvin is the fraction 1/273.16 of the thermodynamic temperature of the triple point of water (XII CGPM, 1967). The temperature of the ice point is 273.15 K. The units of kelvin and Celsius temperature interval are identical.

Candela
The candela is the luminous intensity, in the perpendicular direction, of a surface of 1/600,000 square metre of a black body at the temperature of freezing platinum under a pressure of 101,325 newtons per square metre (XIII CGPM, 1967).

Supplementary units

Radian
The angle between two radii of a circle which cut off on the circumference an arc equal in length to the radius.

Steradian
The solid angle which, having its vertex in the centre of a sphere, cuts off an area of the surface of the sphere equal to that of a square having sides of length equal to the radius of the sphere.

A3.2 Sound

The main units remain unchanged (e.g. 1 hertz = 1 cycle per second; decibels and absorption coefficients are ratios, so they remain unchanged). It is important to note, however, when using premetric tables, that where absorption coefficients are quoted per unit (e.g. person or per seat) an area in square feet is implicit in the figure and a conversion to square metres should be made (square feet $\times$ 0.09 = square metres). The constant for use in the Sabine formula becomes 0.16.

A3.3 Electricity

Units remain unchanged.

A3.4 Conversion factors

	Quantity or application	SI unit, S Description	SI unit, S Symbol	Non-SI unit, N	Conversion factor, C ($S = C \times N$)
Space	length	metre	m	foot	0.31
				inch	0.025
	area	square metre	m^2	square yard	0.84
				square foot	0.09
		square millimetre	mm^2	square inch	0.000 65
		square millimetre	mm^2	square inch	645
	volume	cubic metre	m^3	cubic yard	0.76
				cubic foot	0.028
				gallon (UK)	0.0045
				gallon (US)	0.0038
		litre ($m^3/1000$)	l	gallon	4.55
				pint	0.57
Mass	mass	kilogram	kg	pound	0.45
				ton	1016
	density	kilogram per cubic metre	kg/m^3	pound per cubic foot	16
				pound per gallon	99.78
Motion	velocity	metre per second	m/s	foot per second	0.31
				foot per minute	0.0051
				mile per hour	0.45
Flow rate	volume flow	cubic metre per second	m^3/s	cubic foot per second	0.028
		litre per second	l/s	gallon per minute	0.076
Pressure	pressure	pascal $1\,Pa = 1\,N/m^2$	Pa	foot water gauge	2890
				millibar (mb)	10^5
Temperature	customary temperature (level)	degree Celsius	°C	degree Fahrenheit °F	$\dfrac{5\,(°F - 32)}{9}$
	temperature interval (range or difference)	degree Kelvin	K	degree Fahrenheit	0.56
Heat	quantity (energy)	joule	J	British thermal unit (Btu)	1.055[a]
		kilojoule	kJ	Btu	1.055
		kilojoule	kJ	kilowatt-hour (kWh)	3600
	flow rate (power)	watt	W	Btu per hour	0.29
				ton of refrigeration	3516
	intensity of heat flow rate	watt per square metre	W/m^2	Btu per hour per square foot	3.16
	thermal conductivity	watt per metre per degree Kelvin	$W\,m^{-1}\,K^{-1}$	Btu inch per hour square foot degree F	0.14[b]
	thermal conductance, U value	watt per square metre per degree Kelvin	$W\,m^{-2}\,K^{-1}$	Btu per hour per square foot per degree F	5.68
	thermal resistivity	metre degree Kelvin per watt	$m\,K\,W^{-1}$	square foot hour degree F per Btu inch	6.93[b]
	thermal resistance	square metre degree Kelvin per watt	$m^2\,K\,W^{-1}$	square foot hour degree F per Btu	0.18
	thermal diffusivity	square metre per second	m^2/s	square foot per hour	0.000 026
	thermal capacity per unit mass (specific heat capacity)	kilojoule per kilogram per degree Kelvin	$kJ\,kg^{-1}\,K^{-1}$	Btu per pound degree F	4.19

Quantity or application	SI unit, S		Non-SI unit, N	Conversion factor, C ($S = C \times N$)
	Description	Symbol		
Heat (*cont.*) specific heat (volume basis)	kilojoule per cubic metre per degree Kelvin	kJ m^{-3} K^{-1}	Btu per cubic foot degree F	67.1
calorific value (weight basis)	kilojoule per kilogram	kJ/kg	Btu per pound	2.32
calorific value (volume basis)	kilojoule per cubic metre	kJ/m^3	Btu per cubic foot	37.26
latent heat	kilojoule per kilogram	kJ/kg	Btu per pound	2.32
refrigeration	watt	W	ton	3516
Moisture moisture content	gram per kilogram	g/kg	grain per pound	0.14
vapour permeability	kilogram metre per newton per second	kg m N^{-1} s^{-1}	pound foot per hour pound force	0.000 008 6
Light illumination	lux	lx	foot candle lumen per square foot	10.76
luminance	candela per square metre	cd/m^2	foot lambert	3.43

[a] In practice 1 Btu is taken as equivalent to 1 kJ.
[b] The apparent discrepancy between imperial and SI units may be made resolved by expressing the SI units in basic terms before cancellation of terms, e.g.

$$\frac{\text{J m}}{\text{m}^2\,\text{s K}} \quad \text{becomes} \quad \frac{\text{W}}{\text{m K}} \quad \text{or} \quad \text{W m}^{-1}\,\text{K}^{-1} \quad (\text{m/m}^2 = 1/\text{m} \quad \text{and} \quad \text{J/s} = \text{W})$$

$$\text{similarly} \quad \frac{\text{m}^2\,\text{s K}}{\text{J m}} \quad \text{becomes} \quad \text{m K W}^{-1}$$

Index